Lectures in Applied Mathematics
Volumes in This Series

Proceedings of the Summer Seminar, Boulder, Colorado, 1960

1 **Lectures in Statistical Mechanics,** *G. E. Uhlenbeck and G .W. Ford with E. W. Montroll*
2 **Mathematical Problems of Relativistic Physics,** *I. E. Segal with G. W. Mackey*
3 **Perturbation of Spectra in Hilbert Space,** *K. O. Friedrichs*
4 **Quantum Mechanics,** *R. Jost*

Proceedings of the Summer Seminar, Ithaca, New York, 1963

5 **Space Mathematics. Part 1,** *J. Barkley Rosser, Editor*
6 **Space Mathematics. Part 2,** *J. Barkley Rosser, Editor*
7 **Space Mathematics. Part 3,** *J. Barkley Rosser, Editor*

Proceedings of the Summer Seminar, Ithaca, New York, 1965

8 **Relativity Theory and Astrophysics. 1. Relativity and Cosmology,** *Jürgen Ehlers, Editor*
9 **Relativity Theory and Astrophysics. 2. Galactic Structure,** *Jürgen Ehlers, Editor*
10 **Relativity Theory and Astrophysics. 3. Stellar Structure,** *Jürgen Ehlers, Editor*

Proceedings of the Summer Seminar, Stanford, California, 1967

11 **Mathematics of the Decision Sciences, Part 1,** *George B. Dantzig and Arthur F. Veinott, Jr., Editors*
12 **Mathematics of the Decision Sciences, Part 2,** *George B. Dantzig and Arthur F. Veinott, Jr., Editors*

Proceedings of the Summer Seminar, Troy, New York, 1970

13 Mathematical Problems in the Geophysical Sciences. 1. Geophysical Fluid Dynamics, *William H. Reid, Editor*

14 Mathematical Problems in the Geophysical Sciences. 2. Inverse Problems, Dynamo Theory, and Tides, *William H. Reid, Editor*

Proceedings of the Summer Seminar, Potsdam, New York, 1972

15 Nonlinear Wave Motion, *Alan C. Newell, Editor*

Proceedings of the Summer Seminar, Troy, New York, 1975

16 Modern Modeling of Continuum Phenomena, *Richard C. DiPrima, Editor*

Proceedings of the Summer Seminar, Salt Lake City, Utah, 1978

17 Nonlinear Oscillations in Biology, *Frank C. Hoppensteadt, Editor*

Proceedings of the Summer Seminar, Cambridge, Massachusetts, 1979

18 Algebraic and Geometric Methods in Linear Systems Theory, *Christopher I. Brynes and Clyde F. Martin, Editors*

Proceedings of the Summer Seminar, Salt Lake City, Utah, 1980

19 Mathematical Aspects of Physiology, *Frank C. Hoppensteadt, Editor*

Proceedings of the Summer Seminar, Chicago, Illinois, 1981

20 Fluid Dynamics in Astrophysics and Geophysics, *Norman R. Lebovitz, Editor*

Proceedings of the Summer Seminar, Chicago, Illinois, 1982

21 Applications of Group Theory in Physics and Mathematical Sciences, *Moshe Flato, Paul Sally and Gregg Zuckerman, Editors*

22 Large-Scale Computations in Fluid Mechanics, Parts 1 and 2, *B. Engquist, S. Osher and R. C. J. Somerville, Editors*

23 Nonlinear Systems of Partial Differential Equations in Applied Mathematics, Parts 1 and 2, *Basil Nicolaenko, Darryl D. Holm and James M. Hyman, Editors*

24 Reacting Flows: Combustion and Chemical Reactors, Parts 1 and 2, *G. S. S. Ludford, Editor*

Reacting Flows:

Combustion and Chemical Reactors

Part 1

Volume 24
Lectures in Applied Mathematics

Reacting Flows:
Combustion and Chemical Reactors

Part 1

G.S.S. Ludford, Editor

1986
American Mathematical Society, Providence, Rhode Island

The Proceedings of the 1985 SIAM-AMS Summer Seminar on Reacting Flows: Combustion and Chemical Reactors were prepared by the American Mathematical Society with support from the following sources: Air Force Office of Scientific Research, Army Research Office, Office of Naval Research, and the National Science Foundation through NSF Grant DMS8414688.

1980 *Mathematics Subject Classification* (1985 *Revision*). 76-06, 80-06, 58F40.

Library of Congress Cataloging-in-Publication Data

Main entry under title:

Reacting flows.

(Lectures in applied mathematics; v. 24)

Proceedings of the '85 AMS/SIAM Summer Seminar in Applied Mathematics held at the Cornell University center for Applied Mathematics

Bibliography: p

1. Combustion–Congresses. 2. Thermodynamics–Congresses. 3. Chemical reactors–Congresses. I. Ludford, G. S. S. II. Summer Seminar on Applied Mathematics (17th: 1985: Cornell University) III. American Mathematical Society. IV. Society for Industrial and Applied Mathematics. V. Lectures in applied mathematics (American Mathematical Society); v. 24.

QD516.R418 1986 541.3'61 86-1088

ISBN 0-8218-1124-X (set)

ISBN 0-8218-1127-4 (pt. 1)

ISBN 0-8218-1128-2 (pt. 2)

Printed in the United States of America.

Contents

Part 1

PREFACE

The '85 AMS/SIAM Summer Seminar on Applied Mathematics (Reacting Flows: Combustion and Chemical Reactors) was the culmination of the Special Year '84-85 of the same title held by the Center for Applied Mathematics at Cornell University. The special issue of Physica D containing papers by the main Special-Year speakers (R. Aris, J.D. Buckmaster, D.S. Cohen, H. Dwyer, I. Müller, B. Nicolaenko, H. Rabitz, and F.A. Williams) and another (J. Guckenheimer) forms the background for these Proceedings. In particular, the Preface of that issue spells out the theme of both the Special Year and the Summer Seminar.

Series of introductory lectures at the Seminar were given by R. Aris, H. Dwyer, A. Ghoniem, G.S.S. Ludford, and A. Majda. Written versions are collected at the beginning of this volume in the order of presentation. In addition, the remaining Physica D authors were invited to give Seminar papers, and all but Cohen were able to do so. With the exception of Nicolaenko's, which did not meet the publication deadline, their contributions appear at appropriate places in the following pages.

The remaining 33 lectures were chosen on the basis of four dozen abstracts that were submitted in response to a wide solicitation. Only one of the corresponding papers (by B.T. Chu, K.R. Sreenivasan and S. Raghu) is missing, in spite of the strict deadline. The oral contributions of about 50 other (unnamed) participants should also be mentioned.

All authors were required to produce final typescripts within a few weeks of the Seminar, and they must be commended for meeting the deadlines. The editorial work was finished at the end of September, so that responsibility for the delay beyond then must be placed elsewhere.

In requesting support from agencies we asserted that these Proceedings would provide a solid basis for large-scale computing and motivation for future research; only time will tell whether we delivered on our promise. The editor (over)heard many congratulatory comments about the Seminar, but he is well aware that a meeting is only as good as its participants. The agencies concerned were AFOSR, ARO, NSF, and ONR; the reacting-flow community is indebted to them, in particular J.P. Thomas, J. Chandra, A. Thaler, and T. Mulliken (respectively).

The Organizing Committee of the Seminar - consisting of D.S. Cohen, G.S.S. Ludford (chairman), A. Majda, and F.A. Williams - was ably assisted by an Advisory Committee consisting of the Physica D authors.

G.S.S. Ludford
Ithaca, New York
December 1985

INTRODUCTIONS

Lectures in Applied Mathematics
Volume 24, 1986

LOW MACH NUMBER COMBUSTION*

G.S.S. Ludford

ABSTRACT. This article gives an introduction to combustion theory for low Mach number. The aim is threefold: to set the subject in the context of reacting flows; to illustrate the achievements of asymptotic methods; and to identify the most important directions for future theoretical research to take.

INTRODUCTION

This article is the written version of five introductory lectures given at the Seminar. The original intention was to summarize and complement the combustion lectures delivered by Buckmaster, Müller, Nicolaenko, Rabitz, and Williams during the Special Year '84-'85 at Cornell. The subsequent decision to have them tell their own stories [1,2,3,4,5] in a special issue of Physica D, along with others associated with the Special Year (namely Aris, Cohen, Dwyer, and Guckenheimer), relieved me of a formidable task. The final product is about half as long as originally planned, corresponding to a reduction in the number of lectures from ten to five (a decision of the Organizaing Committee that was welcomed by all). References to the Physica D articles show how summaries of the Special Year lectures would have fitted in.

1980 Mathematics Subject Classification. 35K99, 76 V05.
*Work performed under the auspices of the U.S. Army Research Office and of the U.S. Department of Energy under Contract W-7405-ENG-36 and the Office of Scientific Computing.

Extensive use has been made of my monograph [6] with John Buckmaster; but responsibility for changes lies entirely with me. Even more than there, full treatments of the material and pedantry over original sources have been avoided. Most missing information can be found in [1,2,3,4,5,6]; otherwise try my earlier monograph [7] with John Buckmaster or a standard combustion text. Nevertheless, there is other material; and, wherever it is not fresh, an attempt has been made to view it with a fresh eye. The aim has been threefold: to set the subject in the context of reacting flows; to illustrate the achievements of the asymptotic theory; and to identify the most important directions for future theoretical research to take.

Complexity can rapidly make a mathematical text unattractive; accordingly the notation has been kept as simple as possible:

When part of a multiple display is intended, the letters a,b,... are added to indicate the first, second,... part. To avoid repetition a uniform notation for asymptotic expansions in the small parameter θ^{-1} will be used throughout, namely

$$v = v_o + \theta^{-1} v_1 + \theta^{-2} v_2 + \ldots$$

for the generic dependent variable v. The coefficients v_o, v_1, v_2, etc. will be called the leading term, (first) perturbation, second perturbation, etc. Only the variables on which these coefficients depend need be specified, changing as they do from region to region. In all cases the notation $+\ldots$ is used (when a more accurate estimate is not needed) to denote a remainder that is of smaller order in θ than the preceding term. The same notation will be used for parameters; however, since their expansion coefficients are constants, no specification of variables is needed.

Similar conventions will be used when a small parameter other than θ^{-1} is involved. Except in a few obvious places, numerical subscripts such as 0, 1, 2 will be used exclusively for terms in asymptotic expansions.

1. OVERVIEW OF REACTING FLOWS

In general, two-reactant flames (i.e., zones of reaction) can be classified as diffusion or premixed. In a premixed flame the reactants are mixed and burn when the mixture is raised to a sufficiently high temperature. In a diffusion flame the reactants are of separate origin; burning occurs only at a diffusion-blurred interface.

Both kinds of flames can be produced by a Bunsen burner (Fig. 1). If the air hole is only partly open, a fuel-rich mixture of gas and air passes up the burner tube, and the thin conical sheet of flame standing at the mouth is the premixed flame. The excess gas escaping downstream mixes by diffusion with the surrounding atmosphere and burns as a diffusion flame.

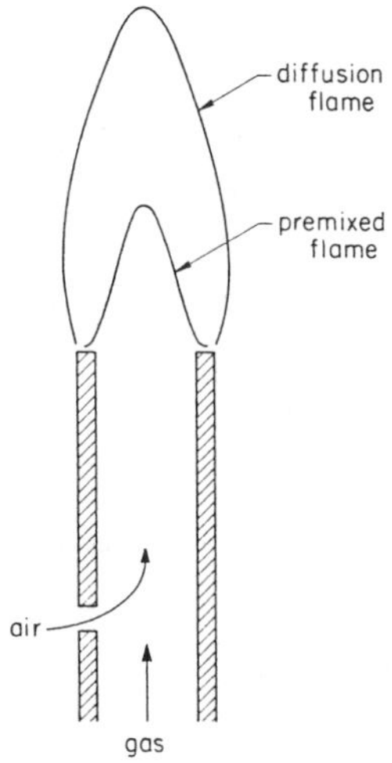

Figure 1. Bunsen burner.

Separate origins do not guarantee a diffusion flame, however. In Fig. 2 the reactants are originally separated by the splitter plate but mix before igniting. The flame spread across the oncoming flow is therfore premixed. Behind this premixed flame the remaining portions of the reactants are separate again, so that a diffusion flame trails downstream.

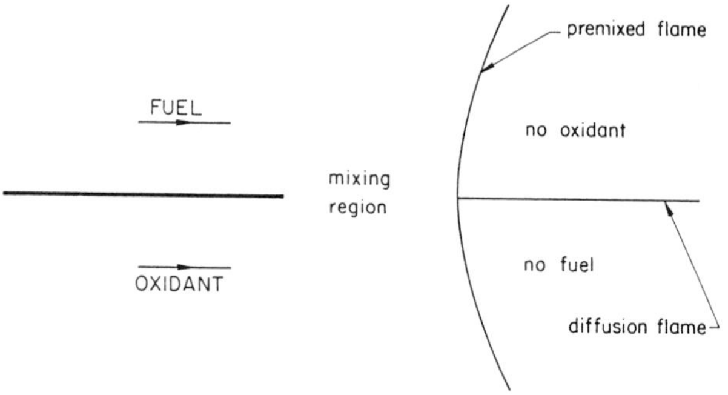

Figure 2. Combustion of initially separated reactants.

There are two types of premixed combustion: deflagrations and detonations. Deflagration is a creature of combustion; there is no counterpart in flow without chemical reaction. By contrast, detonations are modifications of shock waves and must in general be sustained (overdriven) if, like shockwaves, they are to retain their strength. To see the different natures of deflagrations and detonations, it is convenient to introduce the so-called Hugoniot diagram (Fig. 3).

If the deflagration or detonation can be treated as a discontinuity, the density ρ, relative normal velocity u, and pressure p must satisfy the (modified) Rankine-Hugoniot relations

$$\rho_f u_f = \rho_b u_b , \quad \rho_f u_f^2 + p_f = \rho_b u_b^2 + p_b , \tag{1}$$

$$u_f^2/2 + \gamma p_f/(\gamma-1)\rho_f + h = u_b^2/2 + \gamma p_b/(\gamma-1)\rho_b . \tag{2}$$

Here $h > 0$ is the heat released by the reaction at the discontinuity and γ is the specific-heat ratio. When the flow is from side f (fresh) to side b (burned), as we shall suppose, h is positive. If the speeds u_f, u_b are eliminated from these three equations, a relation between ρ_f, p_f, ρ_b, p_b is obtained, namely

$$\frac{\gamma}{\gamma-1}\left(\frac{p_f}{\rho_f} - \frac{p_b}{\rho_b}\right) - \frac{1}{2}\left(\frac{1}{\rho_f} + \frac{1}{\rho_b}\right)(p_f - p_b) + h = 0 . \qquad (3)$$

For given state f, the locus of the state b in the $1/\rho$, p-plane is a hyperbola, as shown in Fig. 3.

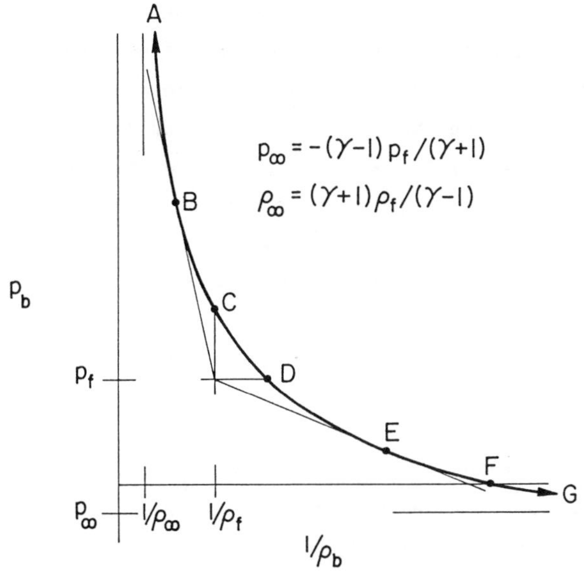

Figure 3. Hugoniot diagram.

The mass flux through the discontinuity is imaginary on the part CD of the hyperbola, so that it and the part FG are ruled out on physical grounds. The remaining four parts lead to the following classification, where M denotes the Mach number $u\sqrt{\rho/\gamma p}$.

AB: (strong) detonations; $M_f > 1$, $M_b < 1$

BC: weak detonations; M_f , $M_b > 1$

DE: (weak) deflagrations; M_f , $M_b < 1$

EF: strong deflagrations; $M_f < 1$, $M_b > 1$

A (strong) detonation of arbitrary (sufficiently large supersonic) speed can be produced by driving a piston at an appropriate speed into the gas mixture. By contrast, for any given mixture, there is only one wave speed at which a weak detonation can propagate, and the piston must be driven at a corresponding speed. Weak detonations are seldom observed, so the part BC is usually ignored.

No structure exists for strong deflagrations, i.e., the equations of steady reacting compressible flow with transport effects have no one-dimensional solution joining the states f and b . On the other hand, there is a unique solution for the (weak) deflagration, giving a definite wave speed for each gas mixture. (There is no question of driving a deflagration with a piston, as there is for a detonation.)

Points on AB correspond to so-called overdriven detonations, ones that have to be maintained by the piston. (We now drop the adjective "strong" since these detonations are the only ones of interest.) As for a shock, if the piston decelerates after having generated the detonation, the rarefaction wave that is emitted overtakes the latter (since M_b is less than 1) and weakens it. A shock can be weakened into a Mach wave, but a detonation cannot be weakened beyond B , the so-called Chapman-Jouget detonation point, at which the state behind is sonic. Thus, the Chapman-Jouget detonation, which is the counterpart of a Mach wave in reactionless flow, is self-sustaining once it has been produced. For this reason, near Chapman-Jouget detonations (corresponding to points on AB close to B) have received the most attention.

As the point D is approached on the deflagration branch DE , the Mach numbers M_f and M_b tend to zero. (We now drop the adjective "weak".) The neighborhood of D is of particular interest because in practice deflagration waves propagate slowly, the Mach numbers being in the range $10^{-3} - 10^{-2}$. Such deflagrations, whose governing equations are obtained by taking the limit of vanishingly small Mach number, will be our sole concern here. Note that in the neighborhood of D the pressures p_f and p_b are almost the same: we shall be dealing with near-isobaric flow.

Combustion at high Mach number is being discussed by Majda elsewhere in these proceedings, but an observation concerning it can be made here. One of the most interesting problems in that area is the deflagration-to-detonation transition, i.e., how a deflagration wave is induced to change into a detonation wave. The Hugoniot diagram gives a graphic impression of the jump required: high Mach-number deflagrations correspond to the neighborhood of E (called the Chapman-Jouget deflagration point), while the weakest detonations correspond to the neighborhood of the point B .

2. Governing Equations for Small Mach Number

The problem of formulating the governing equations of combustion consists, at its simplest, in characterizing the flow of a viscous, heat-conducting mixture of diffusing, reacting gases. This is a formidable task that filled Müller's week [2] at Cornell in October 1984 and, mindful of this, we shall limit ourselves to a description, rather than a derivation, of the simplest equations that can be brought to bear on combustion problems. Only the most important assumptions normally used to justify the equations will be discussed; for a more extensive treatment the reader is referred to Buckmaster & Ludford [6, Chapter 1].

The easiest framework in which to understand the field equations is the "reactant bath". We suppose that most of the mixture consists of a single inert component (e.g. nitrogen), the properties of which determine those of the mixture (e.g., viscosity, specific heat). The reacting components (and their products) are highly diluted by immersion in this bath of inert.

Mass conservation for the mixture is always described by the single-fluid equation

$$\frac{\partial\rho}{\partial t} + \nabla \cdot (\rho\mathbf{v}) = 0 \ , \tag{4}$$

where ρ is the density and $\mathbf{v}$ the velocity. But only for dilute mixtures is the overall momentum balance identical to that for a single homogeneous fluid, namely

$$\rho \frac{D\mathbf{v}}{Dt} = \nabla \cdot \Sigma \ , \tag{5}$$

where

$$\Sigma = -(p + \frac{2}{3} \kappa \nabla \cdot \mathbf{v})\mathbf{I} + \kappa(\nabla\mathbf{v} + (\nabla\mathbf{v})^T) \tag{6}$$

and p is pressure; bulk viscosity has been neglected.

A single-fluid equation for energy balance is also justified, provided account is taken of the release of heat by chemical reaction. But here additional approximations are made, based on the observation that temperatures are high and gas speeds low for a large class of combustion phenomena (excluding detonation); more precisely, a characteristic Mach number is small (typically in the range 10^{-3} -10^{-2}). Then the only significant form of energy, other than that of chemical bonding, is thermal; kinetic energy makes a negligible contribution to the energy balance. For the same reason, the conversion of kinetic energy into thermal energy by way of viscous dissipation can be ignored. Thus, when variations of the specific heat c_p with temperature are neglected, we have

$$\rho c_p \frac{DT}{Dt} - \nabla \cdot (\lambda \nabla T) - \frac{\partial p}{\partial t} = q \ , \tag{7}$$

where q is the heat released per unit volume of the fluid by chemical reaction; the form of q is considered later.

In addition, the assumption of small Mach number implies that spatial variations in pressure are small, so that $\partial p/\partial t$ in the energy balance (7) is due (for unconfined flames) to imposed, uniform pressure variations. We shall assume that the imposed pressure is constant, i.e., the term vanishes. The pressure term in the momentum equation (5) cannot be neglected, however; small spatial variations are needed to account for changes in the weak velocity field. A further consequence of the virtual constancy of the pressure is that the equation of state of the mixture is Charles's law

$$\rho T = \frac{m p_c}{R} \tag{8}$$

if the inert is a perfect gas. Here m is the molecular mass of the inert, p_c the imposed constant pressure, and R the gas constant.

Consider now the individual components of the mixture, denoting the density of the i-component by ρY_i , where Y_i is the mass fraction and $i = 1,2,\ldots,N$. The reactants and their products are convected with the gas velocity $\mathbf{v}$, diffuse relate to the inert diluent, and are consumed or generated by reaction. The diffusion laws of general mixtures are complicated, involving a diffusion matrix (Sec. 3); but for dilute mixtures the matrix is diagonal insofar as the reactants and products are concerned, so that we may write

$$\rho \frac{DY_i}{Dt} - \nabla \cdot (\mu_{ii} \nabla Y_i) = \dot{\rho}_i \quad \text{for} \quad i = 1,2,\ldots,N-1 \ . \tag{9}$$

Here $\dot{\rho}_i$ is the mass production rate per unit volume of the ith component; its precise form is considered below. The equation for the mass fraction Y_N of the inert is more complicated, but it can be obtained from the relation

$$\sum_{i=1}^{N} Y_i = 1 \qquad (10)$$

instead, once the other Y_i's have been determined.

Coupling between the fluid-mechanical equations (4), (5) and the thermal-chemical equations (7), (9) occurs because of density variations. If these variations are ignored, the former may be solved for $\mathbf{v}$, which can then be substituted into the latter, a substantial simplification. Such a procedure is justified if the heat released by the reaction is small, but this is not a characteristic of combustion systems, whose main purpose is to liberate heat from its chemical bonds. For this reason, the simplified system of equations should be thought of as a model in the spirit of Oseen's approximation in hydrodynamics. However, to emphasize the mathematically rational nature of the procedure, we shall refer to the simplified system as the constant-density approximation rather than model. Phenomena whose physical basis is truly fluid-mechanical are not emcompassed by this approximation, but much of importance is; it will play a central role in our discussion.

There remains the question of the contribution of the individual reactions to the heat release q and the production rates $\dot{\rho}_i$. It is possible, in principle, to consider all the reactions that are taking place between the constituents of a mixture. However, this is done but infrequently; often a complete chemical-kinetic description (i.e., how the rates depend on the various concentrations and temperature, or even whether a particular reaction takes place) is not available. Even when it is, its complexity may deter solution by anything short of massive use of computers. For these reasons, simplified kinetic schemes are normally adopted which model, in an overall fashion, the multitude of reactions.

The simplest are the one-step irreversible schemes that account for the consumption of the reactants, here taken to be just a fuel and an oxidant. If the reactants are simply lumped together as a single entity, the scheme is represented by

$$(Y_1) \rightarrow \text{products} , \tag{11}$$

where brackets denote a molecule of the component whose mass fraction is enclosed. On the other hand, if the separate identities of the fuel and oxidant are recognized, we have

$$\nu_1(Y_1) + \nu_2(Y_2) \rightarrow \text{products} ; \tag{12}$$

here the ν_i are stoichiometric coefficients, specifying the molecular proportions in which the two reactants participate. We shall adopt the scheme (11) when discussing premixed combustion and (12) for diffusion flames.

If N_i is the number density of the ith component, so that

$$\dot{\rho}_i = m_i \dot{N}_i , \tag{13}$$

where m_i is the molecular mass of the ith component, the reaction rate ω is defined by the formula

$$\dot{N}_i = -\nu_i \omega \tag{14}$$

It is then common to write

$$\omega = k(T)\ \rho^{\gamma} \prod_j Y_j^{\beta_j} \quad (\gamma, \beta_j \ \text{positive constants}) \tag{15}$$

for the reaction rate, an empirical formula that is suggested by a theoretical treatment of so-called elementary reactions. The product contains a single term for the scheme (11), two terms for (12). The Arrhenius law

$$k = BT^{\alpha}e^{-E/RT} \quad (B, \alpha, E \text{ constants}) , \tag{16}$$

which will be adopted here, is at the heart of the mathematical treatment; E is called the activation energy.

The heat release q is a consequence of the difference between the heats of formation of the products and those of the reactants, so that it is proportional to ω. Combustion is inherently exothermic, so that we shall write

$$q = Q\omega \quad \text{with} \quad Q > 0 , \tag{17}$$

where Q has the dimensions of energy.

3. Nondimensional Equations; Shvab-Zeldovich Formulation; Non-Dilute Mixtures

We shall take units as follows:

$$\text{temperature } \frac{Q}{c_p \Sigma_j \nu_j m_j} \quad (\text{summation over 1 or 2 reactants}), \tag{18}$$

$$\text{pressure } p_c, \text{ density } \rho_r, \text{ mass flux } M_r, \text{ speed } \frac{M_r}{\rho_r}, \tag{19}$$

$$\text{length } \frac{\lambda}{c_p M_r} , \text{ time } \frac{\lambda \rho_r}{c_p M_r^2} , \text{ pressure variations } \frac{M_r^2}{\rho_r} . \tag{20}$$

Appropriate choices for the reference density ρ_r and the reference mass flux M_r are made according to the problem considered. The governing equations in nondimensional form are

$$\rho T = \frac{m p_c c_p \Sigma_j \nu_j m_j}{\rho_r R Q} , \quad \frac{\partial \rho}{\partial t} + \nabla \cdot (\rho \mathbf{v}) = 0 , \tag{21}$$

$$\rho \frac{D\mathbf{v}}{Dt} = -\nabla p + \mathcal{P}\left(\nabla^2 \mathbf{v} + \frac{1}{3}\nabla(\nabla\cdot\mathbf{v})\right) , \tag{22}$$

$$\rho \frac{DT}{Dt} - \nabla^2 T = \Omega , \quad \rho \frac{DY_i}{Dt} - \mathcal{L}_i^{-1} \nabla^2 Y_i = \alpha_i \Omega , \tag{23}$$

where i runs from 1 to $N-1$, and

$$\mathcal{P} = \frac{\kappa c_p}{\lambda} \text{ (Prandtl number)}, \mathcal{L}_i = \frac{\lambda}{\mu_{ii} c_p} \text{ (Lewis numbers)}, \tag{24}$$

$$\alpha_i = -\frac{v_i m_i}{\Sigma_j \nu_j m_j} \quad \text{(with } \sum_i \alpha_i = -1) , \; \Omega = \mathcal{D} e^{-\theta/T} \Pi_j Y_j^{\beta_j} , \tag{25}$$

$$\theta = \frac{E c_p \Sigma_j \nu_j m_j}{QR}, \quad \mathcal{D} = D M_r^{-2}, \; D = \frac{\lambda B Q^{\alpha} \rho_r^{\gamma}}{c_p^{1+\alpha} (\Sigma_j \nu_j m_j)^{\alpha-1}} \rho^{\gamma} T^{\alpha} . \tag{26}$$

Except in the case $\gamma = \alpha$, the Damköhler number $\mathcal{D}$ is variable. (The term is also used for D in spite of its having the dimensions of M_r^2.) In the context of activation-energy asymptotics ($\theta \to \infty$), only the value of D at a fixed temperature T_* plays a role, so that it may be considered an assigned constant.

When one of the Lewis numbers $\mathcal{L}_i$ is equal to 1, the differential operator in its equation (23b) is identical to that in the temperature equation (23a). We may therefore write

$$\left(\rho \frac{D}{Dt} - \nabla^2\right)\left(T - \frac{Y_i}{\alpha_i}\right) = 0 , \tag{27}$$

of which one solution is

$$T - \frac{Y_i}{\alpha_i} \equiv H_i \text{ (const.)} . \tag{28}$$

If this solution is appropriate for the problem at hand, Y_i may be eliminated in favor of T, thereby reducing the number of unknowns. The linear combination $T - Y_i/\alpha_i$ is known as a

Shvab-Zeldovich variable; it is easier to find by virtue of satisfying the reactionless equation (27).

The same equations (21-23) are also used when the mixture is not dilute, justification being based on the continuum theory of mixtures. Buckmaster and Ludford [7] give an outline of their derivation, while Müller [2] provides a deeper treatment of the issues involved. As always in continuum mechanics, the starting point is balance laws for mass, momentum, and energy; in the case of mixtures, these balances must be written for each constituent (here a gaseous species). The question then is to propose constitutive equations that do not violate the general principles of continuum mechanics, and experience with a single fluid is largely the guide for doing so.

The main difficulty is Fick's law, namely

$$\rho Y_i \mathbf{V}_i = - \mu_{ii} \nabla Y_i \quad \text{for} \quad i = 1,2,\ldots,N-1 \tag{29}$$

which was used in writing the mass balances (9); here the $\mathbf{V}_i$ are the velocities of the species (considered as coexistent fluids) relative to the mixture. Experience with a single fluid cannot suggest this law as a constitutive equation, but observations on binary mixtures do. Nevertheless, the proper place to introduce constitutive equations is the interaction forces in the momentum balances of the individual species and, when that is done, the law follows for dilute mixtures (at least in the steady case).

For a general mixture, the corresponding result is the multicomponent diffusion law

$$\nabla X_i = \sum_{j=1}^{N} \frac{X_i X_j}{D_{ij}} (\mathbf{V}_j - \mathbf{V}_i) \quad \text{for} \quad i = 1,2,\ldots N \,, \tag{30}$$

where

$$X_i = \frac{Y_i/m_i}{\sum_{i=1}^{N} Y_i/m_i} \tag{31}$$

is the mole fraction of the ith species. Here D_{ij} is the binary diffusion coefficient for the pair of species i,j and m_i is the molecular mass of the species i. Terms proportional to ∇p and ∇T have been omitted, the former because the process of interest is nearly isobaric and the latter (Soret effect) because of its smallness in practice.

For a dilute mixture we have

$$Y_i, 1-Y_N \ll 1 \quad \text{for} \quad i = 1,2,\ldots,N-1 , \tag{32}$$

so that

$$X_i = m_N Y_i/m_i \quad \text{for all } i , \tag{33}$$

approximately, and we find

$$X_i, 1-X_N, \mathbf{V}_N \ll 1 \quad \text{for} \quad i = 1,2,\ldots,N-1; \tag{34}$$

the last result follows from the requirement $\sum_{i=1}^{N} Y_i \mathbf{V}_i = 0$. As a consequence, the multicomponent diffusion law becomes

$$\nabla Y_i = -Y_i \mathbf{V}_i / D_{iN} \quad \text{for} \quad i = 1,2,\ldots,N-1, \tag{35}$$

i.e., Fick's law with $\mu_{ii} = \rho D_{iN}$. In general, however, we obtain

$$\rho Y_i \mathbf{V}_i = - \sum_{j=1}^{N} \mu_{ij} \nabla Y_j \quad \text{for} \quad i = 1,2,\ldots N \tag{36}$$

on solving the algebraic system (30) for the $\mathbf{V}_i$, the matrix μ_{ij} being related to the inverse of the matrix $X_i X_j (1-\delta_{ij})/D_{ij}$.

In treatments of a general nature there is no difficulty in using a non-diagonal diffusion matrix μ_{ij}, but in specific problems the resulting equations have proved to be intractable in virtually all cases. Fick's law is almost invariably used in solving combustion problems. So long as the purpose is to provide qualitative explanation, there is no harm in doing so; when the essential feature to capture is differential diffusion, that may be done by taking $\mu_{ii} \neq \mu_{jj}$ for $i \neq j$.

When quantitative information is required the multicomponent diffusion law should be used, and many other changes made. (Since the approach is then numerical anyway, no new difficulties are thereby introduced.) Certain changes are obvious: the coefficients κ, λ, c_p should be allowed to vary. Others involve the assumptions that are made in deriving the governing equations, such as equality of specific heats of the species; neglect of radiation and concentration gradients (Dufour effect) in the heat flux; and neglect of pressure gradients and temperature gradients (Soret effect) in the diffusion law. Indeed, in certain circumstances one or other of these assumptions has even been avoided in non-numerical treatments.

4. Basic Problems and Methods

Williams (5) has listed the following areas that have benefited from mathematical analysis in recent years.

1. Homogeneous, isothermal chemical kinetics
2. Homogeneous explosions
3. Ignition
4. Deflagration structure
5. Detonation structure
6. Premixed-flame extinction
7. Diffusion-flame structure and extinction
8. Flame spread
9. Premixed-flame instabilities
10. Instabilities in combustion devices
11. Turbulent diffusion-flame structure
12. Premixed turbulent flame propagation

Many of those who have contributed to these areas gave papers at the Seminar; notable exceptions are Clavin, Joulin, and Sivashinsky. The list is by no means complete; in particular, computational combustion (a special concern of the Seminar) is not listed, no doubt because it is still emerging.

Progress has been achieved largely by numerical and asymptotic methods. The former will be discussed elsewhere in these proceedings; here we shall pursue the latter in the context of low-speed phenomena. (A comprehensive introduction to the limits that play a central role in combustion theory is given by Buckmaster [1].) The limit of vanishingly small Mach number has already been used in deriving the governing equations (21-23), and that of vanishingly small heat release was mentioned in connection with uncoupling these equations. Nothing more need be said about the former, but we shall return to the latter later. (The more realistic limit of infinitely large heat release has received no attention, but should.)

Williams [5] goes on to discuss specific problems in laminar and turbulent combusiotn that could well be solved by numerical and asymptotic methods. We will make general comments about such future prospects in Secs. 13 and 14.

The governing equations (21-23) show that the combustion introduces N+1 parameters, namely $\mathcal{L}_i, \mathcal{D}$, and θ. The limits that have played the most important roles in combustion theory are

$$\mathcal{L}_i \to 1 \, , \quad \mathcal{D} \to \infty \, , \quad \theta \to \infty \, ; \qquad (37)$$

the last of these is preeminent, leading to so-called activation-energy asymptotics, a method that has dominated the subject. The reason for this is the need for an effective tool for dealing with the highly nonlinear reaction term Ω. Activation-energy asymptotics, used in an ad hoc fashion by the Russian school (notably Frank-Kamenetskii and Zeldovich) in the '40s, exploited in the framework of modern singular perturbation theory (but in a very narrow context) by aerothermodynamicists in the '60s, and systematically developed by Western combustion scientists in the '70s, is just such a tool. (For more details, see Buckmaster [1].)

The limit $\theta \to \infty$ is, by itself, of little interest: the definition (25b) shows that Ω vanishes. To preserve the reaction, it is necessary for $\mathcal{D}$ to become unboundedly large; i.e., we must consider a distinguished limit characterized essentially by

$$\mathcal{D} \backsim e^{\theta/T_*} \tag{38}$$

where T_* is a constant that may have to be found. The consequences of this limit then depend on the relative magnitudes of T and T_*.

For $T < T_*$, the reaction term Ω vanishes to all algebraic orders; this is known as the frozen limit. For $T > T_*$, (25b) implies $\Pi_j Y_j^{\beta_j} \to 0$ exponentially rapidly, so that at least one Y_i vanishes in that way, and again Ω vanishes to all orders; the so-called equilibrium limit holds. For T no more than $O(\theta^{-1})$ away from T_*, reaction takes place, usually in a thin layer called a flame sheet. Thus, with a few exceptions, the general feature of high activation energy is the absence of chemical reaction from most of the combustion field, the description of which is thereby simplified. Reaction occurs only in thin layers (spatial or temporal), whose description is also relatively simple.

5. Plane Deflagration Waves

We are now ready to demonstrate the efficacy of the technique that is the main concern of these lectures, by examining the fundamental problem of premixed combustion--the plane unbounded flame.

The plane unbounded flame of premixed combustion, the so-called (plane) deflagration wave, propagates at a well-defined speed through the fresh mixture and, accordingly, can be brought to rest by means of a counterflow. It is natural to take the mass flux of this counterflow as the representative mass flux M_r, which is not known a priori, but is to be determined during the analysis of the combustion field. Indeed, its determination is the main goal of the analysis. A choice must also be made for the reference density ρ_r; we shall take it to be that of the fresh mixture.

The continuity equation (21b) integrates to give

$$\rho v = 1 , \tag{39}$$

so that

$$\frac{dT}{dx} - \frac{d^2T}{dx^2} = \Omega . \tag{40}$$

Since there is only one reactant we shall drop the subscript 1. For $\mathcal{L} = 1$ in the corresponding equation (23b), the Shvab-Zeldovich formulation applies, showing that

$$T + Y \equiv H = T_f + Y_f \equiv T_b , \tag{41}$$

where the subscript f denotes the fresh mixture at $x = -\infty$. (Actually, H is the total enthalpy of the mixture.) Thus,

$$\Omega = \mathcal{D}(T_b - T)e^{-\theta/T} \tag{42}$$

if the most common choice $\beta_1 = 1$ is made (first-order reaction). Equations (40) and (42) form a single equation for T, which must satisfy the boundary condition

$$T \to T_f \quad \text{as} \quad x \to -\infty . \tag{43}$$

The requirement that all the reactant be burnt provides the final boundary condition

$$T \to T_b \quad \text{as} \quad x \to +\infty . \tag{44}$$

Note that neither the equation of state (21a) nor the momentum equation (22) has been used; the former provides ρ once T has been found, and the latter then determines p from $v = 1/\rho$.

It is immediately apparent, since Ω does not vanish for $T = T_f$, that the problem for T cannot have a solution. The mixture at any finite location will have had an infinite time to react and so will be completely burnt. This cold-boundary difficulty, as it is known, is the result of idealizations and can be resolved in a number of ways: the mixture can originate at a finite point; an appropriate initial-value problem can be defined (see Sec. 6), without the solution having a steady limit of the kind originally sought; or a switch-on temperature can be introduced below which Ω vanishes identically. It is one of the virtues of activation-energy asymptotics that it makes such resolutions unnecessary. Reaction at all temperatures below T_* (including T_f) is exponentially small, so that it takes an exponentially large time for it to have a significant effect; in other words, T_* is a switch-on temperature.

We now seek a solution that is valid as $\theta \to \infty$. Our construction will be guided by the assumption that, in the limit, reaction is confined to a thin sheet located at $x = 0$. On either side of this flame sheet, the governing equation (40) simplifies to

$$\frac{dT}{dx} - \frac{d^2T}{dx^2} = 0 , \tag{45}$$

which only has a constant as an acceptable solution behind the flame sheet $(x > 0)$, exponential growth being excluded. The boundary condition (44) then shows that

$$T = T_b \quad \text{for} \quad x > 0 ; \tag{46}$$

T_b is called the adiabatic flame temperature. It follows that the temperature at the flame sheet is T_b, so that this is also the value of T_* needed to specify the distinguished limit (38).

Ahead of the flame sheet, equation (45) has the solution

$$T = T_f + Y_f e^x + C(\theta)e^x \quad \text{for} \quad x < 0 , \tag{47}$$

satisfying the boundary condition (43) and making T continuous at $x = 0$ to leading order, provided C vanishes in the limit $\theta \to \infty$. (No structure would be found for the flame sheet if the temperature were not continuous.) A small displacement of the origin of x can absorb C which can, therefore, be set equal to zero.

Turning now to the structure, which must determine the still-unknown M_r (i.e, $\mathcal{D}$), we note that the form of Ω restricts the variations in T to being $O(\theta^{-1})$. Since the temperature gradient must be $O(1)$ to effect the transition between the profiles (46) and (47), the appropriate layer variable is

$$\xi = \theta x ; \tag{48}$$

coefficients in the layer expansion

$$T = T_b - \theta^{-1}T_b^2\phi + \ldots \text{ with } \quad \phi = (1/T)_1 \tag{49}$$

are now considered to be functions of ξ.

Equation (40) shows that

$$\frac{d^2\phi}{d\xi^2} = \tilde{\mathcal{D}}\,\phi e^{-\phi} \quad \text{with} \quad \tilde{\mathcal{D}} = \frac{\mathcal{D}e^{-\theta/T_b}}{\theta^2} , \tag{50}$$

while matching with the solutions (46), (47) gives the boundary conditions

$$\phi = -\frac{Y_f\xi}{T_b^2} + o(1) \quad \text{as} \quad \xi \to -\infty \, , \quad \phi = o(1) \quad \text{as} \quad \xi \to +\infty \, . \qquad (51)$$

In order for this derivation to be valid, $\tilde{\mathcal{D}}$ must be $O(1)$. Then $\mathcal{D}e^{-\theta/T}$ is $O(\theta^2)$ in $x > 0$, so that equation (40) is unbalanced unless $Y = 0$ (to all orders) behind the flame sheet, consistent with the result (46). Thus, equilibrium prevails in $x > 0$ even though the mixture is no hotter than the flame sheet there.

Integrating (50a) once, using the condition (51b), gives

$$\left(\frac{d\phi}{d\xi}\right)^2 = 2\tilde{\mathcal{D}}\left(1 - (\phi + 1)e^{-\phi}\right) . \qquad (52)$$

The remaining boundary condition will then be satisfied only if

$$\tilde{\mathcal{D}} = \frac{Y_f^2}{2T_b^4} \, , \qquad (53)$$

which corresponds to the determination

$$M_r = \frac{\sqrt{2D}\, T_b^2 e^{-\theta/2T_b}}{Y_f\theta} \qquad (54)$$

of the burning rate. (If D is temperature dependent it must be evaluated at the temperature T_b.)

Determination of the wave speed M_r/ρ_r is the main goal of the analysis, and rightly so. But, at the same time, the structure of the combustion field is obtained (Fig. 4). The reaction zone appears as a discontinuity in the first derivatives of T and Y, a reflection of the delta function nature of Ω in the limit $\theta \to \infty$. Ahead, the temperature rises and the reactant concentration falls as the reaction zone is approached through the so-called preheat zone. It is the preheat zone that delimits the combustion field and, therefore, defines the thickness of the flame. According to the formula (47), more than 99% of the increase in temperature from T_f to T_b is achieved in a distance 5, i.e., $5\lambda/c_pM_r$ in dimensional terms. With this definition of flame thickness, we find that hydrocarbon-air flames are about 0.5 mm thick. The thickness of the reaction zone, which is scaled by θ^{-1}, is typically 10 or 20 times smaller.

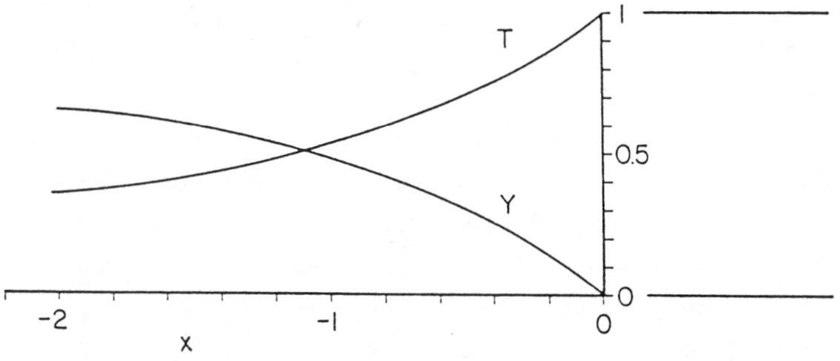

Figure 4. Profiles of T and Y, drawn for $\mathcal{L}=1$ and $T_f=0.25$, $Y_f=0.75$.

If the reaction-zone structure itself is required, the differential equation (52) must be integrated to obtain ϕ as a function of ξ. The constant of integration is fixed by the boundary condition (51a).

The analysis yields a definite value of $\mathcal{D}$, for which the term laminar-flame eigenvalue is often used. A similar analysis for $\mathcal{L} \neq 1$ generalizes the result (54) to

$$M_r = \frac{\sqrt{2\mathcal{L}D}\, T_b^2 e^{-\theta/2T_b}}{Y_f\theta} \quad . \tag{55}$$

The only change is the replacement of $\sqrt{2D}$ by $\sqrt{2\mathcal{L}D}$.

The rate at which the mixture burns is extremely sensitive to the flame temperature. If T_b changes to $T_b - \theta^{-1}T_b^2\phi_*$, i.e., by an $O(\theta^{-1})$ amount, the burning rate changes to

$$M = M_r e^{-\phi_*/2} , \qquad (56)$$

i.e., by an $O(1)$ amount. This is equivalent to a leading-order jump condition on the normal derivative $\partial T/\partial n$, namely

$$\delta\left(\left(\frac{\partial T}{\partial n}\right)^2\right) = Y_f^2 e^{-\phi_*} \qquad (57)$$

when M_r is taken to have the value (54), and in this form has universal validity (i.e., is independent of the way in which ϕ_* is generated) provided the temperature gradient vanishes to leading order behind the flame sheet. The reason is that the perturbation only intrudes through the matching of ϕ at $\xi = +\infty$ (which leads to the exponential factor). This aspect of the structure problem is obscured by the analysis of the steady plane wave given in the last section, where M_r was taken to be the constant (unknown) burning rate. If M_r had been given the value (54) without explanation, and M used to denote the (dimensionless) burning rate, then x would have been replaced by Mx in the formula (47) and the jump condition would have yielded

$$M = 1 . \qquad (58)$$

The condition (57) gives the gradient

$$\frac{\partial T}{\partial n} = Y_f e^{-\phi_*/2} \qquad (59)$$

ahead, a result that will be needed later.

6. The Cold-Boundary Difficulty and Homogeneous Explosion

The source of the cold-boundary difficulty for θ finite is that reaction takes place whatever the temperature, so that even the state of the fresh mixture at $x = -\infty$ is changing in time; seeking a strictly steady solution is futile. To determine this change we shall assume that the fresh mixture is prevented from expanding as its temperature rises due to heat release. According to the perfect gas law (8), p_c must rise with the temperature if the density is to remain constant, so that the energy equation (7) no longer loses the term $\partial p/\partial t$. The term is still spatially constant, however, and hence may be evaluated in the fresh mixture to yield

$$\rho \frac{\partial T}{\partial t} + \rho u \frac{\partial T}{\partial x} - \frac{\partial^2 T}{\partial x^2} = \Omega + \frac{\gamma-1}{\gamma}\frac{dT_f}{dt} \quad \text{with} \quad \Omega = \mathcal{D} Y e^{-\theta/T} \qquad (60)$$

as the modification of equation (40) in the present circumstances; here γ is the ratio of specific heats. As before, the density of the fresh mixture has been taken for ρ_r, so that $\rho_f = 1$. For M_r we shall take the asymptotic burning rate (54), with θ finite and initial values T_f^o, Y_f^o for T_f, Y_f.

The equations governing the heating of the fresh mixture are therefore

$$\frac{1}{\gamma}\frac{dT_f}{dt} = \Omega_f = -\frac{dY_f}{dt}, \qquad (61)$$

to which must be added the initial conditions

$$T_f = T_f^o, \quad Y_f = Y_f^o \quad \text{at} \quad t = 0 . \qquad (62)$$

Equation (23b) has been invoked; and the factor $1/\gamma$ in equation (61a) is due to the change from a constant-pressure to a constant-volume process. The problem reduces to one for T_f alone, namely

$$dT_f/dt = \mathcal{D}(\bar{T}_b - T_b)e^{-\theta/T_f} \ , \ T_f = T_f^o \ \text{ at } \ t = 0 \ , \tag{63}$$

since Y_f may be determined from the integral

$$T_f + \gamma Y_f = T_f^o + \gamma Y_f^o \equiv \bar{T}_b \ . \tag{64}$$

The fresh mixture undergoes the simplest of thermal explosions, a phenomenon that has attracted mathematical analysts ever since it was pointed out by Sememov more than fifty years ago. The solution of the problem (63) can be written in terms of the exponential-integral function, thereby revealing three phases of the process when θ is large (but finite). The induction phase, during which the temperature stays close to T_f^o, is terminated by a sudden increase known as thermal runaway. During the explosion phase that follows, T_f increases rapidly until it approaches $\bar{T}_b$, when depletion of the reactant (i.e., the smallness of Y_f) brings on the relaxation phase: T_f slowly rises to its final value $\bar{T}_b$.

These phases are sharply delineated in the limit $\theta \to \infty$. If we anticipate, on the basis of the analysis in the last section, that

$$\mathcal{D} = 0(\theta^2 e^{\theta/T_b}) \ , \tag{65}$$

the appropriate time variable for the induction phase (to which we limit our attention) is

$$\tau = \varepsilon t \quad \text{with} \quad \varepsilon = \theta^2 e^{\theta/T_b^o - \theta/T_f^o} << 1 \ . \tag{66}$$

The fresh mixture reacts exponentially slowly: at times $o(\varepsilon^{-1})$ its temperature has not changed appreciably from T_f^o.

The behavior of the mixture as a whole is governed by equation (60), which may be written

$$\rho \frac{\partial T}{\partial t} + \rho u \frac{\partial T}{\partial x} - \frac{\partial^2 T}{\partial x^2} = \Omega + (\gamma-1)\ \Omega_f \ , \tag{67}$$

the corresponding equation for the mass fraction (written for unit Lewis number), namely

$$\rho \frac{\partial Y}{\partial t} + \rho u \frac{\partial Y}{\partial x} - \frac{\partial^2 Y}{\partial x^2} = - \Omega , \tag{68}$$

the continuity equation

$$\frac{\partial \rho}{\partial t} + \frac{\partial}{\partial x} (\rho u) = 0 , \tag{69}$$

and

$$\rho T = T_f . \tag{70}$$

The boundary conditions are

$$T, Y, \rho, u \to T_f, Y_f, 1, 0 \quad \text{as} \quad x \to -\infty, \tag{71}$$

$$\partial T/\partial x, Y, \partial\rho/\partial x \to 0, 0, 0 \quad \text{as} \quad x \to +\infty, \tag{72}$$

where T_f and Y_f satisfy equations (61, 62); those on the left no longer violate the differential equations, i.e., the cold-boundary difficulty has disappeared.

Elimination of $T = T_f/\rho$ between equations (67) and (69) yields an equation containing spatial derivatives only. This means that T, u, and Y cannot be arbitrarily prescribed initially, a difficulty that results from having taken the limit of vanishingly small Mach number in Sec. 2. If initial values are taken otherwise, a "Mach" layer is formed; its structure has been considered by Buckmaster [1].

The initial-value problem for the system (67-72) deserves to be studied, with a view to proving what will now be conjectured. Given that a unique solution exists, the aim is to identify the rate at which it propagates into the fresh mixture; but that immediately raises a difficulty. For finite θ the burning rate is defined in a natural way only when the combustion field is steady in some frame of reference; if the structure changes with time in all frames, the location of the flame (and hence its speed) is not a precise concept. It is, therefore, customary to introduce an arbitrary definition, such as the speed with which the inflection point of temperature moves. Since no change in reference frame will remove the unsteadiness in the fresh mixture, some such arbitrary definition has to be made.

The aim is to show that this burning rate, however defined, tends to the value (54) as $\theta \to \infty$; and that might be done as follows. For large θ , the solution (after a short initial development) should be quasi-steady in some reference frame, i.e., the t-derivatives in equations (67-69) should be $O(\varepsilon)$. It is then a matter of showing that the burning rate defined by the speed of this reference frame is close not only to the choice above but also to the asymptotic value $\rho u = 1$. (The latter is suggested by setting $\partial/\partial t = 0$ in equations (67-69), when the steady asymptotic problem of Sec. 5 is obtained.)

To prove (or disprove) this conjecture will not be easy, which just shows that the "cold-boundary difficulty" is not yet dead.

7. General Deflagrations Under the Constant-Density Approximation

In Sec. 5 we examined the plane, steady, premixed flame and deduced an explicit asymptotic formula for its speed. By a judicious choice of parameters this formula can be made to agree roughly with experiment; precision is not a reasonable goal, given the crude nature of our model. Noteworthy is the extreme sensitivity of the speed to variations in the flame temperature: an $O(1)$ change generates an exponentially large change in flame speed. Such variations in speed (caused, for example, by changes in mixture strength) are not excessive numerically (at least for fuels burnt in air), because activation energies and changes in temperature are moderate; but in an asymptotic analysis they present a potential obstacle to discussion of multidimensional and/or unsteady flames. Then significant variations in the flame temperature, spatial and/or temporal, can be expected and, if the sensitivity mentioned above is any guide, there will be correspondingly large variations in the flame speed. A mathematical framework in which to accommodate these is not obvious.

As a consequence, attempts to discuss general deflagrations have, for the most part, been limited to situations where there is an a priori guarantee that variations in the flame temperature are $O(\theta^{-1})$; then flame-speed changes are $O(1)$ and present no mathematical difficulties. Two approaches are known to provide such a guarantee, and this section is largely devoted to their disclosure.

Although the formulation can be carried through for the full equations (21-23), all the essential features are preserved under the assumption that density variations due to the presence of the flame are negligible. If no temperature differences are imposed on the flow, the velocity field is then that of a constant-density fluid and can be calculated in advance; we shall suppose the fluid is at rest. In other words, we shall set

$$\rho = 1 , \quad \mathbf{v} = 0 \tag{73}$$

in the full equations to obtain

$$\frac{\partial T}{\partial t} - \nabla^2 T = \Omega , \quad \frac{\partial Y}{\partial t} - \mathcal{L}^{-1}\nabla^2 Y = -\Omega \tag{74}$$

as those governing the combustion field under the constant-density approximation. (All equations (23b) except the first, corresponding to the single reactant, can be omitted; the subscript 1 can then be dropped.) The pair of equations (74) are known as the diffusional-thermal model.

If the representative mass flux M_r is chosen to be the burning rate (55) of the plane, steady deflagration, then the reaction term becomes

$$\Omega = \mathscr{D} Y e^{-\theta/T} \quad \text{with} \quad \mathscr{D} = \frac{Y_f^2 \theta^2 e^{\theta/T_b}}{2 \mathscr{L} T_b^4} . \tag{75}$$

Note that $\mathscr{L}$ is not necessarily equal to 1 in these equations: the Lewis number plays a very important role in the analysis, especially for unsteady flames.

We shall require that

$$T \to T_f , \quad Y \to Y_f \quad \text{as} \quad x \to -\infty \tag{76}$$

and deal exclusively with situations where equilibrium prevails behind the flame sheet, i.e.

$$Y = 0 \quad \text{in the burnt gas.} \tag{77}$$

The temperature behind the flame will be close to the so-called adiabatic flame temperature (41).

The constant-density approximation, on which most of the premixed flame analysis will be based, clearly provides substantial simplifications. It can be justified as a formal limit in which the heat released by the reaction becomes vanishingly small (compared to the existing thermal energy of the mixture). Small heat release can be due to either a scarcity of reactant $(Y_f \to 0)$ or weak combustion $(T_f \to \infty)$; when confining ourselves to dilute mixtures, we auotmatically assume the former. The relevant parameter is the expansion ratio

$$\sigma = \frac{\rho_f}{\rho_b} = \frac{T_b}{T_f} , \tag{78}$$

where T_b has the definition (41c); asymptotic expansions in $\sigma - 1$ provide a formal basis for the approximation.

Consider now situations in which the flame sheet, in addition to being unsteady, moves in a nonplanar fashion. The goal is to find conditions under which the variations in flame temperature, both temporal and spatial, are $0(\theta^{-1})$ at most; to that end we shall integrate the sum of the basic equations (74).

The x-axis is taken instantaneously along the normal to the flame sheet at the point of interest (pointing into the burnt gas), and a new variable

$$n = x - F(0,0,t) \tag{79}$$

is introduced; here $x = F(y,z,t)$ gives the position of the sheet. Equations (74) then become

$$\frac{\partial T}{\partial t} + V\frac{\partial T}{\partial n} - \frac{\partial^2 T}{\partial n^2} - \nabla_\perp^2 T = -\frac{\partial Y}{\partial t} - V\frac{\partial Y}{\partial n} + \mathcal{L}^{-1}\frac{\partial^2 Y}{\partial n^2} + \mathcal{L}^{-1}\nabla_\perp^2 Y = \Omega \tag{80}$$

where

$$V = -\dot{F}(0,0,t) \tag{81}$$

is the speed of the sheet back along its normal at the instant considered and the subscript $\perp$ denotes the component perpendicular to $\mathbf{n}$.

Equation (80a) is now integrated with respect to n from $-\infty$ to $0+$, thereby yielding

$$\int_{-\infty}^{0+}\frac{\partial}{\partial t}(T+Y)\,dn + \big(V(T+Y)\big)_{-\infty}^{0} = \left(\frac{\partial T}{\partial n} + \mathcal{L}^{-1}\frac{\partial Y}{\partial n}\right)_{-\infty}^{0+} + \int_{-\infty}^{0+}\nabla_\perp^2(T + \mathcal{L}^{-1}Y)\,dn. \tag{82}$$

Certain terms can be evaluated immediately; thus

$$\big(V(T+Y)\big)_{-\infty}^{0+} = V(T_* - T_b)\ , \quad \left(\frac{\partial T}{\partial n}\right)_{-\infty}^{0+} = \left.\frac{\partial T}{\partial n}\right|_{0+}\ , \quad \left(\frac{\partial Y}{\partial n}\right)_{-\infty}^{0+} = 0\ ,$$

so that we may write

$$V(T_*-T_b) = \left.\frac{\partial T}{\partial n}\right|_{0+} + \int_{-\infty}^{0+}\left(\nabla_\perp^2(T+\mathcal{L}^{-1}Y) - \frac{\partial H}{\partial t}\right)dn \text{ with } H \equiv T+Y. \tag{83}$$

This expresses the deviation of the flame temperature T_* from its adiabatic value T_b in terms of the heat lost to the burnt mixture, the transverse diffusion of heat and reactant up to the flame sheet and the temporal variations in enthalpy H of the mixture ahead of the flame sheet.

If deviations of T_* from T_b are to be $O(\theta^{-1})$, the right side of (83) must be of the same order. This is guaranteed when the terms in $\partial/\partial n$, $\nabla_\perp^2$, and $\partial/\partial t$ are made separately small, a step that can be taken in two different ways. One way is to confine attention to disturbances of a steady, plane deflagration that vary over times and distances $O(\theta)$. These are known as slowly varying flames (SVFs). The second way is suggested by the ineffectiveness of the SVF analysis for $\mathcal{L}$ close to 1. For the distinguished limit

$$\mathcal{L}^{-1} = 1 - \ell/\theta \quad \text{with} \quad \ell = O(1), \tag{84}$$

equation (80a) becomes

$$\frac{\partial H}{\partial t} + V\frac{\partial H}{\partial n} - \left(\frac{\partial^2}{\partial n^2} + \nabla_\perp^2\right)H = \theta^{-1}\ell\left(\frac{\partial^2}{\partial n^2} + \nabla_\perp^2\right)Y, \tag{85}$$

of which

$$H = H_f + O(\theta^{-1}) \quad \text{(everywhere)} \tag{86}$$

is one solution. For the corresponding class of solutions, called near-equidiffusion flames (NEFs),

$$\left.\frac{\partial T}{\partial n}\right|_{0+} = \left.\frac{\partial H}{\partial n}\right|_{0+}, \quad \nabla_\perp^2(T+\mathcal{L}^{-1}Y) = \nabla_\perp^2 H + O(\theta^{-1}), \quad \frac{\partial H}{\partial t} \tag{87}$$

are all $O(\theta^{-1})$, so that the right side of equation (83) is also of that order.

It should be emphasized that SVFs and NEFs are restricted classes of solutions, identified by sufficient (but not necessary) conditions fo the flame-temperature variations to be $O(\theta^{-1})$, itself a sufficient condition for the efficacy of the asymptotics. While these classes may be the only general ones, special circumstances make it possible to treat other premixed flames. The most important of these is the spherical (premixed) flame: symmetry ensures that the temperature does not vary at all over its flame sheet, so that it need not be either an SVF or an NEF. (Nevertheless, for certain parameter values it is an SVF and for others an NEF.)

The SVFs have proved to be of considerably less interest than the NEFs, so that we shall restrict attention to the latter.

8. The Basic Equations for NEFs

The NEF is characterized by the requirements (84) and (86), the second of which corresponds to using the expansions

$$T = T_0 + \theta^{-1}T_1 + \ldots , \quad Y = (H_f - T_0) + \theta^{-1}(H_1 - T_1) + \ldots \quad (88)$$

When these are inserted, the basic equations (74) become

$$\frac{\partial T}{\partial t} = \nabla^2 T \quad \text{on either side of the flame sheet,} \quad (89)$$

$$\frac{\partial h}{\partial t} = \nabla^2 h + \ell \nabla^2 T \quad \text{everywhere ;} \quad (90)$$

here T stands for T_0 , and H_1 has been replaced by h . It is important to remember that the use of these equations is restricted to boundary and initial conditions that are consistent with the assumption of H being constant to leading order. This emphasizes once more that NEFs are a restricted class of solutions.

Ahead of the flame sheet the full equations (89, 90) hold; but in the burnt gas the assumption of equilibrium leads to

$$T = T_b \ , \quad \frac{\partial h}{\partial t} = \nabla^2 h \ , \tag{91}$$

the temperature perturbation accounting for the whole of h. The solution on the two sides must be linked by jump conditions, to be derived next.

These conditions are deduced by analysis of the reaction-zone structure, a question that was addressed in Sec. 5. First, the very existence of a structure requires

$$\delta(T) = 0 \quad \text{with} \quad \delta(\cdot) = (\cdot)_{0+} - (\cdot)_{0-} \ ; \tag{92}$$

then, when $\partial T/\partial n = 0$ for $n = 0+$ (as here), the structure gives the result (59):

$$\left.\frac{\partial T}{\partial n}\right|_{0-} = Y_f e^{-\phi_*/2} \ , \tag{93}$$

where $-T_b^2\phi_*$ is the flame-temperature perturbation, i.e., the value of h at the flame sheet. The remaining jump conditions

$$\delta(h) = 0 \ , \quad \delta\left(\frac{\partial h}{\partial n}\right) = \ell \left.\frac{\partial T}{\partial n}\right|_{0-} \tag{94}$$

come from integrating (90) through the reaction zone and matching the result with the combustion fields outside.

The equations (89, 90) governing NEFs have been developed under the assumption (73b), i.e. a quiescent mixture. When the mixture is in motion they must be replaced by

$$\frac{DT}{Dt} = \nabla^2 T \ , \quad \frac{Dh}{Dt} = \nabla^2 h + \ell\nabla^2 T \tag{95}$$

ahead of the flame sheet, and

$$T = T_b , \quad \frac{Dh}{Dt} = \nabla^2 h \tag{96}$$

behind. The system (92-96) defines a free-boundary problem (elliptic if steady) of the fourth order, with the flame sheet as the moving boundary. Solution is a formidable question, tackled in three ways: (i) small perturbations, (ii) numerical integration, (iii) special geometries.

Stability considerations fall under (i). The numerical work under (ii) has dealt only with a parabolic limit of the elliptic problem, resulting in a Stefan problem. An example of (iii) is stagnation-point flow, where the partial differential equations reduce to ordinary ones.

The discussion of general deflagration started (Sec. 1) with those that could be considered discontinuities. To qualify as such, a deflagration must have a length scale L that is large compared to its thickness $(5\lambda/c_p M_r$, see Sec. 5), i.e.,

$$\varepsilon = \lambda c_p / M_r L << 1 \tag{97}$$

must be satisfied. Missing from the early discussion was a determination of the wave speed, witness the use of velocities relative to the discontinuity in the Rankine-Hugoniot conditions (1, 2). The wave speed is determined by the structure of the discontinuity, consideration of which in the context of θ-asymptotics has been the subject of the present section, leading to SVFs and NEFs. The SVF is clearly an acceptable structure for the interior of the discontinuity if

$$\theta = 0(\varepsilon^{-1}) , \tag{98}$$

since then the undulations of the flame sheet follows those of the discontinuity.

No demand of the type (98) is made of NEFs; the activation energy is independently large. If an NEF can be viewed as a discontinuity, it corresponds to a solution with variations on the scale ε^{-1} (other than in the n-direction). To leading order it must, therefore, be a steady, plane deflagration traveling at the adiabatic speed.

Both the SVF and NEF can be generalized to finite heat release. As the structure of a discontinuity, the generalized NEF leads not only to a wave speed but also to jump conditions, which to leading order are the Rankine-Hugoniot relations (1,2). (Details will be given in Sec. 13.) The question is then to determine the flow fields on the two sides of the discontinuity that satisfy the jump conditions for the appropriate wave speed. To emphasize the purely fluid-mechanical nature of the remaining problem the term hydrodynamic discontinuity is used. Apart from the leading-order hydrodynamic discontinuity, the theory has until recently admitted only weak fluid-mechanical effects, a deficiency whose removal should be one of the main thrusts of future research. The introduction of the hydrodynamic discontinuity (beyond the leading order) is an important step in that direction.

NEFs have been most useful when they could not be viewed as discontinuities; witness what we shall have to say about them. Their power is evident in the stability considerations of the next section.

9. Stability of the Plane NEF

Steady, plane deflagration was introduced in Sec. 5; here we shall consider infinitesimal perturbations of it and so examine its stability, restricting attention to the (most important) case of an NEF and adopting the thermal-diffusive model. Without the constant-density approximation the task is not easy, because the perturbation equations (though of course linear) have variable coefficients.

The equations governing NEFs were derived in the preceding section. To obtain the combustion field of a steady, plane deflagration we introduce the coordinate

$$n = x + t \ , \tag{99}$$

based on the flame, and seek a solution of the resulting equations

$$\frac{\partial T}{\partial t} + \frac{\partial T}{\partial n} - \frac{\partial^2 T}{\partial n^2} - \frac{\partial^2 T}{\partial y^2} = \frac{\partial h}{\partial t} + \frac{\partial h}{\partial n} - \frac{\partial^2 h}{\partial n^2} - \frac{\partial^2 h}{\partial y^2} - \ell\left(\frac{\partial^2 T}{\partial n^2} + \frac{\partial^2 T}{\partial y^2}\right) = 0 \tag{100}$$

depending only on n . This yields

$$T_0 = \begin{cases} T_f + Y_f e^n \\ T_b \end{cases} , \quad h_0 = \begin{cases} -\ell Y_f n e^n \\ 0 \end{cases} \quad \text{for} \quad n \lessgtr 0 \tag{101}$$

as the undisturbed temperature and enthalpy (perturbation) profiles, since equilibrium must prevail behind the flame.

If the equation of the disturbed flame sheet is

$$n = F_1(y,t) \ , \tag{102}$$

so that the normal derivative in the jump conditions (92-94) becomes $\partial/\partial n - F_{1y}\ \partial/\partial y$, then these conditions may be written

$$T_1^- = -Y_f F_1 \ , \quad \delta(h_1) = -\ell Y_f F_1 \ , \quad \frac{\partial T_1^-}{\partial n} = -Y_f F_1 + \frac{h_1^+}{\ell_s} \ ,$$

$$\delta\left(\frac{\partial h_1}{\partial n}\right) = \frac{\ell h_1^+}{\ell_s} - 2\ell Y_f F_1 \quad \text{with} \quad \ell_s = \frac{2T_b^2}{Y_f} \ , \tag{103}$$

all quantities being evaluated at the undisturbed flame sheet $n=0$. (We have used the relation $\phi_* = -h_1^+/T_b^2 + \ldots$ and superscripts $\pm$ to denote values at $n = 0\pm$.) The problem is to solve the equations (100) subject to these conditions and the requirement that T_1 , h_1 die out as $n \to \pm\infty$. Arbitrary initial conditions are, as usual, taken into account by considering perturbations for which

$$F_1 = A \exp(iky + \alpha t) \ . \tag{104}$$

The possibility of stability can be seen most easily for $\ell=0$, when the jump conditions

$$\delta(h_1) = \delta(\frac{\partial h_1}{\partial n}) = 0 \tag{105}$$

ensure that

$$h_1 = 0 \quad \text{for all} \quad n \tag{106}$$

is the appropriate solution of the differential equation (100b). The remaining problem is to solve the T_1-equation for $n < 0$ alone, subject to the boundary conditions

$$T_1 = \frac{\partial T_1}{\partial n} = -Y_f A \quad \text{at} \quad n = 0 \ ; \tag{107}$$

the factor $\exp(iky + \alpha t)$ has been omitted. Now, solutions of the perturbation equation are

$$e^{(1\pm\kappa)n/2} \quad \text{with} \quad \kappa = \sqrt{1 + 4\alpha + 4k^2} \quad (-\frac{\pi}{2} < \arg \kappa \leq \frac{\pi}{2}) \tag{108}$$

where, if attention is restricted to unstable modes $\mathrm{Re}(\alpha) > 0$, we have

$$\mathrm{Re}(\kappa) > 1 \ . \tag{109}$$

It follows that the appropriate solution is

$$T_1 = Be^{(1+\kappa)n/2} \quad \text{for} \quad n < 0 \ , \tag{110}$$

and then the boundary condition (107a) requires

$$B = \tfrac{1}{2}(1 + \kappa)B \ , \text{ i.e.,} \quad \kappa = 1 \quad \text{or} \quad \alpha = -k^2 \ . \tag{111}$$

This contradicts our assumption that the mode is unstable, so we must conclude that there are no unstable modes: when $\mathcal{L} = 1$ the flame is stable for all wavenumbers k.

We now consider the jump conditions (103) for $\ell \neq 0$. Solution of the perturbation equations proceeds separately on the two sides of the flame sheet. In front we find

$$T_1 = Be^{(1+\kappa)n/2}, \quad h_1 = \{C + \ell\kappa^{-1}(k^2 - \frac{(1+\kappa)^2}{4})Bn\}e^{(1+\kappa)n/2} \quad \text{for } n<0, \tag{112}$$

where κ has the definition (108b) and again the factor $\exp(iky+\alpha t)$ has been omitted. Behind we have

$$T_1 = 0, \quad h_1 = De^{(1-\kappa)n/2} \quad \text{for} \quad n > 0. \tag{113}$$

The expressions are valid only for $\mathrm{Re}(1+\kappa) > 0$ and $\mathrm{Re}(1-\kappa) < 0$, i.e., when the condition (109) is satisfied, as it is for unstable modes. The jump conditions (103) now yield a homogeneous system for the coefficients A, B, C, D; a nontrivial solution exists only if

$$2\kappa^2(1-\kappa) + \bar{\ell}\left((1-\kappa)^2 - 4k^2\right) = 0 \quad \text{with} \quad \bar{\ell} = \ell/\ell_s, \tag{114}$$

a result due to Sivashinsky.

This dispersion relation should be viewed as determining, for each $\bar{\ell}$, the growth parameter α of any unstable disturbance mode of wave number k. The stability boundaries in the $\bar{\ell}$, k-plane therefore correspond to $\mathrm{Re}(\alpha) = 0$; they are shown in Fig. 5. At any wave number there is a finite band of Lewis numbers, always including $\mathcal{L} = 1$, for which the flame is stable and outside of which it is unstable. The left boundary, on which $\mathrm{Im}(\alpha)$ vanishes also, is a creation of nonplanar disturbances: the dispersion relation (114) does not yield an α with positive real part for $\bar{\ell} < -1$ and $k = 0$. In fact, the unstable mode becomes neutral as this part of the $\bar{\ell}$-axis is approached. On the other hand, the instability predicted by the right boundary, where $\mathrm{Im}(\alpha) \neq 0$, does survive a planar treatment: the dispersion relation (114) with $k = 0$ yields an α with positive real part for $\bar{\ell} > 2(1 + \sqrt{3})$.

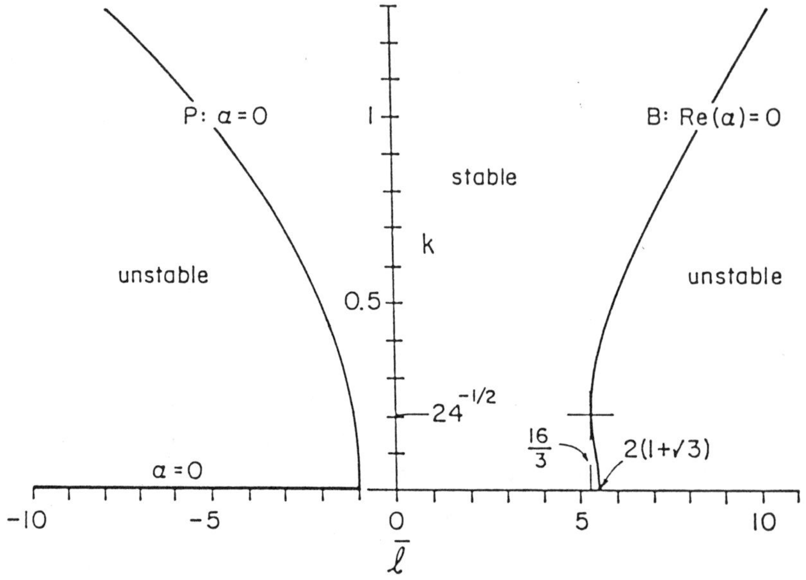

Fig. 5. Linear stability regions for plane NEFs. The boundaries are P: $4k^2=-(\bar{\ell}+1)$ and B: $2(1+8k^2)(1+\sqrt{3(1+8k^2)})/(1+12k^2)$.

There are convincing reasons for believing that the left boundary is associated with cellular flames, a common laboratory phenomenon: it corresponds to $\mathcal{L} < 1$, the Lewis numbers for which such flames are seen; the crests of disturbed flames are cooler than the troughs, this being a characteristic of cellular flames; and, for $\bar{\ell} < -1$, all modes with $k < \frac{1}{2}\sqrt{-(1+\bar{\ell})}$ are unstable, suggesting that the outcome of the instability will not be monochromatic, another characteristic of these flames. In Sec. 10 we shall give a nonlinear theory that arises naturally from the present linear analysis and reinforces this conviction.

The right stability boundary is associated with pulsations or traveling waves; it is relatively inaccessible because $\mathcal{L}$ is rarely much bigger than 1. However similar phenomena may be expected in so-called thermites, for which $\mathcal{L} = \infty$. Otherwise, special means must be devised to make the boundary more accessible. These matters form the subject of Sec. 11.

10. Cellular Flames

We shall now examine the left stability boundary that was uncovered in the last section by our discussion of NEFs (Fig. 5). The boundary is associated with instabilities leading to cellular flames, i.e., flames whose surfaces are broken up into distinct luminous regions (cells) separated by dark lines. Each line is a ridge of high curvature, convex towards the burnt gas. For a nominally flat flame these cells are very unsteady, growing and subdividing in a chaotic fashion; but curvature, for example, can make them stationary.

We shall discuss chaotic and stationary cellular flames in the framework of the weakly nonlinear theory pioneered by Sivashinsky. The constant-density approximation will be used throughout, although perturbations of it will be admitted in two places. (For rigorous results see, Nicolaenko [3].)

The nonlinearity associated with the left stability boundary will be weakest in the neighborhood of

$$\bar{\ell} = -1 \ , \quad k = 0 \ , \tag{115}$$

a possible bifurcation point; accordingly we focus our attention there by taking

$$\bar{\ell} + 1 = O(\varepsilon) \ , \quad k = O(\sqrt{\varepsilon}) \ , \quad \alpha = O(\varepsilon^2) \ , \tag{116}$$

where ε is a small positive parameter that will be found to represent the amplitude of the disturbance. The relative ordering of $\bar{\ell}+1$ and k is suggested by the parabolic shape of the stability boundary, while the order of α follows from the limiting form of the dispersion relation (114) as $\varepsilon \to 0$. This determines the growth rate of the most important Fourier components (the unstable ones) of the disturbance when $\bar{\ell}+1$ is small. In terms of any scalar F that represents the disturbance field, the limiting dispersion relation $\alpha = -(1+\bar{\ell})k^2 - 4k^4$ is equivalent to

$$F_t + 4F_{yyyy} - (1+\bar{\ell})F_{yy} = 0 . \tag{117}$$

For $\bar{\ell}+1 < 0$, this equation predicts unbounded growth. Bifurcation (with weakly nonlinear description) is possible if nonlinear effects, not yet taken into account, limit this growth. We shall first give a heuristic argument to determine these effects and then indicate how to substantiate the result by formal analysis. The argument consists in recognizing that (117) is actually a formula for the wave speed, and modifying it appropriately. In this connection, suppose that F determines the location of the flame sheet as

$$x = -t + \varepsilon F ; \tag{118}$$

then the speed of the sheet is

$$W = 1 + \varepsilon W_1 + \ldots \quad \text{with} \quad W_1 = -F_t , \tag{119}$$

and equation (117) becomes

$$W_1 = 4F_{yyyy} - (1+\bar{\ell})F_{yy} . \tag{120}$$

This formula determines the changes in the flame speed from its adiabatic value of 1 due to the thermal-diffusive effects that are triggered by the distortion of the flame front.

Now, the expression (119b) is mere kinematics, valid in a linear theory only. The exact relation between flame speed and displacement is

$$W = \frac{1 - \varepsilon F_t}{\sqrt{1 + \varepsilon^2 F_y^2}} = 1 - \varepsilon F_t - \tfrac{1}{2}\varepsilon^2 F_y^2 + \ldots \qquad (121)$$

and, for disturbances with wave numbers of the magnitude (116b), the nonlinear term $\frac{1}{2}\varepsilon^2 F_y^2$ is comparable to the linear term εF_t. This suggests that the nonlinear generalization

$$W_1 = - F_t - \tfrac{1}{2}\varepsilon F_y^2 \qquad (122)$$

should be used in the formula (120) and, when ε is purged from the resulting equation by writing

$$\bar{\ell} + 1 = -\varepsilon \; , \quad \eta = \sqrt{\varepsilon} y \; , \quad \tau = \varepsilon^2 t \qquad (123)$$

(in accordance with the ordering (116)), we find

$$F_\tau + \tfrac{1}{2}F_\eta^2 + 4F_{\eta\eta\eta\eta} + F_{\eta\eta} = 0 \, . \qquad (124)$$

Note that this equation holds for $\bar{\ell} < -1$, since ε must be positive.

Substantiation of this result requires a systematic asymptotic development in which x is replaced by the coordinate

$$n = x + t - \varepsilon F(\eta, \tau) \qquad (125)$$

in the governing equations (89, 90); thus,

$$\frac{\partial}{\partial x} = \frac{\partial}{\partial n}, \; \frac{\partial}{\partial y} = -\varepsilon^{3/2} F_\eta \frac{\partial}{\partial n} + \varepsilon^{1/2} \frac{\partial}{\partial \eta}, \; \frac{\partial}{\partial t} = (1 - \varepsilon^3 F_\tau) \frac{\partial}{\partial n} + \varepsilon^2 \frac{\partial}{\partial \tau} \qquad (126)$$

when y, t are replaced by η, τ. The normal derivative, required for the jump conditions (92-94) is

$$(1 + \tfrac{1}{2}\varepsilon^3 F_\eta^2)\frac{\partial}{\partial n} - \varepsilon^3 F_\eta \frac{\partial}{\partial \eta} \tag{127}$$

to sufficient accuracy. Perturbation expansions in ε are now introduced for T, h and F, leading to a sequence of linear problems for the T- and h-coefficients as functions of n, η and τ. These are to be solved under the requirements: $T_1 = T_2 = T_3 = \ldots = 0$ for $n > 0$; conditions as $n \to -\infty$ are undisturbed; and exponential growth as $n \to +\infty$ is disallowed. The problems are overdetermined, but only at the fourth (for T_3, h_3) is a solvability condition required, namely (124) for the leading term in F.

For two-dimensional disturbances of the flame sheet, the basic equation is

$$F_\tau + \tfrac{1}{2}(\nabla F)^2 + 4\nabla^4 F + \nabla^2 F = 0 . \tag{128}$$

Discussion for both one- and two-dimensional disturbances was originally limited to numerical computations. The solutions obtained display chaotic variations in a cellular structure, resembling the behavior of actual flames. Figure 6 clearly shows the ridges that separate the individual cells.

Equation (124) is a balance of small terms; it may be modified to account for any additional physical process whose effect is also small. Hydrodynamic effects can be incorporated, for example, if the density change across the flame is appropriately small (because of small heat release), and this provides important insight into the role of Darrieus-Landau instability (Sec. 13) in actual flames. Equation (124) is replaced by

$$F_\tau + \tfrac{1}{2}F_\eta^2 + 4F_{\eta\eta\eta\eta} + F_{\eta\eta} + \gamma \fint_{-\infty}^{\infty} \frac{F_\eta(\bar{\eta},\tau)}{\bar{\eta} - \eta}\, d\bar{\eta} = 0 \text{ with } \gamma = \frac{\sigma - 1}{2\pi\varepsilon^{3/2}}, \tag{129}$$

the requirement on the heat release being $\gamma = O(1)$: the expansion ratio (78) can only differ from 1 by $O(\varepsilon^{3/2})$. Numerical integration shows that the new (integral) term is destabilizing; an even finer structure is superimposed on the chaotic cellular pattern obtained without it.

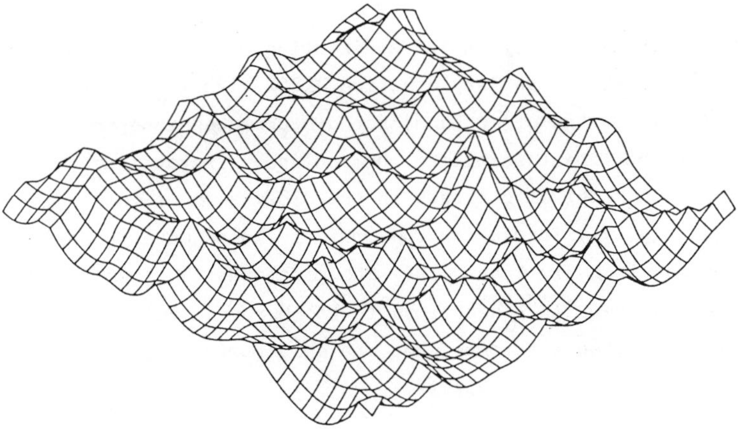

Fig. 6. Numerical calculation of cellular flame. (Courtesy G.I. Sivashinsky).

Weak curvature effects can likewise be incorporated. For a line source of mixture supporting a stationary cylindrical flame of radius $R = O(\varepsilon^{-2})$, the modified equation is

$$F_\tau + \tfrac{1}{2}F_\eta^2 + 4F_{\eta\eta\eta\eta} + F_{\eta\eta} + \gamma F = 0 \quad \text{with} \quad \gamma = \frac{1}{\varepsilon^2 R} . \qquad (130)$$

Comparison with (124) shows that the only new term is γF , and this is found to be stabilizing.

An unusual modification occurs for a flame located in an appropriately weak stagnation-point flow. If the strain rate β is $O(\varepsilon^2)$, the evolution equation becomes

$$F_\tau + \tfrac{1}{2}F_\eta^2 + 4F_{\eta\eta\eta\eta} + F_{\eta\eta} + \gamma(\eta F)_\eta = 0 \quad \text{with } \gamma = B/\varepsilon^2, \tag{131}$$

or

$$F_\tau + \tfrac{1}{2}(\nabla F)^2 + 4\nabla^4 F + \nabla^2 F + \gamma(\eta F)_\eta = 0 \tag{132}$$

when disturbances vary in the z-direction also. Disturbances independent of η satisfy equation (130) with η replaced by $\zeta = \sqrt{\varepsilon}\, z$. On the other hand, for disturbances independent of ζ there may be linear stability but nonlinear instability.

Equation (130) also arises in the discussion of polyhedral flames (Fig. 7), into which the conical flames on a Bunsen burner can suddenly transform. The parameter γ then represents the $0(\varepsilon^2)$ heat loss to the burner. Polyhedral flames are often stationary but can spin rapidly, phenomena that are explained by the analysis.

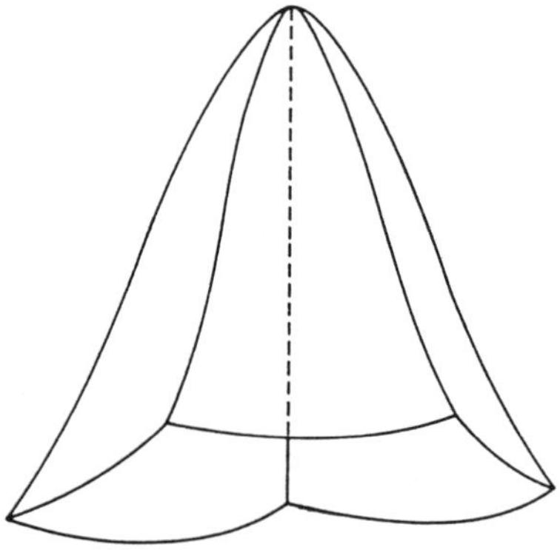

Fig. 7. Five-sided polyhedral flame.

So far we have been concerned with the nonlinear evolution of the linear instabilities associate with values of ℓ slightly less than -1 . Various additional effects were incorporated into the basic nonlinear theory. Our final remarks are concerned with the values of $\bar{\ell}$ slightly greater than -1, where the linear stability of the flame can be destroyed by hydrodynamic effects.

The weakly nonlinear description is now

$$F_\tau+\tfrac{1}{2}F_\eta^2+4F_{\eta\eta\eta\eta}-F_{\eta\eta}+\gamma\,⨍_{-\infty}^{\infty}\frac{F_\eta(\bar{\eta},\tau)}{\bar{\eta}-\eta}\,d\bar{\eta}=0 \quad \text{with } \gamma=\frac{\sigma-.1}{2\pi\varepsilon^{3/2}}\,. \tag{133}$$

Comparison with equation (129) reveals that the sign of $F_{\eta\eta}$ has been changed, because now the definition

$$\varepsilon = 1 + \bar{\ell} \tag{134}$$

is needed to obtain a positive parameter; η and τ still have the definitions (123b,c). Without the integral term, $F \to 0$ as $\tau \to \infty$ whatever the initial conditions are, corresponding to linear stability; but, as before, the hydrodynamic effects (represented by the integral) are destabilizing. Computations show that a progressive wave, consisting of stationary cells, eventually forms provided the flame is not too large. For large flames, the chaotic cellular structure found earlier reasserts itself.

The shape of the progressive wave satisfies a much simpler equation in the limit $\gamma \to 0$, i.e., for significantly larger departures of $\bar{\ell}$ from -1 than of σ from 1. Evolution is then on the scales $\gamma\eta$, $\gamma^2\tau$ rather than η, τ, so that the fourth derivative drops out. If a progressive wave is sought by setting $F_\tau = -V$ and if $F_{\eta\eta}$ is neglected (a valid step where the curvature is not large), the result is

$$\tfrac{1}{2}F_\eta^2+\gamma\,⨍_{-\infty}^{\infty}\left(\frac{1}{\bar{\eta}-\eta}-\frac{1}{\bar{\eta}}\right)F_\eta(\bar{\eta})\,d\bar{\eta}=0 \text{ with } V=\gamma\,⨍^{\infty}\frac{F_{0\eta}(\bar{\eta})}{\bar{\eta}}\,d\bar{\eta}\,, \tag{135}$$

a nonlinear integral equation for the slope F_η. This limiting problem has received a certain amount of attention but deserves more.

11. PULSATING FLAMES AND THE DELTA-FUNCTION MODEL

In Sec. 9 it was found that plane NEFs of sufficiently large Lewis number are unstable. Since $\mathrm{Im}(\alpha) \neq 0$ on the stability boundary, the instability is likely to result in either a pulsating flame or a flame that supports traveling waves. Such flames are the subject of this section.

One difficulty that immediately confronts us is that in contrast to the ubiquitous nature of cellular instabilities, pulsating instabilities are not ordinarily seen. The reason seems to be the large values of $\mathcal{L}$ needed; according to the theory, $Y_f \ell/T_b^2$ must exceed $32/3$ (or $4(1+\sqrt{3})$ if the disturbances are one-dimensional). There is evidence that fuel-rich hydrogen/bromine mixtures attain such values since oscillations have been obtained in a numerical study, but there is no similar evidence for more commonplae gas mixtures.

For this reason we must turn from the commonplace and deal either with unusual combustible materials or else with special configurations, in order to uncver pulsating flames. Our discussion will start with thermites, which are solids that burn to form solids (a phenomenon that is appropriately called gasless combustion). There is no significant diffusion of mass, so that $\mathcal{L}$ is effectively infinite and the equations become

$$\frac{\partial T}{\partial t} - \frac{\partial^2 T}{\partial x^2} = - \frac{\partial Y}{\partial t} = \Omega \ , \qquad (136)$$

since the (constant) density may be given the (dimensionless) value 1 ; here

$$\Omega = \mathcal{D} Y e^{-\theta/T} \quad \text{with} \quad \mathcal{D} = D M_r^{-2} \ . \qquad (137)$$

Numerical solutions uncover a critical value θ_c of the activation energy: for $\theta < \theta_c$ the propagation is steady, but for $\theta > \theta_c$ only pulsating propagation is seen. The prediction of pulsating propagation is consistent with the NEF analysis in Sec. 9, where oscillatory instability was found for sufficiently large $\mathcal{L}$ in the limit $\theta \to \infty$; but activation-energy asymptotics has nothing to say about a phenomenon (here the switch to steady propagation) occurring at some finite value of θ . Even though it is not observed either experimentally or numerically for large enough θ , a steady wave can nevertheless be constructed by means of activation-energy asymptotics; we shall start our discussion by doing so.

The following boundary-value problem presents itself in a frame moving with the flame sheet:

$$\frac{dT}{dx} - \frac{d^2T}{dx^2} = - \frac{dY}{dx} = \mathcal{D} Ye^{-\theta/T} , \tag{138}$$

$$T \to T_F , \quad Y \to Y_f \quad \text{as} \quad x \to - \infty,$$

$$T \text{ bounded}, \quad Y \to 0 \quad \text{as} \quad x \to + \infty. \tag{139}$$

The asymptotic solution follows that for a plane deflagration given in Sec. 5, so that we just quote the replacement

$$M_r = \frac{\sqrt{D}\, T_b e^{-\theta/2T_b}}{\sqrt{Y_f \theta}} \tag{140}$$

for the burning rate (54).

As already noted, there is little point in investigating the stability of this solution using activation-energy asymptotics. Insteady, we introduce a delta-function model suggested by the asymptotics. Thus, the strength of the delta function that replaces the Arrhenius term will be defined so that the mass flux through the flame sheet, in quite general circumstances, is

$$M = \frac{\sqrt{D}\, T_* e^{-\theta/2T_*}}{\sqrt{Y_f \theta}} \quad ; \tag{141}$$

here T_* is the flame temperature. The dimensionless mass flux is then

$$\frac{M}{M_r} = \left(\frac{T_*}{T_b}\right) \exp\left(\frac{\theta(T_* - T_b)}{2T_b T_*}\right) , \tag{142}$$

an expression that will be simplified before use. For θ large (but not necessarily infinite), the preexponential factor is not significant and can be replaced by 1 ; in addition, and consistently, for small deviations of T_* from T_b (such as occur in a linear stability analysis) the exponent can be replaced by $\theta(T_* - T_b)/2T_b^2$. The formula (142) then becomes

$$W = \frac{M}{M_r} = \exp\left(\frac{\theta(T_* - T_b)}{2T_b^2}\right) . \tag{143}$$

since the dimensionless density has been taken to be 1 ; here W is the wave speed.

The Arrenheius term (137) is now replaced locally by

$$\Omega = Y_f W \Delta(n) \quad \text{with} \quad n = x - F(0,0,t) , \tag{144}$$

where Δ is the Dirac delta function and

$$x = F(y,z,t) \tag{145}$$

is the flame sheet, written in a (fixed) coordinate system chosen so that that x-axis coincides with the normal at the point of interest at the instant considered. The equations

$$\frac{\partial T}{\partial t} - \nabla^2 T = - \frac{\partial Y}{\partial t} = \Omega , \tag{146}$$

which generalize the one-dimensional ones (136), then show that the wave speed $-F_t(0,0,t)$ is just W and that

$$\delta(T) = 0 \ , \quad \delta\left(\frac{\partial T}{\partial n}\right) = - Y_f W \ , \quad \delta(Y) = Y_f \ . \tag{147}$$

Precisely the same formulas can be obtained by applying activation-energy asymptotics to the flame sheet in the unsteady case (i.e., with $M \neq M_r$ and a flame temperature T_* within $O(\theta^{-1})$ of T_b) if it is assumed that there is no significant temperature gradient behind the flame sheet. What sets the delta-function method apart from activation-energy asymptotics is that, in solving the governing equations (146), θ is treated as a finite parameter, and there is no requirement for the outer solution to match the inner in the limit $\theta \to \infty$.

Linear stability of the plane wave can now be investigated in the manner of Sec. 9 for a plane NEF, but we omit the details. A critical value of θ is found below which the wave is stable, in accordance with experimental and numerical results. Above the critical value, both pulsations and traveling wave instabilities are possible.

Pulsations can also occur in gaseous deflagrations, and this is of practical interest when there is a mechanism for shifting the right stability boundary in Fig. 5 to the left, so that Lewis numbers of common mixtures are involved. Such a mechanism is heat loss to a burner. There is a difficulty with the boundary conditions since they do not satisfy the NEF requirement (86), but this can be overcome without modifying these conditions artificially or taking recourse to a delta-function model. Another mechanism is the negative strain at a rear stagnation point.

12. Diffusion Flames

The discussion so far has been focused on premixed flames, simply because there is less to say about diffusion flames.

Here we consider the latter and, to keep the discussion simple, select one particular example, the counterflow diffusion flame.

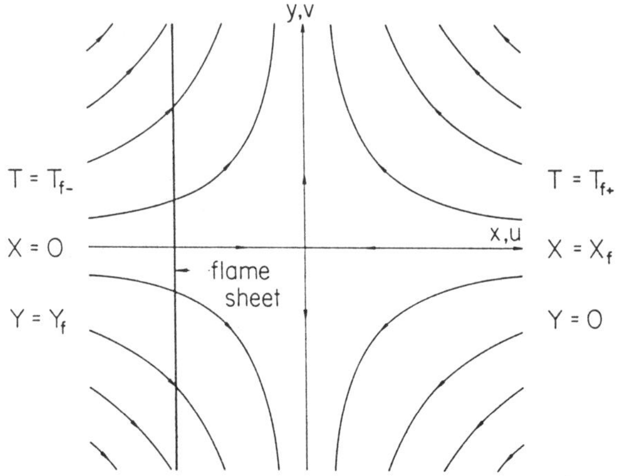

Fig. 8. Notation for the counterflow diffusion flame.

Consider the combustion field sketched in Fig. 8. A stream of gas containing the oxidant $Y_1 = X$ flows to the left and impinges on a stream containing the fuel $Y_2 = Y$ that flows to the right, forming a stagnation point at the origin. The flow field is

$$(u,v) = 2(-x,y) \tag{148}$$

under the constant-density approximation; here a constant of proportionality $\varepsilon/2$, where ε is the straining rate, has been absorbed into the length unit (which is used to define M_r). For such a flow, it is possible for the combustion field to be stratified in the y-direction, with a flat flame sheet at $x = x_*$. The temperature and mass fractions are then functions of x alone satisfying, for unit Lewis numbers,

$$\mathbb{L}(T) = -2\mathbb{L}(X) = -2\mathbb{L}(Y) = -\Omega \quad \text{with} \quad \Omega = \mathcal{D}XYe^{-\theta/T} \; ; \tag{149}$$

here

$$\mathbb{L} \equiv 2x\frac{\partial}{\partial x} + \frac{\partial^2}{\partial x^2} \tag{150}$$

and the boundary conditions are

$$T \to T_{f-} \, , \quad X \to 0 \, , \quad Y \to Y_f \quad \text{as} \quad x \to -\infty \tag{151}$$

$$T \to T_{f+} \, , \quad X \to X_f \, , \quad Y \to 0 \quad \text{as} \quad x \to +\infty \, . \tag{152}$$

The coefficient $\mathcal{D}$ is proportional to $1/\varepsilon$, so that an increase in the straining rate causes a decrease in $\mathcal{D}$. It is the response of the combustion field to variations in $\mathcal{D}$ that is of principal interest.

Since both Lewis numbers have been taken equal to 1, there are two Shvab-Zeldovich variables (Sec. 3), namely

$$T + 2X = S + Z_- \, , \quad T + 2Y = S + Z_+ \, , \tag{153}$$

where

$$S = \tfrac{1}{2}\left(T_{f+}\operatorname{erfc}(-x) + T_{f-}\operatorname{erfc}(x) \right), \quad Z_- = X_f\operatorname{erfc}(-x), \quad Z_+ = Y_f\operatorname{erfc}(x). \tag{154}$$

Both these variables are annihilated by $\mathbb{L}$ and satisfy the boundary conditions (151, 152). We are left with a problem for the temperature alone, when X and Y are suppressed in favor of T.

The usual way of characterizing the solution is to plot variations in some significant parameter, such as the maximum temperature T_m, with the Damkohler number. If $T_{f\pm}$, X_f, Y_f are such that the combustion generates a heat flux to both far fields, this response is found to be S-shaped in the limit $\theta\to\infty$ (Fig. 9). Certain physical conclusions can then be drawn.

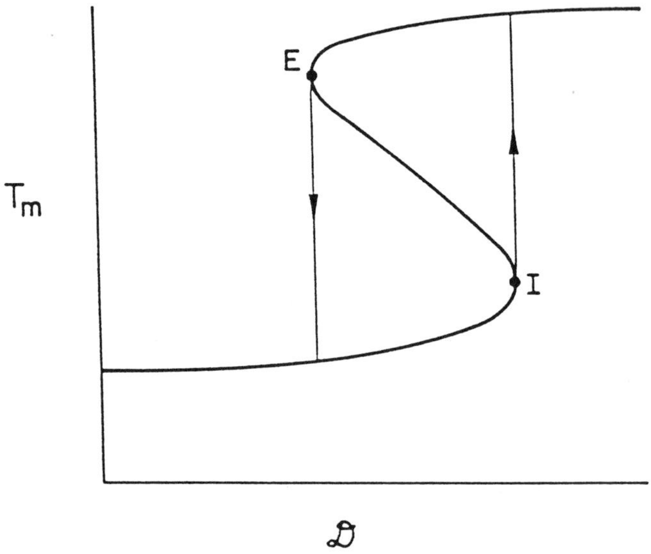

Fig. 9. Ignition and extinction for S-shaped response.

If the system is in a state corresponding to a point on the lower branch, and $\mathcal{D}$ is slowly increased, the solution can be expected to change smoothly until the point I is reached. Rapid transition to the upper branch will then presumably occur, corresponding to ignition. A subsequent slow decrease in $\mathcal{D}$ is likewise anticipated to produce a smooth decrease in burning rate until extinction occurs at E .

If one of the far fields loses heat, the response is monotonic, so that the phenomena of ignition and extinction are absent. For high activation energy the transition from a monotonic to an S-shaped response occurs when the temperature gradient on one side of the flame sheet is small, a case that is not difficult to analyze.

Assume, without loss of generality, that

$$T_{f+} > T_{f-} ; \tag{155}$$

then the small temperature gradient must be on the right of the flame sheet, i.e., on the hotter side. We shall see immediately that T_{f+} is close to the flame temperature, and hence that the temperature gradient is small on the hotter side, if

$$2Y_f = T_{f+} - T_{f-} + k/\theta \quad \text{with} \quad k = \text{const.} \ . \tag{156}$$

In seeking an asymptotic solution as $\theta \to \infty$, we shall assume that equilibrium prevails for $x > x_*$ even though T does not rise above T_* by an $O(1)$ amount there, and check a posteriori that the solution thereby constructed is self-consistent.

In view of the assumption (156), the Shvab-Zeldovich relations (153b) becomes

$$T + 2Y = T_{f+} + \left(\frac{k}{2\theta}\right)\operatorname{erfc}(x) \tag{157}$$

and, hence,

$$T = T_{f+} + \left(\frac{k}{2\theta}\right)\operatorname{erfc}(x) \quad \text{for} \quad x > x_* \ , \tag{158}$$

since $Y = 0$ there. Thus, the temperature gradient is indeed small, namely $O(\theta^{-1})$, on the hotter side. To complete the description of the combustion field outside the reaction zone, we need the temperature in the frozen region ahead of the flame sheet, i.e., the linear combination of 1 and $\operatorname{erf}(x)$ that takes on the values T_{f-} at $x = -\infty$ and T_* at $x = x_*$; clearly

$$T = \frac{T_{f-}\left(\operatorname{erfc}(-x_*)-\operatorname{erfc}(-x)\right) + T_*\operatorname{erfc}(-x)}{\operatorname{erfc}(-x_*)} \quad \text{for} \quad x < x_* \ . \tag{159}$$

This, like the result (158), is correct to any order in θ^{-1}, provided T_* is determined to the same order. We shall only need leading-order accuracy in the result (159), so that taking

$$T_* = T_{f+} \tag{160}$$

is good enough. Determination of x_* (which need not be expanded) comes from analysis of the reaction zone, for which the leading-order result

$$X_* = \tfrac{1}{4}(T_{f-}-T_{f+})\operatorname{erfc}(x_*)+\tfrac{1}{2}X_f\operatorname{erfc}(-x_*) \quad , \tag{161}$$

a consequence of the Shvab-Zeldovich relation (153a), is needed.

The appropriate variable in the reaction zone is

$$\xi = \theta(x - x_*) \quad , \tag{162}$$

so that coefficients in the layer expansion

$$T = T_{f+} - \theta^{-1}T_{f+}^2\phi + \ldots \quad \text{with} \quad \phi = (1/T)_1 \tag{163}$$

are considered to be functions of ξ. The Shvab-Zeldovich relation (153b) gives

$$Y = \tfrac{1}{2}\theta^{-1}T_{f+}^2(\phi-\phi_*) + \ldots \quad \text{with} \quad \phi_* = -\frac{k\operatorname{erfc}(x_*)}{2T_{f+}^2} \quad , \tag{164}$$

so that the structure equation is

$$\frac{d^2\phi}{d\xi^2} = \tilde{\mathcal{D}}(\phi - \phi_*)e^{-\phi} \quad \text{with} \quad \tilde{\mathcal{D}} = \frac{\mathcal{D}X_*e^{-\theta/T_{f+}}}{2\theta^2} \; . \tag{165}$$

Note that $\tilde{\mathcal{D}} = O(1)$ implies $\mathcal{D}e^{-\theta/T} = O(\theta^2)$ for $x > x_*$: there must be the equilibrium $Y = 0$ on the right of the flame sheet since otherwise the reaction term in the Y-equation could not be balanced there. In other words, our solution is self-consistent.

Equation (165) is precisely the structure equation for the premixed flames discussed in Sec. 5. Since the gradient dT/dx vanishes on the right of the flame sheet, the gradient on the left is determined (as was explained at the end of Sec. 5). But the latter is known in terms of x_* from the expression (159), so the result is an equation for x_*, namely

$$\frac{2(T_{f+} - T_{f-})}{\sqrt{\pi}} \cdot \frac{e^{-x_*^2}}{\operatorname{erfc}(-x_*)} = \frac{T_{f+}^2 \sqrt{\mathcal{D} X_*}\, e^{-\phi_*/2}}{\theta e^{\theta/2T_{f+}}} . \tag{166}$$

When the definition (161) is used to eliminate X_*, equations (164b) and (166) give ϕ_* (representing, when it is negative, the maximum temperature in the combustion field) and $\mathcal{D}$ as functions of x_*, i.e., the required relation between the maximum temperature and the Damköhler number.

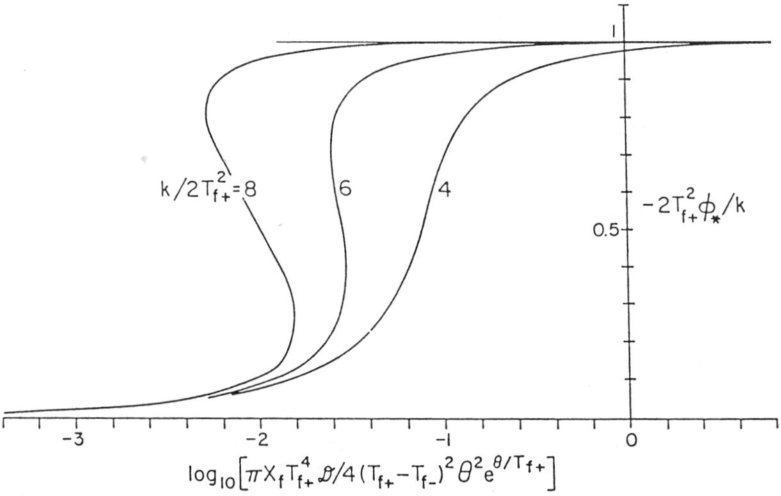

Fig. 10. Steady-state responses when the temperature gradient is positive for $x < 0$ and small for $x > 0$, the latter being represented by k. Drawn for $X_f = \frac{1}{2}(T_{f+} - T_{f-})$.

The corresponding response curve is shown in Fig. 10 for several positive values of the constant k ; when k is negative, the maximum temperature is T_{f+} and not $T_{f+} - \theta^{-1}T_{f+}^2\phi_*$. For k sufficiently small, the response is monotonic; otherwise it is S-shaped. Responses in the shape of an S are quite generally found to be associated with a flux of heat away from the flame sheet on both sides, but this has never been proved.

Consideration of the S-shaped response in general, i.e., when there is an $O(1)$ heat flux away from the flame sheet in both directions, is complicated and will not be given here. The three branches have to be considered separately, and various boundary-value problems arise that were originally solved numerically but have since been treated analytically.

The stability of the three branches of the S has also received attention, and it is now known that exchange of stability takes place exactly at the turning point E (Figure 9). Whether this is also true of the turning point I is not yet known for counterflow diffusion flames.

Other types of diffusion flames, e.g., those in droplet burning and those confined to chambers, lead to the same problem near extinction but to different problems near ignition. As for premixed flames, there are Lewis-number effects, and these have recently been worked out by Choi, Laine-Schmidt and Ludford [8] in the context of chambered diffusion flames.

13. Hydrodynamic Effects

The two assumptions, small heat release and one-step kinetics, have taken the theory a long way in view of their unrealistic nature. (It is interesting to note that the former was introduced by the combustion engineers Burke and Schumann, and that the latter was invariably used by combustion engineers in pre-asymptotic analysis; but criticism started only when applied mathematicians adopted these assumptions.) Only a few of the achievements have been mentioned in the present lectures, but Williams' list (Sec. 4) gives an idea of their extent. Nevertheless, the need to remove these assumptions has long been recognized, and this is arguably the most important task for future research in combustion theory. (In the special case of one-dimensional unsteady flow, finite density variations may be incorporated by means of von Mises' transformation, i.e., by introducing a density-weighted spatial coordinate.) In recent years there have been two promising developments in this respect, and we shall devote the last two sections to sketching the main ideas involved.

We start with Clavin's [9] theory of large premixed flames, as formulated by Matalon and Matkowsky [10], where the heat release is no longer assumed small. Nevertheless, the fluid mechanics is still uncoupled from the thermal-diffusive processes, now by using the small parameter implied by the term "large". If L is a characteristic length of the flame and the streamlines in its neighborhood (e.g., the minimum radius of curvature), the requirement is

$$\varepsilon = \lambda/c_p M_r L << 1 , \tag{167}$$

where $\lambda/c_p M_r$ represents the thickness of the flame.

On either side of the reaction zone, where the length scale is $\lambda/c_p M_r$ (with M_r the burning rate (55)), the appropriate variables are

$$(\bar{x}, \bar{y}, \bar{z}, \bar{t}) = \varepsilon(x, y, z, t) ; \tag{168}$$

so that the governing equations (21b-23) become

$$\frac{\partial\rho}{\partial\bar{t}} + \bar{\nabla} \cdot (\rho\mathbf{v}) = 0, \quad \rho\frac{D\mathbf{v}}{D\bar{t}} = -\bar{\nabla}p + \varepsilon\mathcal{P}(\bar{\nabla}^2\mathbf{v} + \frac{1}{3}\nabla(\nabla\cdot\mathbf{v})), \tag{169}$$

$$\rho\frac{DT}{D\bar{t}} - \varepsilon\bar{\nabla}^2 T = 0 , \quad \rho\frac{DY}{D\bar{t}} - \varepsilon\mathcal{L}^{-1}\bar{\nabla}^2 Y = 0 . \tag{170}$$

(We have not written the equations for components other than the single reactant $i = 1$, and the subscript 1 has been dropped.)

As $\varepsilon \to 0$, we have

$$\frac{DT_o}{D\bar{t}} = \frac{DY_o}{D\bar{t}} = 0 \ , \tag{171}$$

where now $D/D\bar{t} \equiv \partial/\partial\bar{t} + \mathbf{v}_o \cdot \bar{\nabla}$; so that constant values of T_o and Y_o are carried by the fluid particles. We conclude that

$$T_o = T_f \ , \quad Y_o = Y_f \tag{172}$$

everywhere ahead of the discontinuity surface if, as we shall suppose, these constant values are assumed by each particle at its point of origin. Likewise

$$T_o = T_b \ , \quad Y_o = 0 \tag{173}$$

everywhere behind the discontinuity since, as we shall see presently, these values are assumed by each particle as it leaves the flame. Equations (170) then show that

$$T_1 = Y_1 = 0 \tag{174}$$

on both sides of the reaction zone. Charles' law (21a) now shows that ρ has the constant values ρ_f , $\rho_b = \rho_f/\sigma$ on the two sides of the discontinuity, where σ is the expansion ratio (78). We are left with Euler's equations

$$\bar{\nabla} \cdot \mathbf{v}_o = 0 \ , \quad \rho \frac{D\mathbf{v}_o}{D\bar{t}} = -\bar{\nabla}p_o \tag{175}$$

for the leading-order flow field, and with

$$\bar{\nabla} \cdot \mathbf{v}_1 = 0 \ , \quad \rho\left(\frac{D\mathbf{v}_1}{D\bar{t}} + (\mathbf{v}_1 \cdot \bar{\nabla})\mathbf{v}_o\right) = -\bar{\nabla}p_1 + \varepsilon \ \bar{\nabla}^2\mathbf{v}_o \tag{176}$$

for its perturbation.

The two ideal-fluid regions are coupled through jump conditions that, to leading order, are just the Rankine-Hugoniot conditions (1-3), the last being modified to

$$T_f + Y_f = T_b \,, \tag{177}$$

continuity of tangential velocity being added, and the velocity u_f being that of a steady plane wave, i.e., $1/\rho_f$ (as is appropriate for the chosen M_r in the limit $\theta \to \infty$).

These conditions are derived in the same way as for a shock wave in reactionless gasdynamics, i.e., by integrating the basic equations (21b-23) through the reaction zone. Indeed, the conditions (1,2) are identical to those for shocks since they follow from the same continuity and momentum equations. The requirement (177) can also be recognized as a Rankine-Hugoniot condition, but with kinetic energy neglected and a heat-release term (Y_f) added. It follows from the combination

$$\rho \frac{D(T + Y)}{Dt} = \nabla^2(T + \mathcal{L}^{-1}Y) \tag{178}$$

of the basic equations (23).

As in the case of the shock wave, these jump conditions are insufficient. If the state f immediately ahead of the wave is given, there are three equations for the four unknowns $\rho_b(=\rho_f T_f/T_b)$, $v_{nb}(=u_b+V)$, p_b and V (the velocity of the discontinuity). In the case of a shock, another condition is imposed from outside (such as the deflection of the streamlines at a sharp body or the pressure p_b behind the wave in a shock tube). Here there is no external condition; the deficiency arises from using only the combination (178) of the basic equations (23). The reaction then plays no role in the derivation of the jump conditions other than implying that Y vanishes for the burnt gas. Otherwise stated, the combustion inside the wave provides information about the burning rate $\rho_f u_f$, i.e., about the wave speed

$$u_f = v_{nf} - V \; . \tag{179}$$

Evaluation of u_f from a combustion analysis has often been sidestepped. Instead, hypotheses are introduced; the simplest is that it is a constant, given by the burning-rate formula of steady, plane deflagrations. This hypothesis turns out to be correct in the limit $\varepsilon \to 0$, as might be expected.

Obtaining jump conditions for the perturbed hydrodynamic field is one of the recent developments alluded to at the beginning of this section. The equations are integrated through the reaction zone in the limit $\theta \to \infty$ under the assumption (84) of near equidiffusion, to yield

$$\delta(\rho \mathbf{u}\cdot\mathbf{n}) = \varepsilon\kappa \ln \sigma \; , \tag{180}$$

$$\delta(\mathbf{u} \times \mathbf{n}) = -\varepsilon(\mathcal{P} + \alpha)\left[\delta(\boldsymbol{\nabla} \times \mathbf{v}) + 2(\sigma - 1)\boldsymbol{\nabla} \times \mathbf{n}\right] \; , \tag{181}$$

$$\delta(p + \rho(\mathbf{u} \cdot \mathbf{n})^2) = \varepsilon\left[\alpha\delta(\boldsymbol{\nabla} p \cdot \mathbf{n}) - (\sigma-1)(\alpha+1)\boldsymbol{\nabla} \cdot \mathbf{n}\right] \tag{182}$$

correct to order ε , where $\delta(\;)$ denotes the jump of the quantity parenthesized from the fresh to the burnt side and all quantities are evaluated at the discontinuity

$$F(\bar{x}, \bar{y}, \bar{z}, \bar{t}) = 0 \tag{183}$$

separating fresh $(F < 0)$ from burnt $(F > 0)$ mixture. Here

$$\mathbf{u} = \mathbf{v} - V\mathbf{n} \tag{184}$$

is the velocity relative to the (moving) discontinuity and

$$\kappa = -\left(V\,\bar{\nabla}\cdot\mathbf{n} + \mathbf{n}\cdot\bar{\nabla}\mathrm{x}(\mathbf{u}\mathrm{x}\mathbf{n})\right), \quad \alpha = (\sigma \ell n\, \sigma)/(\sigma-1) \;, \tag{185}$$

$$\mathbf{n} = \bar{\nabla}\mathbf{F}/|\bar{\nabla}\mathrm{F}| \;, \quad V = F_t/|\bar{\nabla}\mathrm{F}| \;. \tag{186}$$

When evaluated at the discontinuity, $\mathbf{n}$ and V are the normal (pointing into the burnt mixture) and speed of the discontinuity; κ is known as the stretch of the flame, a concept that is used extensively in discussing the behavior of flames, especially heuristically. Finally, the wave speed is found to be

$$u_f = 1 - \varepsilon\beta\kappa \quad \text{with} \quad \beta = \alpha + \tfrac{1}{2}\ell \int_0^\infty \ell n \left(1+(\sigma-1)e^{-s}\right)ds \;, \tag{187}$$

when the reference density ρ_r is taken to be that of the fresh mixture (so that $\rho_f = 1$).

As $\varepsilon \to 0$, equations (180-182) reduce to the corresponding Rankine-Hugoniot conditions linking the leading-order flow fields on the two sides of the discontinuity. No useful purpose is served here by writing out the jump conditions for the perturbations (including that of the discontinuity).

Darrieus and later Landau considered the stability of a plane flame by solving Euler's equations (175) under the jump conditions (180-182) with $\varepsilon = 0$. For the flame speed they took the value (187a) with $\varepsilon = 0$, thereby assuming that the disturbed flame had the same speed as the undisturbed. They were led to the conclusion that plane flames are unstable, a result that defied experience and plagued the subject for many years. It is now clear that they were considering only disturbances of long wavelength, ones that are screened out by the finiteness of an apparatus in practice. The effect of disturbances of shorter wavelength was considered, at least for small heat release, in Sec. 9.

The system of equations (175, 176) and jump conditions (180-182) with speed determination (187) form a Stefan problem which must in general be solved numerically. In spite of its limitation to large flames, this problem may tell us a great deal about hydrodynamic effects.

14. Complex Kinetics

In practice, no combustion process consists of a single irreversible step, as we have assumed throughout these lectures. Some processes, such as hydrocarbon oxidations, involve dozens of steps, their number and nature often being matters of controversy for chemical kineticists. What is important for us is to determine which steps are significant in describing flame behavior, and the approach depends on whether a quantitative or qualitative description is desired.

At the quantitative level, sensitivity analysis is the right tool, and Rabitz [4] has been prominent in applying it to combustion problems. Some recent developments are described by him elsewhere in this volume. At its simplest, sensitivity analysis consists in determining the effect of variations in an input parameter (e.g., one of the reaction rates) on an output parameter (e.g., the burning rate). In "local" analysis, on which attention has been largely focused, this involves the partial derivatives of the dependent variables with respect to the input parameter and hence differentiation of the governing equations. Simultaneous solution of the differentiated system and the original system can be simplified by noting that the key terms in the former make up the Jacobian matrix of the latter, and this is automatically calculated by some numerical schemes.

In combustion, the analysis can clearly be used to identify those reaction parameters whose values need to be determined more accurately because of the sensitivity of the output parameter to them. It can also be used to find those reactions that can safely be ignored, by determining derivatives with respect to reaction rates at zero values of the latter.

These brief remarks about sensitivity analysis could well be expanded, in particular into "global" questions, but that is best left for others (e.g., Rabitz). Suffice it to say that, as with all numerical methods, when used as a research tool rather than as a means of answering specific questions (typically relating to design for safety), sensitivity analysis can make fundamental contributions to our understanding; that is now the case in combustion.

At the qualitative level, the approach starts at the other end. The one-step irreversible reaction is modified by a second step which models some feature of a complex kinetic scheme previously neglected. When the two-step model is found to be inadequate for the purpose at hand, a third step is added, and so on. Implicit in such an approach is a means of discriminating between the possible steps that could be added, i.e., a way of evaluating a proposed scheme, so as to avoid the mathematical game of exploiting a scheme because of its tractability rather than its physical significance. Such a systematic evaluation was sketched by Nicolaenko [3] in a summary of his work with Fife. We shall explain the main idea by means of a certain two-step reaction.

It has long been known that free radicals can exert considerable influence on combustion processes. A simple two-step model that takes the production and recombination of radicals into account is

$$X + Y \to 2X \ , \quad 2X + M \to X_2 + M \qquad (188)$$

consisting of a chain-branching (production) step and a chain-breaking (recombination) step. Here X is the radical, Y the reactant, and M a third body. The activation energy of the production step is very large, while that of the recombination step is small and taken to be zero. The reaction was suggested by Zeldovich, who considered an isothermal production step; but it may be generalized by allowing this step to be exothermic or even endothermic. (Endothermic recombination is excluded on physical grounds.) The model was later discussed by Liñán using activation-energy asymptotics; his results will be recovered in a systematic fashion.

For the reaction (188) the dimensionless equations of a steady plane flame are

$$\frac{dT}{dx} - \frac{d^2T}{dx^2} = \mathcal{D}_1 q_1 XYe^{-\theta/T} + \mathcal{D}_2 q_2 X^2 , \tag{189}$$

$$\frac{dX}{dx} - \mathcal{K}^{-1}\frac{d^2X}{dx^2} = \mathcal{D}_1 XYe^{-\theta/T} - \mathcal{D}_2 X^2 , \tag{190}$$

$$\frac{dY}{dx} - \mathcal{L}^{-1}\frac{d^2Y}{dx^2} = -\mathcal{D}_1 XYe^{-\theta/T} . \tag{191}$$

Here T is the temperature while X and Y are the mass fractions of radical and reactant (respectively); $\mathcal{K}$ and $\mathcal{L}$ are the Lewis numbers of radical and reactant;

$$\mathcal{D}_1 = D_1/M_r^2 \quad \text{and} \quad \mathcal{D}_2 = D_2/M_r^2 , \tag{192}$$

where D_1 and D_2 are the rate constants of the reaction steps; q_1 and q_2 are the proportions of the total heat released in the first and second steps of the reaction, so that

$$q_1 + q_2 = 1 ; \tag{193}$$

and θ is the activation energy of the first step, that of the second step being taken zero. Given the parameters $\mathcal{K}$, $\mathcal{L}$, D_1, D_2, q_1/q_2, and θ, the problem is to determine the burning rate M for which these differential equations have a solution satisfying the boundary conditions

$$X,Y,T \to 0,\ Y_f,T_f \text{ as } x \to -\infty,\quad X,Y,dT/dx \to 0 \text{ as } x \to +\infty. \tag{194}$$

The solution is sought in the limit $\theta \to \infty$ and, in general, involves numerical integration with jump conditions at the flame sheet (x = 0) derived from the cooresponding structure problem. The model proves to be remarkably rich, the various types of flame depending on the state of the fresh mixture. (Recall that for the one-step model there is just one type of flame.)

We integrate the equations (189-191) with respect to x from $-\infty$ to $+\infty$ and note that derivatives tend to zero at both limits. Then

$$X_b = \alpha_1 - \alpha_2 , \quad Y_f = \alpha_1 , \quad T_b - T_f = q_1\alpha_1 + q_2\alpha_2 , \tag{195}$$

where

$$M_r^2\alpha_1 = \int_{-\infty}^{\infty} XYH_1 dx, \; M_r^2\alpha_2 = \int_{-\infty}^{\infty} X^2H_2 dx \quad \text{with } H_1 = D_1 e^{-\theta/T}, \; H_2 = D_2 . \tag{196}$$

Note that X_f and Y_b have been set zero, in accordance with the boundary conditions (194), but that X_b has not. When the recombination reaction is too slow, X does not decay to zero behind the flame on the scale of the preheat zone, i.e., the x-scale, though of course recombination is eventually completed.

In equations (195b,c) Y_f and hence α_1 is known but X_b and hence α_2 is not. Nevertheless, the non-negativity of X_b and α_2 places restrictions on them, namely

$$0 \le X_b \le Y_f , \quad 0 \le \alpha_2 \le Y_f \; (= \alpha_1) . \tag{197}$$

It follows that the burned temperature is likewise bounded, i.e.,

$$T_\ell \equiv T_f + q_1 Y_f \le T_b \le T_f + Y_f \equiv T_u . \tag{198}$$

The lower bound corresponds to heat being released in the first reaction but not in the second, and the upper bound to heat being released in both reactions. Both bounds depend on the state of the fresh mixture, unlike the temperature T_c defined below, which is a property of the mixture.

The actual temperature of the burnt mixture depends on α_2 and hence the reaction rates XYH_1 and X^2H_2, since

$$\frac{\alpha_2}{\alpha_1} = \frac{\int_{-\infty}^{\infty} X^2H_2dx}{\int_{-\infty}^{\infty} XYH_1dx} . \tag{199}$$

To determine α_2 which, according to the result (197b), is never bigger than α_1, it is convenient to sketch the so-called power functions H_1 and H_2, and that has been done in Fig. 11 for $D_1 > D_2$. (We consider the less interesting case $D_1 < D_2$ later.) The graphs intersect and thereby define a crossover temperature T_c, which is a property of the mixture:

$$T_c = \theta/(\ell n\ D_1 - \ell n\ D_2) . \tag{200}$$

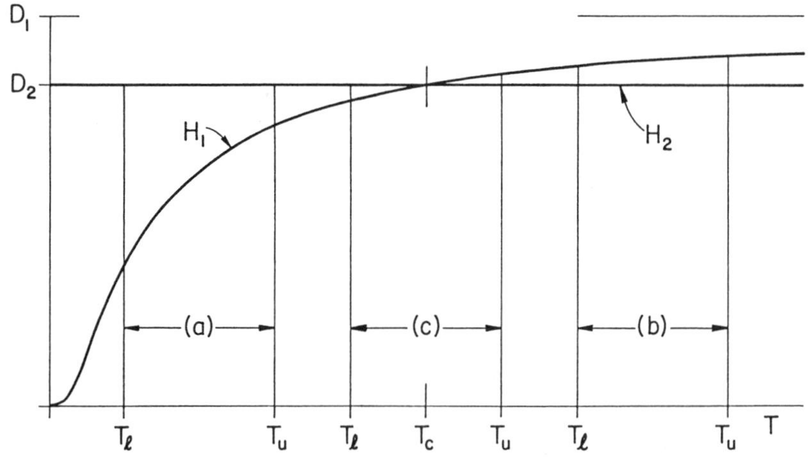

Fig. 11. The power functions H_1 and H_2 for (a) fast, (b) slow, and (c) intermediate recombination (see text).

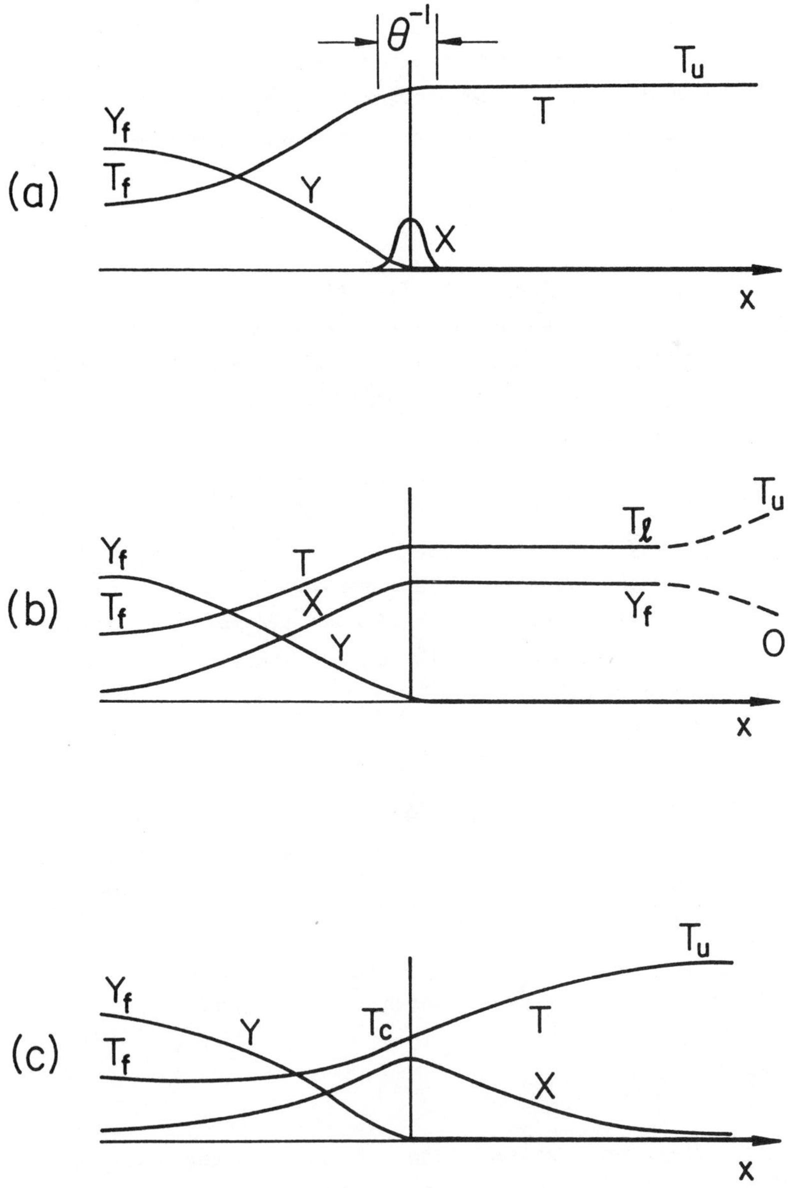

Fig. 12. Profiles for (a) fast, (b) slow, and (c) intermediate recombination.

We now consider the limit $\theta \to \infty$, for which all of the first reaction is concentrated at a flame sheet, and we need only consider H_1 there since the intrgrand of α_1 is zero elsewhere. There are three possibilities for the flame temperature T_* , in the discussion of which we shall use $<<$ to denote "exponentially smaller than".

(i) $T_* < T_c$: Then $H_1 << H_2$ and the only way to prevent α_2 from exceeding α_1 is to take $X << 1$ everywhere. This implies $X_b = 0$ and $\alpha_2 = Y_f$, so that $T_* = T_b = T_u$. Recombination is completed within the flame sheet (see Fig. 12a).

(ii) $T_* > T_c$: Then $H_2 << H_1$ and hence $\alpha_2 << \alpha_1$, i.e., $\alpha_2 = 0$. It follows that $T_* = T_b = T_\ell$, and (correspondingly) X_b has the nonzero value Y_f . Radicals remain at $x = +\infty$, and recombination takes place over exponentially long distances, i.e., downstream of what may be called the flame proper (see Fig. 12b).

(iii) $T_* = T_c$: H_2 is (exponentially) comparable to H_1 , so that the recombination reaction must go to completion on the x-scale, i.e., $X_b = 0$ (see Fig. 12c). Hence $\alpha_2 = Y_f$ and $T_* < T_b = T_u$.

The subsequent discussion depends on the bounds

$$T_\ell \leq T_* \leq T_u \tag{201}$$

for the flame temperature. (These follow from the observation that the mixture immediately behind the flame sheet has benefited at least by the release of heat from the first reaction and at most by that from both reactions.) The aim is to show that the state of the fresh mixture determines the type of flame obtained. There are three main possibilites, separated by two limiting cases, giving five types of flame; which of these occurs depends on the temperature and reactant concentration in the fresh mixture.

The three possibilities may be characterized by the terms (a) fast, (b) slow, and (c) intermediate recombination. They occur in the following circumstances (Fig. 11).

(a) $T_\ell < T_u < T_c$. Case (i) applies and the flame temperature has to be at its upper bound.

(b) $T_c < T_\ell < T_u$. Case (ii) applies and the flame temperature has to be at its lower bound.

(c) $T_\ell < T_c < T_u$. Assuming either case (i) or case (ii) leads to a contradiction; so case (iii) applies and the flame has to be at the crossover temperature.

These are summarized by saying that T_* is driven as close as it can be to T_c . Otherwise stated, as T_ℓ , T_u increase, T_* accompanies T_u as far as T_c where it waits until T_ℓ arrives for it to accompany thereafter. Determination of the flame temperature is the crucial step in solving the problem (189-194). The (asymptotic) analysis is now similar to that in Sec. 5 for the one-step reaction, and leads to the T,X,Y-profiles sketched in Figure 12 and to formulas for the burning rate. We shall not consider these matters here.

There are two limiting cases as follows; in both, the flame temperature is T_c .

(d) $T_\ell < T_u = T_c$. For this fast/intermediate recombination the profiles are similar to those in Fig. 12a, except that the radical concentration is much larger (but still confined to the flame sheet).

(e) $T_c = T_\ell < T_u$. For this intermediate/slow recombination the profiles are similar to those in Fig. 12b, except that recombination is completed by $x = +\infty$.

It is now clear that for $D_1 < D_2$ only (a) can occur. But, for $D_1 > D_2$ (as we have supposed above) there are five possibilities depending (we emphasize again) on the state of the fresh mixture. The four-step model of Peters and Smooke [11] has many more possibilities; to begin with there are 24 orderings of the rate constants D_1, D_2, D_3, D_4, and then there will be several possibilities for each ordering. Complicating the kinetics therefore leads to innumerable types of flame and, while these could keep a mathematician happy for years, a mathematical scientist would want to concentrate on those offering some physical insight.

The promising few have yet to be identified.

REFERENCES

1* J. Buckmaster. The contribution of asymptotics to combustion, to appear in Physica D (1985).

2* I. Müller. Thermodynamics of diffusive and reacting mixtures of fluids, to appear in Physica D (1985).

3* B. Nicolaenko. Some mathematical aspects of flame chaos and flame multiplicity, to appear in Physica D (1985).

4* H. Rabitz. Local and global parametric analysis of reacting flows, to appear in Physica D (1985).

5* F.A. Williams. Lectures on applied mathematics in combustion. (Past contributions and future problems in laminar and turbulent combustion.), to appear in Physica D (1985).

6 J.D. Buckmaster and G.S.S. Ludford. Lectures on Mathematical Combustion (CBMS-NSF Regional Conference Series in Applied Mathematics), SIAM Publications, 1983.

7 J.D. Buckmaster and G.S.S. Ludford. Theory of Laminar Flames, Cambridge University Press, 1982.

8 Y.S. Choi, G.S.S. Ludford, and C. Schmidt-Lainé. Stability of chambered diffusion flames, to appear (1986).

9 P. Pelcé and P. Clavin. Influence of hydrodynamics and diffusion upon the stability limits of laminar premixed flames, J. Fluid Mech. **124** (1982), 219-237.

10 M. Matalon and B.J. Matkowsky. Flames as gasdynamic discontinuities, J. Fluid Mech. **124** (1982), 239-259.

11 N. Peters and M.D. Smooke. Fluiddynamic - chemical interactions at the lean flammability limit, Combust. Flame **60** (1985), 171-182.

*Special-Year lecture series.

Center for Applied Mathematics
Cornell University
Ithaca, NY 14853

Lectures in Applied Mathematics
Volume 24, 1986

The Mathematical Background of Chemical Reactor Analysis II The Stirred Tank

Rutherford Aris
Department of Chemical Engineering and Materials Science
University of Minnesota
Minneapolis, Minnesota 55455

ABSTRACT

The continuous flow stirred tank reactor has long been the system in which the methods of analysis of dynamical systems have found ready application and to which they have brought much insight. Some of the results of stability analysis, singularity theory, bifurcation and periodic forcing are surveyed.

0075-8485/86 $1.00 + $.25 per page

1. Introduction

The form of reactor into which reactants flow continuously, from which unreacted feed and products emerge and in which the contents are mixed has long been the work-horse of both practical and theoretical reactor analysis. The quality of the mixing is a fluid mechanical problem, which has been much discussed experimentally, but not so much theoretically. We will assume, as is commonly done, that the contents of the reactor are uniform so that its state is represented by $S + 1$ functions of time $c^j(t), T(t)$. For the jth species the net flux is $F = q(c_f^j - c^j)$, where q is the volumetric flow rate and c_f^j the feed concentration. The amount of A_j is $H = Vc^j$ and, if R reactions $\alpha_i^j A_j = 0$, $i = 1,...R$, are taking place the rate of generation of A_j is $G = \alpha_i^j r^i$. Then, if $\Theta = V/q$,

$$\Theta \dot{c}^j = c_f^j - c^j + \Theta \alpha_i^j r^i(c1,...c^S,T). \tag{1}$$

If $c^j(0) = c_o^j$ is stoichiometrically compatible with c_f^j, i.e., $c_o^j = c_f^j + \alpha_i^j \xi_o^i$ for some ξ_o^i, then these S equations may be reduced to R equations for the $\xi^i(t)$ defined by

$$\dot{c}^j(t) = c_f^j + \alpha_i^j \xi^i(t) \tag{2}$$

$$\Theta \xi^i = -\xi^i + \Theta r^i(c^1 + \alpha_k^1 \xi^k,....,T),\ \xi^i(0) = \xi_o^i. \tag{3}$$

If the c_o^j are not compatible with the c_f^j a further variable must be added and eqn. (44) replaced by

$$c^j(t) = c_f^j + \alpha_i^j \xi^i(t) + (c_o^j - c_f^j)\varsigma(t) \tag{4}$$

Then equations (45) still hold but with r^i functions of ς as well as of ξ^i. ς itself satisfies

$$\Theta\dot{\varsigma} = -\varsigma, \ \varsigma(0) = 1 \qquad (5)$$

so that ς = exp $-t/\Theta$ and it is clear that the trajectories subside quickly into the [R] stoichiometric subspace of the feed composition. (For stoichiometric background see [1]).

The energy balance has to be done with care [2] but yields an equation for T,

$$\Theta\dot{T} = T_f - T + \Theta J_i r^i - h(T - T_{cf}) \qquad (6)$$

where T_f = feed temperature, T_{cf} = coolant feed temperature, h = dimensionless time constant for heat transfer = UA/C_p, $J_i = (-\Delta H_i)/C_p$, heat of reaction parameter ΔH_i= actual heat of reaction.

2. A Single, Isothermal Reaction

For one reaction at constant temperature a single equation in the extent suffices

$$\Theta\dot{\xi} = -\xi + \Theta r(\xi) \qquad (7)$$

There is one distinguished parameter Θ and there may be others in the expression for the reaction rate. For example the reaction rate expression

$$r(\xi) = \frac{k(c_o\xi)}{1 + (c_o - \xi)^2} \tag{8}$$

allows us to put

$$x = \xi/c_o,\ \tau = kt,\ \alpha = k\Theta,\ \beta = c_o$$

and write

$$\dot{x} = \frac{-x}{\alpha} + \frac{x\,(1 - x)}{1 + \beta(1 - x)^2} = f(x) \tag{9}$$

in which α, the Damköhler number is the distinguished parameter and β is a kinetic parameter. The solution of such an equation can always be accomplished by quadratures. We notice however, that for $\beta > 27$ there is an interval of α for which there are three critical points at which the right hand side vanishes. The middle one is always unstable and divides the real line $0 \leq x \leq 1$ into the two regions of attraction of the two stable states. When the steady state is unique it is often attractive but need not be so. Such kinetic expressions as (8) usually conceal a number of mechanistic assumptions as described in [1] and care must be taken with these, for at some point these assumptions may have to be questioned. But the most complicated reaction rate expression can only result in the division of the interval $0 \leq \xi \leq 1$ into a finite number of regions of attraction to n stable steady states. The (N-1) internal end points of these intervals are the unstable steady states.

3. Systems With Two Equations

The simplest example of a non-isothermal reaction is the irreversible exothermic reaction A $\rightarrow$ B, for which, if c be the concentration of A, we have the equations

$$\Theta\dot{c} = c_f - c - \Theta A e^{-E/RT} c \tag{10}$$

$$\Theta\dot{T} = T_f - T + J\Theta A e^{-E/RT} c - h(T - T_{cf}) \tag{11}$$

This system has long history stretching from the prehistory of empirical discussion before any equations were formulated to the treatment of its steady states by van Heerden [3] and its recognition as a dynamical system by Bilous and Amundson [4]. Denbigh had earlier treated some isothermal cases [5].

The discovery of multiple steady states was first made experimentally through the phenomena of ignition and extinction and theoretically through the balance of heat generation and removal. Thus in the steady state (10) gives

$$c = c_f/(1 + \Theta A e^{-E/RT}).$$

and substituting in (11) we have

$$T - T_f + h(T - T_{cf}) = Jc_f\Theta A e^{-E/RT}/(1 + \Theta A e^{-E/RT}). \tag{12}$$

The left hand side is proportional to the heat removal rate and is clearly linear, being

$$(1 + h)T - (T_f + hT_{cf}) = (1 + h)(T - T) \tag{13}$$

The right hand side is sigmoidal and can clearly be intersected three times by the straight line if its slope and intersection are appropriate (Fig. 1a). Thus the locus of steady state temperature T_s as a function of $(1 + h)$ and T is a folded surface (Fig. 1b) with a cusp at C, such that for values of h and T lying within the arms of the bifurcation curve ACB there are three steady states whereas the steady state is unique elsewhere.

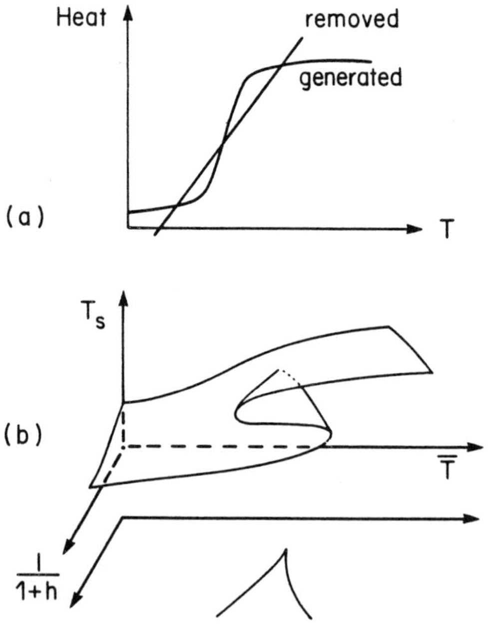

Figure 1a. The basic heat balance at steady state.

Figure 1b. The steady state temperature as a function of T and h.

This example of a first-order, non-isothermal reaction is historically so important and so rich in behavior that it is worth an acronym - FONI is suggested despite the unfortunate homonym. Apart from the fact that no reaction is truly unreversible it is a thoroughly realistic model and is structurally stable to the introduction of reversibility which only adds two further constants and unnecesary complexity. It was this system that Amundson and Bilous took up in the 1950's when they showed how clear a picture of its behavior could be a obtained by the methods of ordinary differential equations. They introduced the phase plane to chemical engineers showing how the whole behavior for a given set of parameters could be presented graphically and simply. The changes that might take place during the attempt to gain control of the intermediate steady state were discussed by Amundson and Aris [6] in a paper which introduced limit cycles and Hopf-bifurcation to the chemical engineering literature. These were often thought to be artefacts of control until Uppal, Ray and Poore's papers in 1974 and 1976. [7,8] They showed, by a study of the Hopf-bifurcations that stable and unstable periodic solutions were quite common and mapped on the parameter space to a large extent. Refinements to their work by Vaganov et al. [9], Calo et al. [10] and Tsotsis [11] have probably covered the field pretty completely - it is necessary to put it somewhat guardedly as in these high dimensional parameter spaces there may yet lurk peculiar forms of behavior. In particular Vaganov's description of the many possible phase portraits was so cryptic that it is only by looking at some of the degenerate Hopf bifurcations in great detail that they can be understood.[12] The multiplicity question will be dealt with in the next section.

An alternative system of two equations which have many of the same properties as the FONI system is an autocatalytic isothermal scheme of reactions developed by Gray and Scott [13, 14, 15]. This is $A + 2B \rightarrow 3B$, $B \rightarrow C$. Writing down equations for the concentrations a and b gives

$$\Theta \dot{a} = a_f - a - \Theta k_1 a b^2 \tag{14}$$

$$\Theta \dot{b} = b_f - b + \Theta k_1 a b^2 - \Theta k_2 b \tag{15}$$

It should be no surprise that these have some of the same properties as eqns. (10) and (11) for the heat generated by the exothermic reaction is the autocatalytic agent in the FONI case whilst the degenerating reaction $B \rightarrow C$ of the Gray-Scott autocatalator ((GSAC) removes the catalytic agent just as the cooling term in eqn. (11) removes the heat. The nonlinear function ab^2 is notably simpler than the Arrhenius temperature dependence making the algebraic labors considerably lighter. However, the transcendental subtlety of the Arrhenius function is needed for the full richness of FONI behavior.

The reaction scheme known as the Brusselator has been extensively exploited by Nicolis, Prigogine and colleagues and has proved to be a very suggestive model. It involves two more reactions than GSAC and is usually written $A \rightarrow X$, $B + X \rightarrow Y + D$, $Y + 2X \rightarrow 3X$, $X \rightarrow E$ in which it is clear that the last two steps are GSAC with Gray's A, B and C playing the roles of Nicolis' Y, X and E. Most who have used this scheme have assumed that B can be prescribed, as might be the case if they were gasses above a sparsely covered catalytic surface and the second reaction were of Eley-Rideal type. Then equations for x and y are

$$\dot{x} = k_1 a - k_2 bx + k_3 x^2 y - k_4 x \tag{16}$$

$$\dot{y} = k_2 bx - k_3 x^2 y \tag{17}$$

If in fact this were the case and the reactants A and B were introduced into a well mixed space above the catalytic surface we would need two further equations of the form

$$\Theta \dot{a} = a_f - a - \alpha\Theta k_1 a \tag{18}$$

$$\Theta \dot{b} = b_f - b - \alpha\Theta k_2 bx \tag{19}$$

and we see that, though the first would justify the prescription of A, there is no way in which the second will allow B to remain constant unless the last term is insignificant. This could only be through the smallness of α which represents the ratio of catalytic surface to the volume of the reactor. More seriously as Wake shows elsewhere in this volume any attempt to inform this system of the principles of chemistry turns it into a very dull dog indeed without any oscillatory behavior at all. This is because the second and third steps of the Brusselator are essentially X $\rightarrow$ Y and Y $\rightarrow$ X and there is not way of making these only very slightly reversible since the principle of "microscopic reversibility" forces the product of the reverse constants to equal the product of the forward constants. The GSAC is more robust and can be made as slightly reversible as we please.

In fact the autocatalytic kinetics of GSAC can be justified by looking at a mechanism A + B = C, B + C $\rightarrow$ 3B for the autocatalytic step A + 2B $\rightarrow$ 3B. Setting

$$x = a/a_f \ , \ y = b/a_f \ , \ \tau = k_2 t \ , \ \alpha = k_2 \Theta \ ,$$
$$\beta = k_1 a_f^2 / k_2 \ , \ \gamma = b_f / a_f \tag{20}$$

gives the dimensionless equations

$$\alpha \dot{x} = 1 - x - \alpha\beta x y^2 = f(x,y) \tag{21}$$
$$\alpha \dot{y} = \gamma - y + \alpha\beta x y^2 - \alpha y = g(x,y) \tag{22}$$

On the other hand, the proposed mechanism would give rise to the equations

$$\begin{aligned} \Theta \dot{a} &= a_f - a - \Theta k_3(ab - c/K_3) \\ \Theta \dot{b} &= b_f - b - \Theta k_3(ab - c/K_3) + 2\Theta k_4 bc - \Theta k_2 b.sp2If \\ \Theta \dot{c} &= - c + \Theta k_3(ab - c/K_3) - \Theta k_4 bc \end{aligned} \tag{23}$$

$$x = a/a_f \ , \ y = b/a_f \ , \ z = c/a_f \ , \ \tau = k_2 t \ , \ \alpha = k_2 \Theta \ ,$$

$$\beta = k_4 K_3 a_f^2 / k_2 \ , \ \gamma = b_f/a_f \ , \ \kappa = K_3 a_f \ , \ \epsilon = k_2 / k_3 a_f \ . \tag{24}$$

The parameter ϵ has obviously been designed to be small and it will turn out that κ must also be small. With this notation the equations (23) become

$$\alpha \dot{x} = 1 - x - (\alpha/\epsilon)(xy - z/\kappa) \tag{25}$$
$$\alpha \dot{y} = \gamma - y - (\alpha/\epsilon)(xy - z/\kappa) + 2\alpha\beta yz/\kappa - \alpha y \tag{26}$$
$$\alpha \dot{z} = -z + (\alpha/\epsilon)(xy - z/\kappa) - \alpha\beta yz/\kappa \tag{27}$$

The last equation may be written

$$\kappa xy = z\left\{1 + (\kappa\epsilon/\alpha)[1 + \alpha\beta y/\kappa + \alpha\dot{z}/z]\right\}$$

whence

$$(\alpha/\epsilon)(xy - z/\kappa) =$$

$$\kappa xy\left\{1 + \frac{\alpha\beta y}{\kappa} + \frac{\alpha\dot{z}}{z}\right\}\left\{1 - \frac{\kappa\epsilon}{\alpha}\left[1 - \frac{\alpha\beta y}{\kappa} + \frac{\alpha\dot{z}}{z}\right]\right\}$$

$$= \alpha\beta xy^2 + \kappa xy + \alpha\kappa(\dot{x}y + \dot{y}x) + \text{h.o.t.}$$

The term $\kappa(\dot{x}y + \dot{y}x)$ comes from the fact that κ is a sufficient approximation to z in this position and the higher order terms involve $\epsilon\kappa$ and κ^2. Substituting this relation in (25) and (26) gives

$$\alpha\dot{x} = f(x,y) - \kappa h(x,y) \tag{28}$$

$$\alpha\dot{y} = \xi(x,y) - \kappa h(x,y) \tag{29}$$

where

$$h(x,y) = \gamma x + y - xy + \alpha\beta xy^2(x - y) \tag{30}$$

Thus in addition to ϵ being small we must insist that κ should also be small, i.e., the equilibrium be unfavorable to the intermediate C. However in contrast to the case of the Brusselator a small perturbation is chemically meaningful and the GSAC equations are structurally stable.

4. Systems With Three Equations

Just as two equations are needed to describe the non-isothermal reactor with only a single reaction so three will suffice for two independent reactions. If the reactor is isothermal, three equations permit us to describe those independent reactions. Care must be taken not to multiply equations without necessity and the argument in Sec. 1 above may be used to show that though the inclusion of eqn. (5) may permit a more accurate computation of numerical values it adds nothing to the understanding of the structure of the process since it does not change the stability in any way.

Though many systems of reactions have been considered it is probably the non-isothermal system of two consecutive reactions (CRNI) which has received the most widespread attention. This is A → B → C, a natural extension of FONI. It is commonly assumed that the reactions are of the first order and so the equations

$$\begin{aligned} \Theta\dot{a} &= a_f - a - \Theta k_1(T)a \\ \Theta\dot{b} &= b_f - b + \Theta k_1(T)a - \Theta k_2(T)b \\ \Theta\dot{T} &= T_f - T + J_1\Theta k_1(T)a + J_2\Theta k_w(T)b - h(T - T_{cf}) \end{aligned} \tag{31}$$

Even if b_f is taken to be zero there are still seven dimensionless parameters to content with. For, let

$$\tau = t/\Theta\ ,\ u = a/a_f\ ,\ v = b/a_f\ ,\ w = \left(\frac{E}{RT^2}\right)(T - T),$$
$$T = \frac{T_f + hT_{cf}}{1 + h} \tag{32}$$

and

$$\alpha = \Theta k_1(T) \ , \ \beta = J_1 a_f E_1/RT^2 \ , \ \gamma = E_1/RT$$

$$\sigma = k_2(T)/k_1(T) \ , \ \rho = J_2/J_1 \ , \ \nu = E_2/E_{12} \tag{33}$$

then

$$\dot{u} = 1 - u[1 + \alpha E(w)] \tag{34}$$

$$\dot{v} = \alpha u E(w) - v[1 + \sigma\alpha E^{\nu}(w)] \tag{35}$$

$$\dot{w} = -(1 + h)w + \alpha\beta u E(w) + \sigma\alpha\rho\beta v E^{\nu}(w) \tag{36}$$

$$E(w) = \exp \gamma w/(w + \gamma) \tag{37}$$

There are seven parameters here and it is clearly impossible to explore the whole space. Most of the work that has been done has used the simplifying assumption of $\nu = 1$ and allowed γ to be very large so that E(w) can be taken to be the exponential function.

The FONI reactor with a term for the heat capacity of the vessel gives a three-dimensional system of immense interest as the work of Planeaux, Jensen and Farr in this volume shows. The justification of the GSAC mechanism given above requires three equations though the intention there is to return to two. An isothermal analogue of CRNI can be made by taking an autocatalyst C in the set $A + 2C \rightarrow B + mC$, $B + 2C \rightarrow D + nC$, $C \rightarrow E$. By taking m and n as 3 exothermic reactions can be emulated while "endothermicity" comes from taking these coefficients to be 1. This will be pursued elsewhere [27].

5. Multiplicity Questions and Singularity Theory

As mentioned above the question of multiplicity of steady states arose in the earliest work or the stirred tank, whilst the extended series of papers by Luss and his co-workers, most notably

Balakotaish, is one of the glories of the recent literature of chemical engineering [16-22]. Its springs are in the pioneering paper of Golubitsky and Keyfitz [23] who used the stirred tank reactor as an example of their elegant method of applying singularity theory with a distinguished parameter. If eqn. (10) is solved for c and this is substituted in (11) then it can be rearranged in the form (using their notation excepting λ here = their ϵ)

$$G(y,\lambda;B,\delta,\eta) \equiv \eta - (1+\lambda)y + B\lambda/[1+\lambda\delta A(y)] \tag{38}$$

where

$$y = (T - T_f)/T_f \ , \ \lambda = 1/h \ , \ B = J_{cf}/T_f \ , \ \gamma = E/RT_f$$
$$\eta = (T_{cf} - T_f)/T_f \ , \ \delta = h/\Theta k(T_f) \ , \ A(y) = k(T_f)/k(T).$$

They showed that G was contact equivalent to the canonical form of what they called winged cusp

$$g(x,\lambda) = x^3 + \lambda^2,$$

a singularity of codimension three whose unfolding is

$$x^3 + \alpha_2 x\lambda + \lambda^2 + \alpha_3 x + \alpha_1 \ .$$

In the neighborhood of the organizing center where $G = G_y = G_{yy} = G_\lambda = G_{\lambda y} = 0$, $G_{yyy} \neq 0$,$G_{\lambda\lambda} \neq 0$ every form of x, λ-diagram can be found. The surfaces H (the hysteresis variety on which inflection points of vertical inflection are to be found) and B (on which isolas form either in isolation or as pinched-off mushrooms)

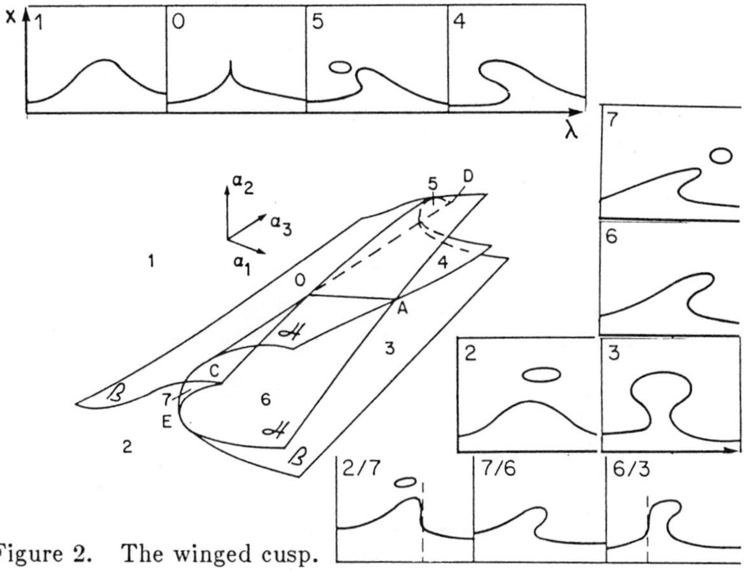

Figure 2. The winged cusp.

are shown in Fig. 2. The point 0 is the organizing center (where $x^3 + \lambda^2 = 0$ is obviously the cusp shown in the second box at the top. Other possible x,λ-diagrams are shown as 1 - 7 and correspond to parameters α_1 , α_2 , α_3 in the regions between the various leaves of the surfaces H and B as shown. Three of the transitional diagrams corresponding to points on these surfaces are shown at the bottom. It was Balakotaiah and Luss [18] who showed that η, B and ς could be used as the unfolding parameters in the orientation shown in Fig. 3. This figure may be further explained by Fig. 4 which represents a ς, B-plane of constant η. The typical λ, y-diagram showing the steady state temperature as a function of flow rate has the form shown in the various regions. A broken line box encloses the typical diagram for a whole region; a solid box interrupting a line shows the transitional forms; while the three points J, L and M give rise to diagrams linked to values with arrows.

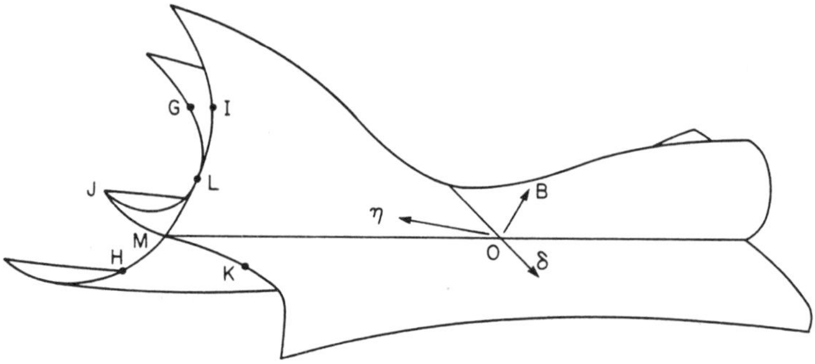

Figure 3. The organizing center of the stirred tank.

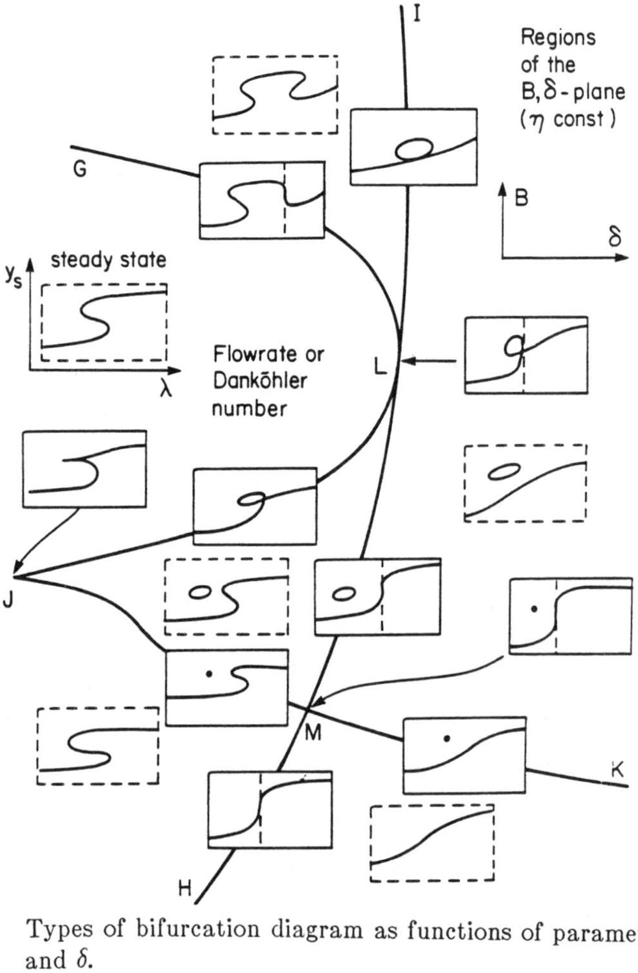

Figure 4. Types of bifurcation diagram as functions of parameters B and δ.

The key to Golubitsky and Keyfitz method is the recognition problem and in their paper and, more fully, in an excellent exposition with Schaeffer [24], Golubitsky gives the criteria for all the low order singularities. It is easier to see the mechanics of this process in the GSAC where the elimination of x from the steady state equations gives a cubic F = 0. Treating α as the distinguished parameter we have:

$$F \equiv \alpha(1+\alpha)\beta y^3 - \alpha\beta(1+\gamma)y^2 + (1+\alpha)y - \gamma$$

$$F_y \equiv 3\alpha(1+\alpha)\beta y^2 - 2\alpha\beta(1+\gamma)y + (1+\alpha)$$

$$F_{yy} \equiv 6\alpha(1+\alpha)\beta y - 2\alpha\beta(1+\gamma)$$

$$F_{yyy} \equiv 6\alpha(1+\alpha)\beta$$

$$F_\alpha = (1+2\alpha)\beta y^3 - \beta(1+\gamma)y^2 + y$$

$$F_{\alpha y} = 3(1+2\alpha)\beta y^2 - 2\beta(1+\gamma)y + 1$$

$$F_{\alpha\alpha} = 2\beta y^3$$

$$F_\beta = \alpha(1+\alpha)y^3 - \alpha(1+\gamma)y^2$$

$$F_\gamma = -\alpha\beta y^2 - 1$$

Were we to eliminate y between F = 0 and $F_y = 0$ we would have a complicated equation between α, β and γ representing the surfaces in α,β,γ-space that separate regions of one steady state from regions of three. Steady state diagrams giving y as a function of, say α could be constructed by passing through this space on a line of constant β and γ.

However Golubitsky's method calls for going to the highest order singularity we can find. Clearly we could have $F = F_y = F_{yy} = 0$ simultaneously and this would be a line in α,β,γ-space. We cannot make $F_{yyy} = 0$ except by putting α or $\beta = 0$ and this is pointless.

But it is possible to find a point where $F = F_y = F_{yy} = F_\alpha = 0$, namely $\alpha = 1$, $\beta = 256/27$, $\gamma = 1/8$, $y = 3/16$, and we find that, though we had no right to expect it, $F_{\alpha y}$ also vanishes here, but $F_{\alpha\alpha}$ does not. These are precisely the conditions for the "winged cusp" singularity.

It is however, a degeneracy for we have no right to expect a codimension three problem to be solved with only two undistinguished parameters. Balakotaiah has found that to remove this degeneracy he has to make both reactions reversible and feed the product c of the now reversible B = C, but he then has much more than a winged cusp on his hands [25]. (I owe this information to Balakotiaiah who was at this conference and will by now, I hope, have published his result in full.)

The application of singularity theory to multiple reactions has been fruitful in the work of Balakotaiah and Luss (see references already given) and in particular applies to the non-isothermal A $\rightarrow$ B $\rightarrow$ C [25]. Here, for reasons similar to those given for the extended GSAC, we might suppose there to be no more than five steady states. By careful search however, Farr has found conditions for which seven may be present [27]. The challenge of the future is to apply singularity theory to distributed systems but this takes us beyond the bounds of the stirred tank.

6. Phase Portraiture

When Amundson and Bilous [4] introduced the phase phane to chemical reaction engineering, they both used simple geometrical arguments to indicate the nature of the flow and also used it as a convenient representation of the solution. In work which followed this up a few years later [6] the evolution of the phase plane under the increase of a proportional control parameter was studied and a Hopf bifurcation discussed for the first time in this context. The developments that took place in the next few years were largely associated with Amundson's work and may be followed in the selection of his papers that was published in 1980. The big advance in the systematic presentation of possible phase portraits was made when Uppal, Ray and Poore used the locus of Hopf bifurcations and degenerate Hopf bifurcation to divide the parameter space up into regions of qualitatively similar phase planes [7,8], and, as has been remarked, this has been refined by further work [9,10,11,12]. With the current refinements of singularity theory and bifurcation analysis progress is beingmade in the systematic description of the possible phase portraits and is well represented in this conference by the contributions of Pismen (see also [29,30]), Guckenheimer and Planeaux, Jensen and Farr.

With only two equations the systematic evolution of the phase plane can be presented more vividly by thinking of them as sections in the state-distinguished parameter space. If necessary a topologically equivalent distortion should be used for clarity. As an example of the sort of thing we have in mind the parameter α, the residence time, might be distinguished in the GSAC model. Let us take equations (21) and (22) and fix $\beta = 40$, $\gamma = 1/15$, so that

$$\begin{aligned} \dot{x} &= (1-x)/\alpha - 40\,xy^2 \\ \dot{y} &= 1/15\alpha - (1+\alpha)y/\alpha + 40xy^2. \end{aligned} \tag{39}$$

The steady states all lie in the ruled surface

$$x + (1 + \alpha)y = 16/15$$

which is indicated as the line of short and long dashes UVWZ in Fig. 5.

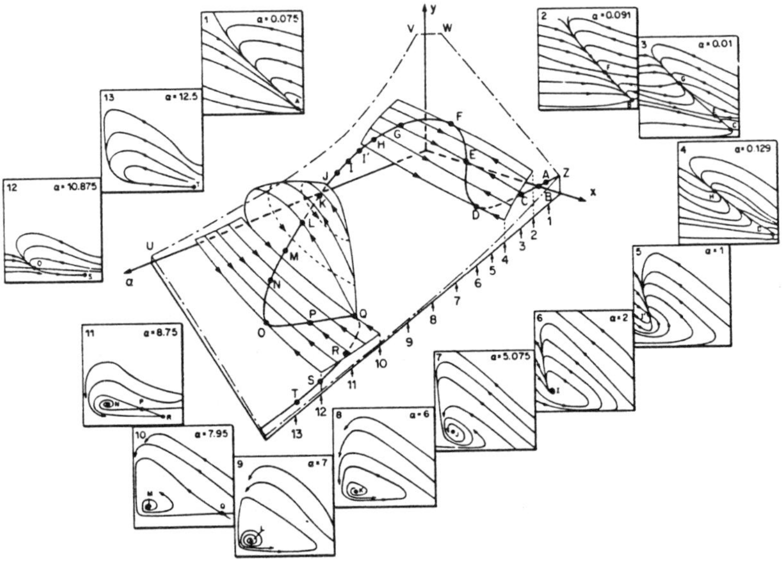

Figure 5. Variation of phase portrait with residence time for a case of GSAC.

The locus of steady states starts from Z (where $\alpha = 0$) and runs throught the alphabet to T as α increases to 12·5. (The parameters are the same as those for Fig. 9c, in [13] p.31. Both figures are distorted in scale as otherwise the first region of multiplicity would be very slight.) The two sets of five curves through five points of the curves FED and OPQ respectively are the stable manifolds of these two branches of saddle-points. These are in the intervals $0.091 \leq \alpha \leq 0.129$ and $7.95 \leq \alpha \leq 10.875$. The steady states B and C lie below and G and H lie above the first of these separatrix surfaces and R, S, and N, M, are similarly disposed below and above the second. To avoid confusion the only other surface of trajectories which we have attempted to show is generated by the family of limit cycles born at $\alpha = 5.075$ and concluding with a homoclinic orbit at $\alpha = 7.95$. Instead the phase planes are numbered with an encircled number marked on the edge parallel to his α-axis and on the small phase planes encircling the diagram clockwise. Each phase plane is the square $0 \leq x, y, \leq 1$.

For $\alpha \leq 0.091$, there is but a single steady state close to the bottom right corner of the square. In fact it travels from (1, 0.66) at $\alpha = 0$ to (0.975, 0.084) at $\alpha = 0.091$ and is a stable node. It therefore attracts all points in the plane and since its slow eigentrajectory has a slope close to -1 the trajectories from most of the plane either go directly to the slow eigentrajectory of the node from the left or swing up to a higher value of y and then tuck themselves in to the eigentrajectory with a quick turn.

At $\alpha = 0.091$, approximately, a node-saddle pair is born at (0.579, 0.447), the point F on the curve of steady states and in phase plane 2. This is a turning point where $F_y = 0$. But $F(y, \alpha)$ is obtained by eliminating x between the two equations for the steady state $f(x, y, \alpha) = g(x, y, \alpha) = 0$. Thus $f_x g_y - f_y g_x = 0$ at this point an so one of the eigenvalues is zero. Thus at the moment of birth of the saddle-node pair there is a trajectory that is one-sidedly stable for though in phase plane 2 trajectories from above go to f they would not stay there as the slightest perturbation would send them down to the stable node at b. This trajectory is the leading edge of the separatrix surface betweén sections 2 and 4. Figure 6 shows the way in which the eigenvalues vary with α and the transgression of the branch DEF into the positive real-half plane corresponds to the first branch of saddles. A similar phenomenon occurs at the further edge where the saddle has moved down FED to meet the lower node on BCD at D, the node having moved up to (0.921, 0.129). In between, say at $\alpha = 0.1\Theta$, = 4, as in phase plane 3 we have the familiar picture of a separatrix dividing the plane into two open regions of attraction to their respective stable steady states.

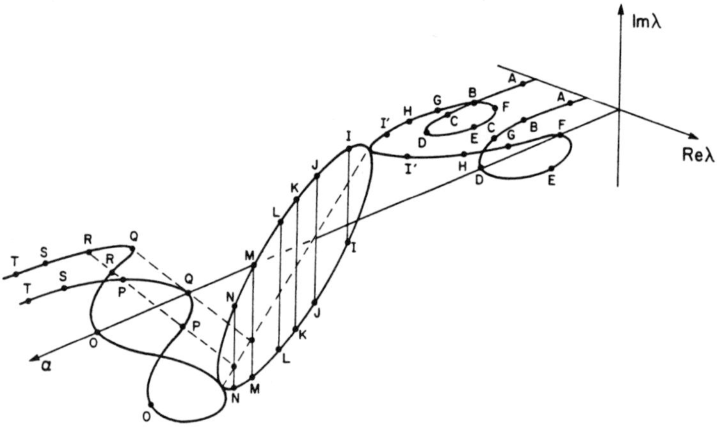

Figure 6. Variation of eigenvalues with residence time.

After the first multiplicity region the now unique steady state remains stable as α increases to 5.075. Sometimes it is a node and sometimes a focus but in either case it is globally attractive. From α = 1.75 onwards it is a stable focus but its eigenvalues have an increasing real part which eventually crosses the real axis at α = r 5.075. In Fig. 6, the segment I,J,K,L,M,N shows pairs of complex conjugate eigenvalues whose real part is shown as a broken curve. This real part crosses into the unstable half-plane at J, $\alpha = 5.075$ and an analysis shows that this is a non-degenerate Hopf bifurcation which sheds a stable limit cycle.

As α increases through the range $5.075 \leq \alpha \leq 7.95$ the limit cycle grows and develops a very spiky character. This is because the lower right hand corner is approaching the point (0.915, 0.017) where another node saddle pair will be born. In fact it will become a homoclinic orbit consisting of the union of one outward branch of the eigentrajectory throught he saddle point with an inward branch. Thus shortly before the birth of the saddle-node the trajectory is spending enormous lengths of time in the neighborhood of the birth point.

Once the saddle and node separate slightly both of the outgoing branches from the saddle go to the stable new born node, one directly and the other by a counter clockwise circuit around the still unstable node as in 11. Of the incoming branches to the saddle, one comes from the unstable node and one passes above the stable node to divide between trajectories that lead immediately to the node and those that swing up and around the unstable state.

We are now back in a region of multiplicity and the saddle moves up until it coaleses with the still unstable nodes at 0 in 12. They disappear and leave a single node in the corner near (1,0). Notice however that the character of the phase plane is rather different from what it was when α was small. The slow dominant diagonal is gone and now the trajectories pass slowly along near the x-axis to the steady state. Almost all of them swing up and round the region where the saddle-node pair "out the black west went".

In two dimensions these kinds of evolutions can be followed in their full unfolding. Even with GSAC it would be a formidable, though quite possible, task to give such a portrait for each region of qualitatively different behavior.

7. Forced Oscillations

There is growing evidence that many two-dimensional non-linear systems with an autonomous periodic solution show a qualitatively similar behavior when forced by an oscillatory input. A general picture of the response, which might be called an excitation diagram, can be given by finding the boundaries of regions of qualitatively similar behavior in the frequency-amplitude plane. This is the quadrant whose axes are the frequency of excitation, ω, as a multiple of the natural frequency, ω_0, and the amplitude of the forcing oscillation, a. The latter will normally be limited by some physical constraint which may not be transgressed either at $\overline{A} + a$ or $\overline{A} - a$, where $\overline{A}$ is the mean value of the perturbed variable. By qualitative behavior we mean that the response of the system is either a harmonic (i.e., periodic with the same period as the forcing), p-subharmonic (..e., periodic with a period p times that of the forcing), quasi-periodic or chaotic. A good indication of this kind of portraiture this involves is provided by the work on Duffing's equations reported by Ueda [31] though he chose to keep the forcing frequency fixed and vary the natural frequency of the system. Tomita's work on the forced Brusselator [32] has the most extensive connection to chemical systems though Bailey and Sincic [33] applied non-linear analysis to FONI. The brief description which follows is more extensively treated in [34-

38].

In the work of Kevrekidis [37] Tomita's forced Brusselator

I:

$$\begin{aligned} \dot{x} &= x^2y - Bx - x + A + a \cos \omega t \\ \dot{y} &= Bx - x^2y , \end{aligned} \tag{40}$$

the dimensionless FONI system in the form

II:

$$\begin{aligned} \dot{x} &= -x + D(1 - x)e^y \\ \dot{y} &= -y + DB(1 - x)e^y - C(y - A - a \cos \omega t), \end{aligned} \tag{41}$$

and a catalytic mechanism due to Schmidt and Takoudis [38]

III:

$$\begin{aligned} \dot{x} &= (A + a \cos \omega t)(1 - x - y) - Bx - xy(1 - x - y)^2 \\ \dot{y} &= C(1 - x - y) - Dy - xy(1 - x - y)^2 \end{aligned} \tag{42}$$

are all considered and the same qualitative picture emerges. In the above A, B. C and D are parameters, A being the mean value about which the forcing perturbation of amplitude a and frequency ω is made.

At low amplitudes the forcing produces either a p-harmonic or quasi-periodictity. Some regions of entrainment for which p-harmonic responses are otherwise (also known as a resonance horn or Arnol'd tongue) are as shown in Fig. 7. It is impossible to show them all as there is p-horn touching down at each rational number p/q, where p and q have no common divisor. The larger the integers p and q the narrower and smaller are the tongues. Between them are regions of quasiperiodicity where the frequency of forcing is incommensurable with the natural frequency and the amplitude is insufficient to ensure entrainment.

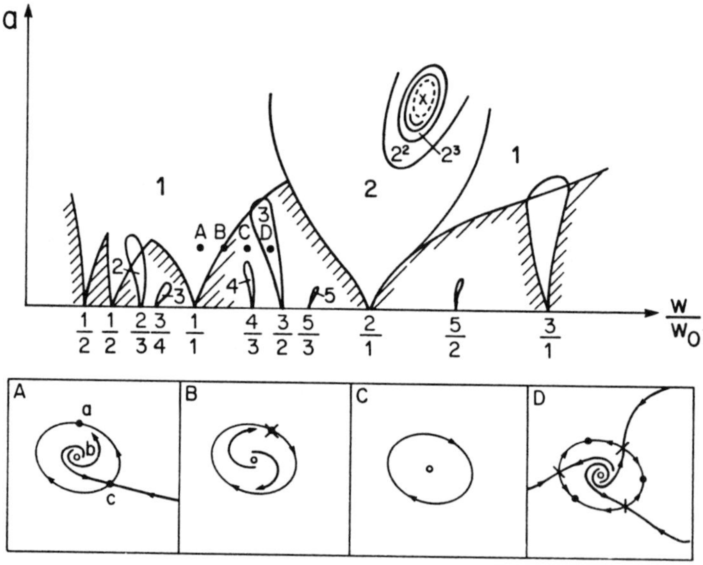

Figure 7. The generic excitation diagram. (Based on Tomita's work and Kevrekidis' extension of it).

In a stroboscopic map, the state of the system is recorded at intervals of $2\pi/\omega$, and a response of the same period as the forcing appears as a fixed point of period 1. Its stability as a point of the map $(x_n, y_n) \rightarrow (x_{n+1}, y_{n+1}, x_n = x(2\pi n/\omega), y_n = y(2\pi n/\omega)$, will be the stability of the periodic solution it represents. For example the natural limit cycle perturbed by the forcing will still persist for small a, but has become unstable. Though the stroboscopic map is a point map it is given convenient to make it continuous by considering its invariant curves. Thus there is a curve passing through any sequence of points $x_0,x_1,x_2,...$ such that if a starting point x_o^1 not identical with any x_n were chosen, then the sequence $x_o^1,x_1^1,x_2^1,...$ would lie on the same curve. This device may be used to summarise the behavior as is done in some of the smaller figures bordering the excitation diagram in Fig. 7.

For example, at the values of a and ω represented by the point A. There is only one stable periodic response which will be of the same period as this forcing a. This lies on a closed curve which is invariant as a whole and is often known as an invariant circle. From any point on it the stroboscopic point will move through a sequence of points to "a" the stable period-one response. But this is not possible without there being also an unstable point on the invariant circle "c" which represents an unstable period-one response. This is not the relict of the natural oscillation for that is to be found within the invariant circle, b. One trajectory of the continuous stroboscopic plane gives from this remnant oscillation to the unstable point on the invariant circle, for this is a saddle. The other branch of the stable manifold of the saddle is merely a curve leaving the invariant circle. Thus for every starting point from within the invariant circle the stroboscopic point moves outwards and gradually approaches the node "a" and the same is true for starting points without the invariant circle.

The quasiperiodic response (as at C) is represented in the continuous stroboscopic plane by an invariant circle surrounding an unstable point. Thus once on the invariant circle the strobscopic point hops around it never quite landing on the same place twice. If N_n is the number of times the typical point goes round the invariant circle in time nT, the rotation number is the limit of N_n/n as $n \rightarrow \infty$. The rotation number is clearly a multiple of 1/p when the response of the system is entrained as a p-subarominic. For a quasiperiodic response it must have irrational values but set of points on which it takes rational values has the structure of a Cantor set.

We can now see how the transition from A to C is made, for at such a point as B, very close to the quasiperiodic border, the saddle and node have moved around the circle and are about to coalesce. When they do so they evaporate and leave behind the quasiperiodic trajectory. As the frequency is increased and the resonance horn 3/2 is invaded, three pairs of saddle-nodes are born. Typically these move around the invariant circle and either return to their cognates or move on to the next. Some of the subtleties of this behavior are discussed elsewhere [36]. Figure 8 shows an actual stroboscopic plane at the top of resonance horn 3, where the saddles and nodes are no longer on the invariant circle which is itself stable. Thus an unstable period 1, a stable and unstable period 3 and a stable quasi-periodic response all coexist, the basin of attraction of the latter being the triangular region between the stable manifolds of the saddles.

Much more remains to be said of the fine structure of the stroboscopic plane and its bifurcations but the subject is unfolding in as fascinating and exciting a way as did the understanding of the more elementary aspects of the stirred tank thirty years ago.

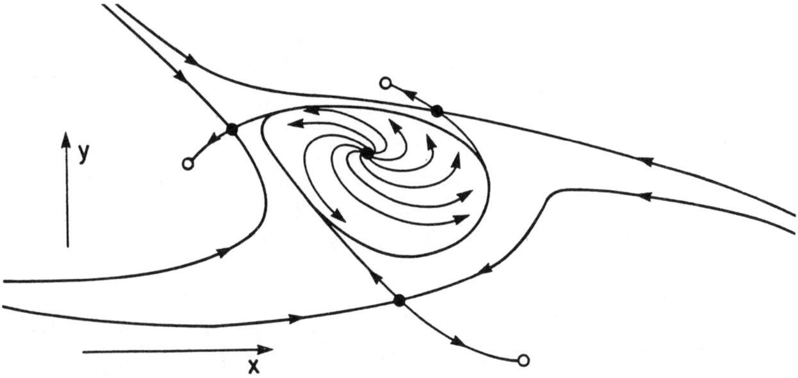

Figure 8. Stroboscopic map of period-3 forced oscillations formed away from the invariant circle. Two period-3 (one stable and one unstable), an unstable period-1 and a stable quasi-periodic response all coexist.

Bibliography

1. Aris, R. "The mathematical background of chemical reactor analysis. I. Preliminaries." Physica D. (1986).

2. Denn, M. M. "Process Modeling" Pitman, Marshfield, Mass., 1985.

3. Van Heerden, C. "Autothermic processes, Properties and reactor design", Ind. Eng. Chem. **45**, (1953) 1242.

4. Bilous, O. and N. R. Amundson, "Chemical reactor stability and sensitivity", AIChE J. **1**, (1955) 513.

5. Denbigh, K. G. "Velocity and yield in continuous reaction systems". Trans. Faraday Soc. **40**, (1944) 352.

6. Amundson, N. R. and R. Aris. "Chemical reactor stability and control", Chem. Engng. Sci. **7**, (1958) 121-160.

7. Uppal, A., W. H. Ray and A. B. Poore, "On the dynamic behavior of continuous stirred tank reactors". Chem. Engng. Sci. **29**, (1974) 967.

8. Uppal, A., W. H. Ray and A. B. Poore, "The classification of the dynamic behavior of stirred tank reactors . Influence of reactor residence time", Chem. Engng. Sci. **31**, (1976) 205.

9. Vaganov, D. A., N. G. Samoilenko and V. G. Abramov, "Periodic regimes of continuous stirred tank reactors", Chem. Engng. Sci. **33**, (1978) 1133.

10. Williams, D. C. and J. M. Calo, "Fine structure of the CSTR parameter space", AIChE Journal **27**, (1981) 514.
Kwong, V. K. and T. T. Tsotsis, "Fine structure of the CSTR parameter space", AIChE Journal **29**, (1983) 343.

11. Tsotsis, T. T., A. E. Haderi and R. A Schmitz, "Exact uniqueness and multiplicity criteria for a class of lumped reaction systems" Chem. Engng. Sci. **37**, (1982) 1235.

12. Farr, W. W., Ph.D. Dissertation, University of Minnesota 1985.

13. Gray, P. and S. K. Scott, "Autocatalytic reactions in the isothermal continuous stirred tank reactor. Isolas and other forms of multistability", Chem. Engng. Sci. **38**, (1983) 29.

14. Gray, P. and S. K. Scott, "Autocatalytic reactions in the isothermal continuous, stirred tank reactor. Oscillations and instabilities in the system $A + 2B \rightarrow 3B$; $B \rightarrow C$". Chem. Engng. Sci. **39**, (1984) 1087.

15. Gray, P. and S. K. Scott. "Sustained oscillations and other exotic patterns of behavior in isothermal reactions", J. Phys. Chem. **89**, (1985) 22.

16. van den Bosch, B. and D. Luss, "Uniqueness and multiplicity criteria for an n-th order chemical reaction". Chem. Engng. Sci. **32**, (1977) 203.

17. Luss, D., "Uniqueness criteria for lumped and distributed chemically reacting systems". Chem. Engng. Sci **26**, (1971) 1713.

18. Balakotaiah, V. and D. Luss, "Analysis of multiplicity patterns of a CSTR". Chem. Eng. Commun. **13**, (1981) 111; **19**, (1982) 185.

19. Balakotaiah, V. and D. Luss. "Exact steady state multiplicity criteria for two consecutive or parallel reaction in lumped parameter systems." Chem. Engng. Sci. **37**, (1982) 433.

20. Balakotaiah, V. and D. Luss, "Structure of the steady state solutions lumped-parameter chemically reacting systems". Chem. Engng. Sci. **37**, (1982) 1611.

21. Balakotaiah, V. and D. Luss, "Multiplicity features of reacting systems. dependence of the steady-states of a ESTR on the residence time", Chem. Engng. Sci. **38**, (1983) 1709.

22. Balakotaiah, V. and D. Luss, "Global analysis of multiplicity features of multi-reaction lumped-parameter systems", Chem. Engng. Sci. **38**, (1984) 865.

23. Golubitsky, M. and B. L. Keyfitz, "A qualitative study of the steady state solutions for a continuous flow stirred tank chemical reactor", SIAM J. Math Anal. **11**, (1980) 316.

24. Golubitsky, M. and D. Schaeffer, "Singularities and groups in "bifurcation theory", (1985) Springer Verag, New York.

25. Balakotaiah, V., Personal communication.

26. Jorgensen, D. V., W. W. Farr and R. Aris, "More on the dynamics of a stirred tank with consecutive reactions", Chem. Engng. Sci. **34**, (1984) 1741.

27. Farr, W. W. and R. Aris, "Yet who would have thought the old man to have had so much blood in him?" Chem. Engng. Sci. **41**, (1986) 000.

28. Amundson, N. R., "The mathematical understanding of chemical engineering systems". Selected papers of N. R. Amundson", Ed. R. Aris and A. Varma. (1980) Pergamon Press. Oxford.

29. Pismen, L. M., "Dynamics of lumped chemically reacting systems near singular bifurcation points", Chem. Engng. Sci **39**, (1984) 1063.

30. Pisner, L. M., "Dynamics of lumped chemically reactine systems near singular bifurcation points II. Almost Hamiltonian dynamics", Chem. Engng. Sci. **40**, (1985) 905.

31. Ueda, Y., "Steady motions exhibited by Duffing's equation: A picture book of regular and chaotic motions: in "New approaches to nonlinear problems in dynamics". Ed. P. J. Holmes. (1980) 311. SIAM, Philadelphia.

32. Tomita, K.,"Chaotic response of non-linear oscillators", Physics Report, **86** (1982) 113.

33. Sincic, D. and J. E. Bailey, "Pathological dynamic behavior of forced periodic chemical processes". Chem. Engng. Sci. **32** (1977) 281.

34. Kevrekidis, I. G., R. Aris and L. D. Schmidt, "Some common features of periodically forced reacting systems". Chem. Engng. Sci. **40**, (1985).

35. Kevrekidis, I. G., R. Aris, L. D. Schmidt and S. Pelican, "Numerical computation of invariant circles of maps". Physica **16D**, (1985) 243.

36. Kevrekidis, I. G. , R. Aris and L. D. Schmidt, "Phenomena and algorithms in the numerical study of peridically forced systems".

37. Kevrekidis, I. G. (1985) Ph.D. Dissertation, University of Minnesota.

38. Takoudis, C. G., L. D. Schmidt and R. Aris, "Isothermal oscillations in surface reactions with coverage independent parameters". Chem. Engng. Sci. **37** (1982) 69.

Department of Chemical Engineering and Materials Science
University of Minnesota
Minneapolis, Minnesota 55455

Lectures in Applied Mathematics
Volume 24, 1986

HIGH MACH NUMBER COMBUSTION

Andrew Majda*

INTRODUCTION. In this paper, the author presents recent developments in the mathematical theory of high Mach number combustion. The phenomena in this regime are enormously complex and the mechanisms responsible for the variety of physical phenomena are poorly understood. The current advances in the mathematical theory combine asymptotic methods, careful numerics, qualitative modelling, and rigorous proofs for important model problems as well as an interplay with the documented experimental literature. This paper attempts to retain the flavor of these research efforts.

Table of Contents

*Partially supported by grants from: U.S. Army Research Office, Office of Naval Research, and National Science Foundation.

I. The Theory and Structure for Planar Detonation Waves and an Instructive Qualitative Model

In this chapter, we discuss the qualitative and quantitative properties of planar reacting shock waves. Unlike the material discussed in subsequent chapters, here we only discuss unidirectional wave propagation. Thus, the basic equations which we use are the one-dimensional compressible Navier Stokes equations for a reacting mixture. We assume a standard simplified form for the reacting mixture here and throughout this paper. Thus, there are only two species present, the reactant and the product, and we postulate that the reactant is converted to the product by a one-step irreversible chemical reaction. Under the above hypothesis, the compressible Navier Stokes equations for the reacting mixture are the system of four equations,

The Compressible Navier Stokes Equations for a Reacting Mixture

$$\begin{aligned}
&\rho_t + (\rho v)_x = 0\\
&(\rho v)_t + (\rho v^2 + p)_x = \mu v_{xx} \qquad (1.1)\\
&(\rho E)_t + (\rho vE + vp)_x = (\mu (v^2/2)_x)_x + (\rho q_o d Z_x)_x + c_p(\lambda T_x)_x\\
&(\rho Z)_t + (\rho v Z)_x = -\rho K_o \phi(T) Z + (d\rho Z_x)_x
\end{aligned}$$

Here ρ is the density, $\tau = 1/\rho$ is the specific volume, v is the fluid velocity, E is the total specific energy, and Z is the mass fraction of the reactant. The total specific energy has the form

$$E = e + q_o Z + \frac{v^2}{2} \qquad (1.2)$$

with e the specific internal energy and q_o the amount of heat released by the given chemical reaction. The internal energy, e, and the temperature T are given via thermodynamics through an equation of state as well-defined functions, $e(\tau,p)$, $T(\tau,p)$. The rate function, $\phi(T)$ typically has the form,

$$\phi(T) = T^{\alpha}e^{-A/T}, \quad T > T_i \tag{1.3}$$

with T_i an ignition temperature or often $\phi(T)$ can be given by some other simplified approximation to (1.3).

For an ideal gas mixture with the same γ-gas laws, which we often use below as a simple illustrative example, we have the explicit formulae

Ideal Gas Mixture

$$\begin{aligned} p\tau &= (\gamma - 1)e \\ T &= p\tau/R \times M \end{aligned} \tag{1.4}$$

with R, Boltzmann's gas constant, M the molecular weight, c_p the specific heat and γ with $1<\gamma<2$ defined by $c_p(\gamma-1)=R$.

The coefficients μ, γ, and d in (1.1) are coefficients of viscosity, heat conduction, and species diffusion respectively. The compressible Euler equations for the reacting mixture are the special case of (1.1) with $\lambda = \mu = d = 0$. In this chapter our discussion will be largely qualitative so it will not be important to discuss a non-dimensional form of (1.1) for the rest of this chapter.

Qualitative-Quantitative Simplified Model Equations

The equations in (1.1) have extremely complex solutions with many peculiar effects produced through the nonlinear interaction of the sound waves and the chemical reactions. It is not surprising that simpler qualitative-quantitative models have been developed ([1], [2], [3]) in an attempt to explain the essential feature of this wave interaction. Independently, Fickett ([1]) and the author ([2]) introduced models of this sort and here we follow the ideas from [2]. These simplified model equations are a coupled 2x2 system given by a Burgers equation coupled to a chemical kinetics equation. These equations have the form

The Simplified Model Equations

$$u_t + (\tfrac{1}{2}u^2 - q_0 Z)_x = \beta u_{xx}$$
$$Z_x = K_0 \phi(u) Z \ , \quad t > 0 \ , \quad -\infty < x < \infty \tag{1.5}$$

where u is a lumped variable with some features of pressure or temperature, Z is the mass fraction of reactant, $q_o > 0$ is the heat release, $\beta \geq 0$ is a lumped diffusion coefficient, K_0 is the reaction rate, and $\phi(u)$ is a structure function for the kinetics and might have the form in (1.3) or some other dependence derived through asymptotic considerations. At the present moment, the reader can regard the equations in (1.5) as a qualitiative model for the equations in (1.1). However, we will derive the model equations in (1.5) as a distinguished asymptotic limit of the equations in (1.1) in the beginning of Chapter 2 with a simplified version of the methods used in [3]. The reader should not be confused by the appearance of Z_x on the left hand side of (1.5) rather than Z_t. The coordinate x in (1.5) is not the space coordinate but is determined through the asymptotics as an appropriately scaled space-time coordinate representing distance to the reaction zone - the x differentiation in (1.5) occurs because Z in (1.1) is convected at the much slower fluid velocity rather than the much faster reacting shock speed. This creates one important difference in the prescription of well-posed data for (1.5) when compared with (1.1) for time-dependent problems.

For the time-dependent reacting Navier Stokes equations in (1.1), we give the initial data,

$$ {}^t(\rho(x,t),\ v(x,t),\ p(x,t),\ Z(x,t))\Big|_{t=0} = {}^t(\rho_0(x), v_0(x), p_0(x), Z_0(x)) = \vec{u}_0(x) \qquad (1.6a)$$

with $\vec{u}_0(x)$ a prescribed function. On the other hand, for (1.5) we give initial data for $u(x,t)$ and signalling data for $Z(x,t)$ at $x = +\infty$, i.e.,

$$ u(x,t)\Big|_{t=0} = u_0(x), \quad -\infty < x < \infty $$
$$ \lim_{x\to\infty} Z(x,t) = Z_0(t), \quad t \geq 0 \qquad (1.6b)$$

with $u_0(x)$ and $Z_0(t)$ prescribed functions. The success of simplified qualitative models rests on their ability to produce analogues in a transparent fashion of the complex phenomena which they model. Below, we indicate that this is the case for the equations in (1.5) as regards the elementary theory of detonation waves for (1.1).

Ia. The Chapman-Jouget Theory - The Simplest Theory for Reacting Shock Waves

The tacit assumptions of the Chapman-Jouget theory are the following:

Chapman-Jouget Assumptions

(1.7)

a) Reacting shock fronts are infinitely thin and are idealized as discontinuities.

b) The gas particles crossing the front instantaneously adjust from thermodynamic equilibrium in the reactant ahead of the front to thermodynamic equilibrium in the product behind the front.

As regards the equations in (1.1), these assumptions are equivalent quantitatively to requiring that

a) All transport effects from viscosity, heat conduction, and species diffusion are ignored, i.e., $\lambda = \mu = d = 0$. (1.8)

b) The reaction rate, K_o, in (1.1) is infinitely large for $T > T_i$ and zero for $T < T_i$ with T_i an ignition temperature.

The Chapman-Jouget theory attempts to mimic the standard non-reactive inviscid shock wave theory to as large an extent as possible by replacing (1.1) by inviscid gas dynamics with a general discontinuous equation of state.

Next, with these assumptions, we briefly discuss the travelling wave solutions moving with velocity, D. With $\vec{u} = {}^t(\tau, v, p, Z)$ and the Chapman-Jouget assumptions enforced, we seek piecewise constant travelling wave solutions,

$$\vec{u}(x-Dt) = \begin{cases} {}^t(\tau_o, 0, p_o, 1), & x-Dt \geq 0 \\ {}^t(\tau_f, v_f, p_f, 0), & x-Dt < 0 \end{cases} . \tag{1.9}$$

Such solutions are discussed in detail in all elementary books on combustion theory ([4], [5], [6]) so we will be terse - now we will describe waves propagating into undisturbed reactant at rest so that of course, $D > 0$. The mechanical shock conditions expressing conservation of mass and momentum yield the familiar jump relations,

$$-m = \rho_o D = \rho_f (D - v_f) \tag{1.10a}$$

$$m^2 = \frac{p_f - p_o}{\tau_f - \tau_o} \tag{1.10b}$$

while, by utilizing the form in (1.2), we compute that conservation of energy yields

$$e(\tau_f,p_f) - e(\tau_o,p_o) - \tfrac{1}{2}(p_f+p_o)/(\tau_f-\tau_o) = q_o \tag{1.10c}$$

where given (τ_o,p_o), the Hugoniot function in the (p,τ) plane, $H(\tau_f,p_f)$ is defined by the left hand side of (1.10c).

Next, we summarize several of the well known peculiar dynamical features of such travelling waves when compared with ordinary inviscid shock wave theory.

There is a critical wave velocity, D_{CJ}, so that

1. No compressive wave solutions (with $\tau_f < \tau_o$) satisfy (1.9), (1.10) for $D < D_{CJ}$.
2. For $D > D_{CJ}$, there are always two compressive travelling wave solutions satisfying (1.9), (1.10); a strong detonation wave defined by (τ_f^*, v_f^*, p_f^*) and a weak detonation wave defined by $(\tau_{f*}, v_{f*}, p_{f*})$. The strong detonation wave is more compressive with a higher post pressure, i.e., $\tau_f^* < \tau_{f*}$ and $p_f^* > p_{f*}$. The wave is subsonic (supersonic) from behind for a strong (weak) detonation. (1.11)
3. For $D = D_{CJ}$, there is a unique travelling wave, the Chapman-Jouget wave, and D_{CJ} is exactly the sound speed in the burnt gas.

We refer the reader to any of the previously mentioned texts or Figure 1.1 for the well-known diagrams. Strictly speaking the facts in (1.11) only hold for appropriate general equations of state but are always satisfied for the ideal gas mixture in (1.4).

Next, we indicate that the travelling waves for the qualitative model under the Chapman-Jouget assumptions from (1.8) corresponding to $\beta = 0$ and K_o satisfying (1.8b) have all the qualitative features mentioned in (1.11) in a transparent and analogous fashion.

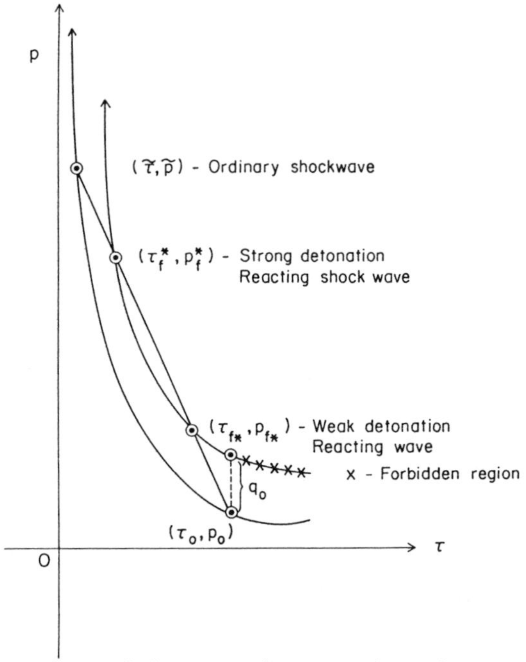

Figure 1.1: Reacting Hugoniot Diagram

Chapman-Jouget Theory for the Qualitative Model

Under the assumptions in (1.8), we construct piecewise constant travelling wave solutions for the equations in (1.5) with the form

$$(u(x-Dt), Z(x-Dt)) = \begin{cases} (0,1), & x-Dt > 0 \\ (u_f, 0), & x-Dt \leq 0 \end{cases} \tag{1.12}$$

The analogue of the three gas dynamic jump conditions in (1.10) is the single jump relation,

$$-D(u_0 - u_f) + \tfrac{1}{2}(u_0^2 - u_f^2) = q_0(Z_0 - Z_f) \tag{1.13}$$

and given u_o , the analogue of the Hugoniot function is defined by the left hand side of (1.13). With the data, (0,1) , for $x-Dt > 0$, it is a trivial matter to solve the quadratic equation and compute the following explicit formulae:

1. For $D < (2q_o)^{\frac{1}{2}}$, there are no travelling waves moving with speed D .
2. For $D > (2q_o)^{\frac{1}{2}}$, there are two travelling waves moving with speed D . The strong detonation wave with u_f^* defined by

$$u_f^* = D + \sqrt{D^2 - 2q_o}$$

and the weak detonation wave, u_{f*} defined by

$$u_{f*} = D - \sqrt{D^2 - 2q_o} \qquad (1.14)$$

3. For $D_{CJ} = \sqrt{2q_o}$, $u_f^{CJ} = \sqrt{2q_o}$ and since u is the nonlinear wave speed for the inviscid Burgers equation, u_f^{CJ} is the analogue of the Chapman-Jouget point where the wave speed equals the speed of propagation in the burnt gas.
4. $u_f^* > D > u_{f*}$ - this is the analogue for the model of the fact that strong detonations, like ordinary fluid shocks, are subsonic when viewed from behind the wave while weak detonations are supersonic when viewed from behind.

Thus, we have an extremely simple and transpartent analogue of the Chapman-Jouget theory in the qualitative model. All of these results are easily explained through the following graph:

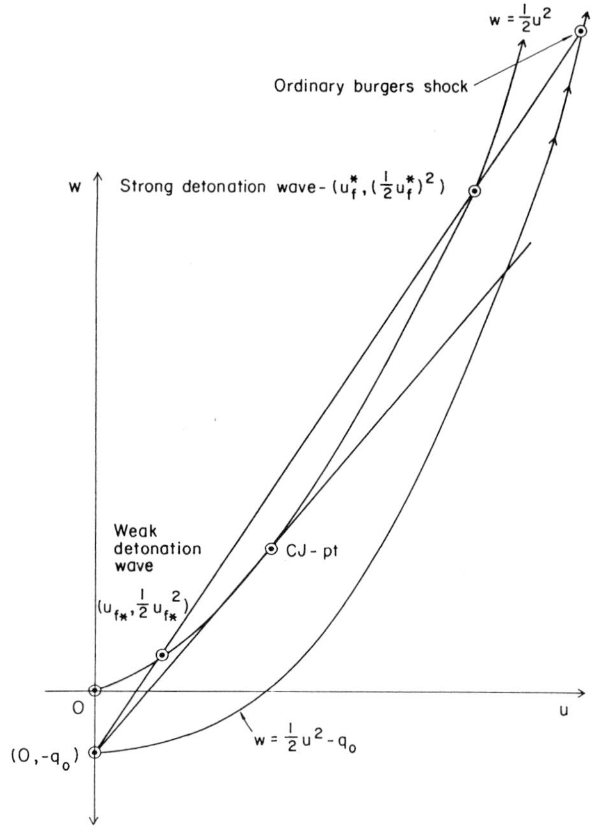

Figure 1.2: Reacting Hugoniot Diagram for Qualitative Model

Ib. The Z-N-D Theory and the Structure of Detonation Waves with Finite Reaction Rates

The Chapman-Jouget theory of wave structure involves both assumptions of zero diffusion coefficients and infinite reaction rates. The requirement of infinite reaction rates is the more severe assumption since diffusive length scales are determined by molecular collision theory and the chemical reactions proceed as a consequence of these collisions; thus, the reaction zone is typically much larger than the diffusive length scale. One consequence of the Chapman-Jouget theory is that two distinct reacting shock waves, strong detonations and weak detonations, can propagate with a given wave speed. The Z-N-D theory for reacting gases developed independently by Zeldovich, von Neumann, and Döring is a theory motivated both by the attempt to resolve the ambiguity of the C-J theory and also to develop a more realistic theory of the physical structure of detonations. This theory is forty years old and one of the modern byproducts of this theory is to provide researchers with relatively explicit realistic steady wave profiles to investigate in studying the multi-dimensional instabilities which occur in detonation waves. The main postulate of the Z-N-D theory is that

All diffusion coefficients are zero but all reaction rates are finite. (1.15)

Next, we present the analogue of the Z-N-D structure for reacting shock waves in the qualitative model from (1.5).

Z-N-D Theory in the Qualitative Model

Under the Z-N-D postulate from (1.15) the simplified model equations have the form

$$\begin{aligned} &u_t + (\tfrac{1}{2}u^2 - q_o Z)_x = 0 \\ &Z_x = K_o \phi(u) Z \end{aligned} \tag{1.16}$$

In the Z-N-D theory, we attempt to construct explicit travelling waves for the equations in (1.16) with K_o finite for a given constant preshock state. With the preshock state, $(0,1)$, from (1.12) and $\phi \geq 0$, $\phi(0) = 0$, we look for travelling wave solutions of (1.16) with $D > 0$ given by $(\tilde{u}(x-Dt), \tilde{Z}(x-Dt))$ with the form

$$(\tilde{u}(x-Dt),\tilde{Z}(x-Dt)) = \begin{cases} (0,1) \ , & x-Dt > 0 \\ (\tilde{u}(x-Dt),\tilde{Z}(x-Dt)) \ , & x-Dt \le 0 \end{cases} . \quad (1.17)$$

Here $\tilde{Z}(0) = 1$ and $\lim_{\xi\to-\infty} \tilde{Z}(\xi) = 0$ while $\lim_{\xi\to-\infty} \tilde{u}(\xi) = u_f$ remains to be determined where $\xi = x-Dt$; of course, to generate a nontrivial wave, we need to assume that $\tilde{u}(0) \neq 0$. How do we determine $\tilde{u}(0)$? Since $\tilde{Z}(0) = 1$ and $\tilde{Z}$ is continuous across $\xi = 0$, according to the jump relations in (1.13),

$$2D = \tilde{u}(0) \ , \quad D > 0 \quad (1.18)$$

and we have an ordinary non-reactive shock discontinuity for the inviscid Burgers equation at $\xi = 0$ - see Figure 1. for $\xi<0$, by substituting the ansatz from (1.17) into (1.16) we obtain the ODE's

$$\begin{aligned} (-D\tilde{u} + \tfrac{1}{2}\tilde{u}^2)'(\xi) &= q_o \tilde{Z}'(\xi) \ , \quad \xi < 0 \\ \tilde{Z}'(\xi) &= K\phi(\tilde{u}(\xi))\tilde{Z} \end{aligned} \quad (1.19a)$$

with the initial conditions at $\xi = 0$ given by

$$\begin{aligned} \tilde{u}(0) &= 2D \\ \tilde{Z}(0) &= 1 \end{aligned} \quad (1.19b)$$

The first equation in (1.19a) can be integrated through (1.13) resulting in the formula,

$$D\tilde{u}(\xi) - \tfrac{1}{2}\tilde{u}^2(\xi) = q_o(1 - \tilde{Z}(\xi))$$

so that

$$\tilde{u}(\tilde{Z}(\xi)) = D + (D^2 - 2q_o(1 - \tilde{Z}(\xi)))^{\frac{1}{2}} . \tag{1.20}$$

We remark that we have chosen the plus sign in the square root factor in (1.20) to guarantee the assumed continuity of the profile from the left, i.e., so that $\lim_{\xi \uparrow 0} \tilde{u}(\xi) = 2D$. The entire structure of the wave is determined now as the solution of the scalar nonlinear ODE

$$\frac{d\tilde{Z}}{d\xi} = K\phi(\tilde{u}(\tilde{Z}))\, \tilde{Z} \quad , \quad \xi < 0$$
$$\tilde{Z}(0) = 1 \quad . \tag{1.21}$$

In particular, if we make a mild assumption on the kinetic structure function requiring that $\phi(u)>0$ on $[D+(D^2-2q_o)^{\frac{1}{2}},2D]$, then it is easy to see that necessarily

$$\lim_{\xi \to -\infty} \tilde{Z}(\xi) = 0$$

and from the formula in (1.20), we deduce that

$$\lim_{\xi \to -\infty} \tilde{u}(\xi) = D + (D^2 - 2q_o)^{\frac{1}{2}} = u_f^*$$

where u_f^* is the appropriate strong or Chapman-Jouget detonation value for the qualitative model.

Let's summarize our <u>conclusions for travelling wave structure for the qualitative model under the postulate of the Z-N-D theory</u>:

1. The structure of a detonation wave is an ordinary non-reactive shock wave followed by a chemical reaction zone.

2. Only strong or C-J detonation waves are possible under the Z-N-D hypothesis; in particular, weak detonations cannot occur. (1.22)

3. For any given kinetics structure function, the Z-N-D profile can be calculated through quadrature of a single scalar nonlinear ODE. The details of the profile depend strongly on the nature of this kinetics structure function.

The reader already familiar with the Z-N-D theory of reacting gases will observe that we have recovered the main aspects of this theory in (1.22) for the qualitative model in a straightforward fashion.

Z-N-D Structure for the Reacting Compressible Euler Equations

The ideas and conclusions of the Z-N-D theory for the reacting compressible Euler equations coincide with those given in (1.22) and the discussion parallels that in (1.17)-(1.21) although the algebra is very messy. Since much of the first few chapters of the lucid book by Fickett and Davis is devoted to this theory with generalizations to more complex chemistry, we will not present any details here. We only mention here that the equations for conservation of mass, momentum, and energy can be explicitly integrated to determine the temperature, $T(Z(\xi))$; once this function is determined explicitly, the detailed structure of the detonation wave is determined by the scalar nonlinear ODE,

$$\frac{dZ}{d\xi} = \frac{K_o}{m} \phi(T(Z))Z, \quad \xi < 0$$

$$Z(0) = 1 \tag{1.23}$$

with m the mass flux and $\xi = x-Dt$. We would also like to mention the graphical interpretation of the structure of the Z-

N-D wave in the (p,τ) plane - the analogous structure in the qualitative model is presented in Figure 1.3. Given the pre-shock specific volume and pressure (τ_o,p_o), we find the ordinary fluid dynamic shock moving with speed D and with value $(\tilde{\tau}, \tilde{p})$; we then slide along the Rayleigh line monotonically to the strong detonation values, (τ_f^*,p_f^*), as $\xi \to -\infty$. The rate of the slide down this straight line is governed by the specific kinetics structure function in (1.23). Thus, the pressure is highest at $\xi = 0$ and drops monotonically through the wave with the same behavior for the density - this is the celebrated von Neumann spike with

$$p(\xi_1) < p(\xi_2) \;, \quad \xi_1 < \xi_2 < 0$$

$$\rho(\xi_1) < \rho(\xi_2) \;, \quad \xi_1 < \xi_2 < 0 \,.$$

Below, in Figure 1.3, we give the graph of the Z-N-D detonation with a pressure spike which emerged from a resolved time dependent numerical calculation with initial data given by the piecewise constant Chapman-Jouget state. The detonation has fairly small heat release with one-step chemical kinetics using the dominant rates of the Ozone decomposition detonation (complete details can be found in [7]). Such time dependent calculations confirm the dynamic stability of such waves and can be used to predict the onset of instability.

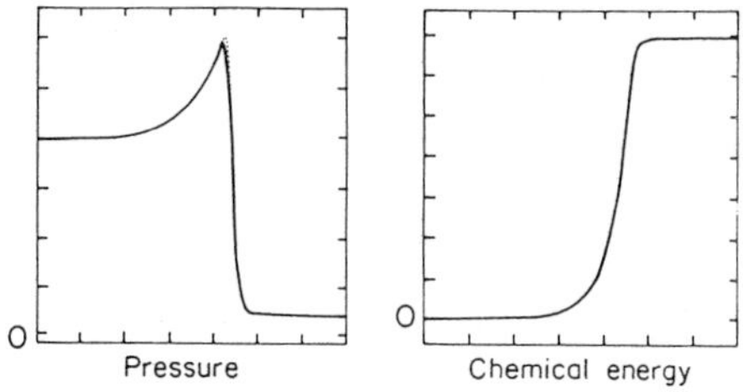

Figure 1.3: Pressure and Chemical Energy Profiles in Z-N-D Wave

These steady wave structures emerged from a time-dependent calculation.

The Structure of Z-N-D Waves at Large Activation Energies

For the case of Arrhenius kinetics and the ideal gas mixture from (1.4) with identical γ-gas laws, here we will present several comments regarding the detailed Z-N-D structure. At high activation energy, we point out the analogy with the asymptotic problem of homogeneous explosions studied by Kassoy.

Given the preshock state (τ_o, p_o), we define the Mach number of the shock by

$$M_0^2 = (D/c_o)^2$$

with c_o the sound speed in the unburnt gas. We have $M_0^2 > 1$. With $(\tilde{\tau}, \tilde{p})$, the specific volume and pressure for the post-shock state of the fluid dynamic shock wave moving at speed D, we have the corresponding temperature, T. A lengthy algebraic calculation establishes that in this special case, $T(Z)$ is given explicitly by

$$\frac{T(Z)}{T} = \frac{1 + \omega(\gamma-1)\theta(Z) - \omega^2\gamma\theta^2(Z)}{(1-\omega)(1+\omega\gamma)}$$

with

$$\text{a)} \quad \theta(Z) = \left(1 - \frac{(1-Z)}{\Omega} Q\right)^{\frac{1}{2}}$$

$$\text{b)} \quad \omega = \frac{M_o^2 - 1}{\gamma M_0^2 + 1} \tag{1.25}$$

$$\text{c)} \quad \Omega = \frac{\gamma(M_o^2 - 1)^2}{2(\gamma^2 - 1)M_o^2}$$

d) $Q = q_0/RT_0$

For this special case, the Mach number, M_0^{CJ}, corresponding to the C-J detonation is given in terms of the heat released by

$$(M_0^{CJ})^2 = 1 + h(1 + \tfrac{1}{4}h^2)^{\frac{1}{2}} + \tfrac{1}{2}h^2 \tag{1.26}$$

with $h^2 = 2(\gamma^2-1)Q/\gamma$. The qualitative structure of the temperature profile in the Z-N-D wave can have a completely different character depending on a second critical Mach number $\hat{M}_0$ defined for fixed Q by the formula,

$$(\gamma-1) - 2\omega(\hat{M}_0)\gamma\left(1 - \frac{Q}{\Omega(M_0)}\right)^{\frac{1}{2}} = 0 \tag{1.27}$$

(The equation in (1.27) is a quadratic equation for $\hat{M}_0$).

Case I: In the first case, if the Mach number M_0 satisfies $M_0 > \hat{M}_0$, the temperature function $T(Z)$ from (1.24) is always a monotone decreasing function of Z for Z with $0 \le Z \le 1$. In particular, sufficiently overdriven detonations satisfy this structure.

On the other hand, it is possible for $\hat{M}_0$ to satisfy $\hat{M}_0 > M_0^{CJ}$ and we have

Case II: For Mach numbers, M_0, satisfying $\hat{M}_0 > M_0 \ge M_0^{CJ}$, the temperature function $T(Z)$ from (1.24) has a maximum at a point Z^* with $0 < Z^* < 1$ with $T(Z)$ monotone increasing for $0 < Z < Z^*$ and $T(Z)$ monotone decreasing for $Z^* < Z < 1$. For Arrhenius kinetics and $T(Z)$ determine explicitly from (1.24), the scalar nonlinear ODE describing the detailed structure of the Z-N-D wave reduces to

$$\frac{dZ}{d\xi} = \frac{K_o}{m} e^{\frac{-A}{T(Z)}} Z \, , \quad \xi < 0$$
$$Z(0) = 1 \, . \tag{1.28}$$

At large activation energies, $A \uparrow \infty$, in Case I, this ODE behaves in a similar asymptotic fashion as the nonlinear ODE

$$\frac{dZ}{dt} = - K e^{\frac{-A}{T_f - q_o Z}} A \, , \quad t > 0$$
$$Z(0) = 1 \tag{1.29}$$

The equation in (1.29) arises in the theory of homogeneous explosions in a familiar fashion as a reduction of the problem

$$\frac{dZ}{dt} = - K e^{-A/T} Z$$
$$\frac{dT}{dt} = q_o K e^{-A/T} Z \qquad t > 0 \tag{1.30}$$

with $Z(0) = 1$, $T(0) = T_o$ and $T_f = T_o + q_o$. In both situations, T is a monotone decreasing function of Z and as $A \to \infty$, there is an induction zone where Z changes negligibly followed by an extremely thin layer where rapid burning occurs and most of the reactant is consumed - this thin layer is followed by a "tail" zone where Z slowly adjusts to zero. For the Z-N-D waves from Case II, a similar scenario is obtained as described above except that in the "tail" zone the reaction is proceeding at a slower relative rate and there is a longer "tail" for Z to adjust to zero. Scott Stewart has commented that Erpenbeck in [8], did the high activation energy asymptotics for (1.28) much earlier.

In either case, as $A \to \infty$, the Z-N-D wave approaches the square wave shape and for sufficiently high activation energies the following picture emerges from the Z-N-D theory. A fluid dynamic shock heats the gas to a temperature where nontrivial reaction occurs; after the induction period the gas burns extremely rapidly in a narrow zone because the temperature also increases causing the reaction to go increasingly faster at high activation energies. This idea was essentially von Neumann's initial physical one which motivated his development of the Z-N-D theory. In the early sixties, it was common for engineers to use the discontinuous square-wave model approximation in studying detonations (see the reveiw article by Erpenbeck [9] from 1969 for a critique of this discontinuous square-wave model including many of the undesirable artifacts of such a simplified approach in discussing detonation stability). The conditions in Case II also play a role in the theory of detonation stability and this was one of our reasons for pointing out this separate case - it is often ignored in the literature.

Ic. The Structure of Detonation Waves with Finite Reaction Rates and Nonzero Diffusion

We have already seen that the qualitative model from (1.5) gives an extremely simple caricature for both the Chapman-Jouget and Z-N-D structure theories for detonation waves. One way to assess the validity of such approximate theories is to study the corresponding travelling waves when both reaction rates are finite and diffusion coefficients are non-zero. This corresponds to the usual shock-layer theory for inviscid compressible flow which is used to justify and assess the validity of the inviscid discontinuous shock wave solutions. The existence and qualitative structure of diffusive travelling wave solutions to the reacting compressible Navier-Stokes equations in (1.1) is a highly nontrivial problem involving the connection of two critical points by an orbit for an autonomous system of four nonlinear ODE's. This problem has recently been solved with complete mathematical rigor by Gardner ([10]). Gardner's approach to this problem is quite interesting. He uses the structure of travelling waves for the qualitative model in (1.5) first established by the author in [2] and deforms the travelling wave problem for (1.1) to the one for the qualitative model to obtain the existence and properties of the diffusive travelling waves for the reacting Navier Stokes equations - this is an additional remarkable justification of the model! A heuristic treatment of the connection problem for (1.1) is contained in Williams' treatise ([5]) and the formal structure at high activation energies in various regimes has been discussed in [11] - a recent rigorous paper by Holmes and Stewart [12] also treats the connection problem at high activation energies. At the symposium, David Wagner has pointed out that Gardner ignores the contribution form $(p_o q_o dZ_x)_x$ in the energy equation in his analysis -Gardner's approach certainly should apply when this term is retained. However, it is important to prove this is the case, to add this term and also to study, more explicitly, the structure of the waves in Gardner's theorem as regards dependence on activation energy and qualitative wave shape - the possibility of no connection for $q_o < q_{CR}$ should also be clarified explicitly in the analysis. The calculations from [7] to be reported in Figure 1.6 provide overwhelming numerical evidence that no connection occurs in a variety of wave regimes. Here first we describe the structure of reacting shock waves with nonzero diffusion and finite reaction rates in detail for the model system in (1.5) - the analysis can be done using only phase plane techniques and still has a remarkable number of subtle features when compared with the standard diffusive shock layer problems of standard gas dynamics and as we have mentioned earlier, the same phenomena occur in an analogous fashion for the reacting compressible Navier Stokes equations. Then we will give several comments regarding structure for the reacting Navier-Stokes equations including recent fully resolved dynamic stability calculations given in a paper by Colella, Roytburd, and the author which demonstrate the existence and one-dimensional dynamic stability

<u>of stable weak detonations for the reacting compressible Navier Stokes equations!</u>

The Theory of Reacting Shock Layers for the Qualitative Model

For simplicity in exposition, we only consider the explicit pre-shock constant state for $\vec{w} = (u,Z)$ with the form $(0,1)$ - this is the special state where we have previously discussed the Chapman-Jouget theory with extremely simple algebra in Section Ia). We seek special travelling wave solutions of (1.5) with the form

$$\vec{w} = \vec{w}\left(\frac{x - Dt}{\beta}\right) \tag{1.30a}$$

so that with $\xi = (x-Dt)/\beta$,

$$\lim_{\xi\to\infty} \vec{w}(\xi) = (0,1) \tag{1.30b}$$

and

$$\lim_{\xi\to-\infty} \vec{w}(\xi) = (u_f,0) \tag{1.30c}$$

where u_f needs to be determined. This is the basic problem of existence of reacting shock layers for the qualitative model. With $\tilde{Z} = q_0 Z$ and $K = \beta K$, we substitute the ansatz from (1.30a) into (1.5) and obtain the autonomous system of 2 non-linear ODE's,

$$\begin{aligned} u' &= \tfrac{1}{2}u^2 - Du - \tilde{Z} + q_0 \\ \tilde{Z}' &= \tilde{K}\,\phi(u)\,\tilde{Z} \quad . \end{aligned} \tag{1.31}$$

The integration constant, q_0 , for the first equation in (1.31) has been chosen to guarantee that the requirement in (1.30b) is satisfied - the coordinate, $\tilde{Z}$, is the heat release and we consider the phase plane analysis of the autonomous system of ODE's from (1.31) in the $u - \tilde{Z}$ plane. We assume that $\phi(0) = 0$ and that the velocity, D, satisfies $D > (2q_0)^{1/2}$ so that the speed exceeds the Chapman-Jouget speed from (1.14). What are the critical points of the above system?

a) Since $\phi(0) = 0$, there are an infinite number of critical points coinciding with the $\tilde{Z}$ axis, i.e., $(0,\tilde{Z})$ with $\tilde{Z}$ arbitrary. (1.32)

b) Since D satisfies $D > D_{CJ}$, there are two isolated critical points along the u-axis. From (1.13) and (1.14) these are defined by the corresponding strong and weak detonation values, u_f^*, u_{f*} and are given by $(u_f^*,0)$, $(u_{f*},0)$.

Furthermore, it is easy to check from (1.14-4) that

c) $(u_f^*,0)$ is an unstable node
$(u_{f*},0)$ is a saddle. (1.32)

We fix the wave speed D and u_f^*, u_{f*} satisfying $\frac{1}{2}(u^*_f)^2 - Du^*_f = \frac{1}{2}(u_{f*})^2 - Du_{f*}$ but vary the location of the critical point $(0,\tilde{Z})$ with $\tilde{Z} > 0$. Then, for a given heat release q_0, the existence of a travelling wave satisfying (1.30) becomes a problem of "shooting" from the critical point $(0,q_0)$ to hit one of the critical points, either $(u_f^*,0)$ or $(u_{f*},0)$, as the parameter $\xi \to -\infty$. Since $(u_f^*,0)$ is an unstable node, trajectories are more likely to arrive at the strong detonation value; on the other hand, since $(u_{f*},0)$ is a saddle point, there is at most a single value, q_{CR}, where the travelling wave structure defines a weak detonation. Of course, for a fixed D, only those values of the critical point $(0,Z)$ with $\tilde{Z} < D^2/2$ define solutions satisfying the original problem in (1.30).

With the above comments as motivation, it is not difficult to anticipate the following results which are proved by the author in [2] (see Figure 1.4).

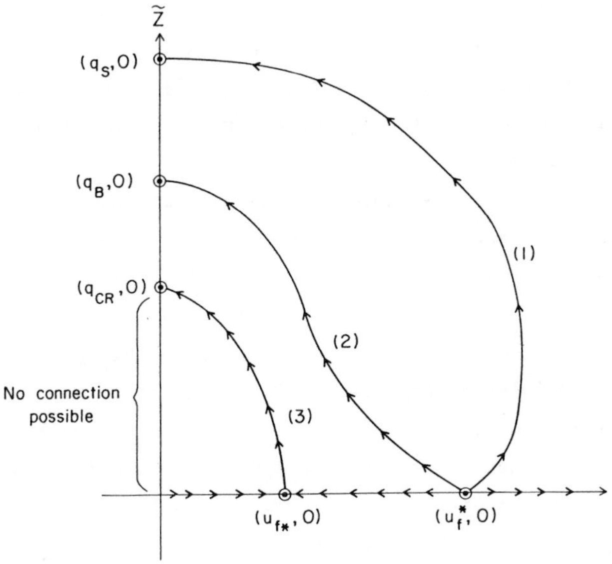

Figure 1.4: Phase Portrait With Three Connecting Profiles.

<u>Wave Structure for the Model</u>

For fixed values u_f^*, u_{f*} and a fixed positive value for $\tilde{K} = \beta K_o$, as the heat release varies, there is a <u>critical heat release</u>, q_{CR}, so that

a) For $q_o > q_{CR}$, a <u>strong detonation profile</u> with speed D <u>exists</u> connecting $(0,1)$ with $(u_f^*,0)$.

b) For $q_o = q_{CR}$, a <u>weak detonation profile</u> with speed D <u>exists</u> connecting $(0,1)$ with $(u_{f*},0)$. (1.33)

c) For $q_o < q_{CR}$, <u>no combustion profile</u> moving with speed D is possible.

This theorem can be deduced from the five phase portrait pictures on pages 83 and 84 of [2]. The parameter $\tilde{K} = \beta K$ is a rough measure of the width of the shock layer to the width of the reaction zone. Furthermore, for $\tilde{K} << 1$ the Z-N-D theory is approached while for $\tilde{K} \geq 0(1)$ the detonation profile can be monotone with a step-structure - see Figure 1 on page 73 of [2].

One consequence of the above theorem is that the predictions of the Chapman-Jouget and Z-N-D theories are completely wrong for $q_o \leq q_{CR}$. What happens to the initial value problem when piecewise constant strong detonation initial data is given but $q_o < q_{CR}$ is an amusing theoretical question. This problem has been solved by Colella, Roytburd, and the author in [7] - a bifurcating wave pattern with a faster moving dynamically stable weak detonation emerges as the time-dependent solution - a complete discussion is given in V. Roytburd's lecture at this meeting. In recent work, Larrouturou ([13]) has generalized the above results to the qualitative model from (1.6) with species diffusion added - both models can occur in asymptotic limits.

<u>Reacting Shock Layers for the Compressible Navier Stokes Equations</u>

We have already mentioned that the exact analogue of the wave structure in the model as presented above in (1.33) has been rigorously proved by Gardner ([10]) for the reacting compressible Navier Stokes equations. Here we present the results of time dependent numerical calculations which show that <u>surprisingly small increases in the reaction prefactor are sufficient to change the reacting wave pattern from</u> nearly a <u>classical Z-N-D wave to</u> a completely different <u>dynamically stable bifurcating wave</u> pattern <u>with</u> a stable <u>precursor weak detonation</u> - in fact, an increase in K_o by a factor of only five is sufficient to create this in the problem we have considered.

We used the same piecewise constant Chapman-Jouget state which produced an inviscid dynamically stable emerging Z-N-D wave discussed below (1.23). We recall that this problem had fairly small heat release and a one-step reaction rate crudely modelled by the Ozone detonation. In the first reported calculation, we included realistic values of the diffusion coefficients (once again modelled by Ozone), we numerically resolved all diffusive scales, and integrated the reacting compressible Navier-Stokes equations - the expected nearly Z-N-D pressure profile from Figure 1.5 emerged dynamically after some time. In the second reported calculation, we kept all other parameters fixed as in the first calculation except the reaction prefactor which we increased by a factor of five. The sequence in Figure 1.6 describes successive pictures of the wave patterns which emerged and the graphics ultimately focuses on the fastest moving wave. The plots include both pressure and chemical energy - the initial growing hump which emerges is a dynamically stable weak detonation - one way to confirm this fact is to observe that all of the chemical energy is released in the leading hump of the wave which emerges and moves at the faster speed. A detailed description is given in [7].

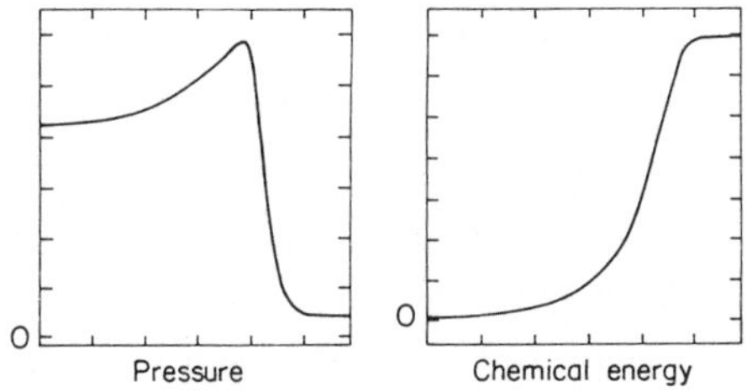

Figure 1.5: Nearly Z-N-d Reacting Shock Profile that Emerged from Time-Dependent Numerical Simulation of Reacting Compressible Navier Stokes Equation

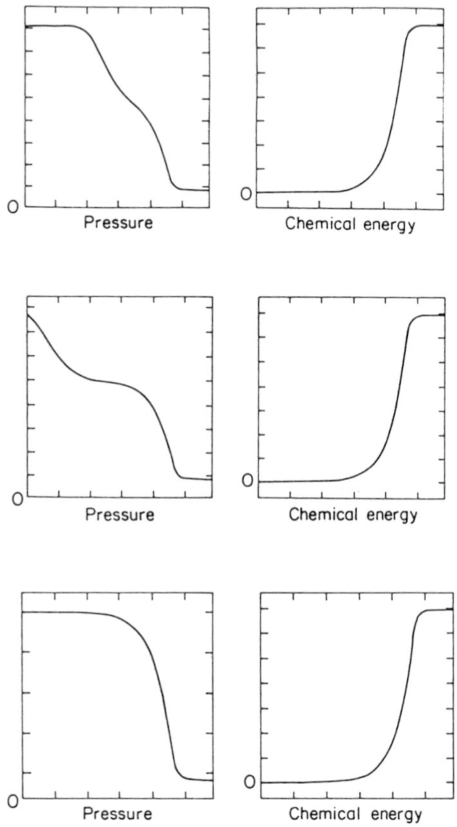

Figure 1.6: Dynamically Stable Weak Detonation
All parameters same as in Figure 1.5 except K_0 increased by factor of five successive snapshots of the emerging wave. Note: Observe that all chemical energy is released in percursor weak detonation wave.

How accurate is the Z-N-D theory for giving the reacting shock structure in real reacting gases? For many systems such as the hydrogen-oxygen system, the Z-N-D theory is an excellent approximation (see Williams [5]). On the other hand, Oppenheim and Rosciszewski ([14]) have established that the Z-N-D theory is extremely poor approximation to the reacting shock structure for detonations based on the Ozone decomposition reaction.

Finally, we conclude this Chapter with a few comments about weak detonations. Weak detonations are observed by experimentalists but not in shock tube experiments - an external means of initiation is required. Weak detonations are supersonic analogues of highly subsonic deflagration waves. Both waves are dynamically unstable in the inviscid theory - weak detonations are supersonic from both sides while deflagrations are subsonic from both sides; furthermore, the speeds of the weak detonation waves which emerge from a given time-dependent setting cannot be predicted in advance without detailed knowledge of the heat release, diffusion coefficients, and reaction rates - this is a well known feature of deflagration waves. In most problems of practical interest in combustion, weak detonations are only a pathological artifact; however, weak detonations might play a significant role in some situations involving supersonic initiation of detonation - we have just illustrated that they are dynamically stable solutions of the compressible Navier Stokes equations under appropriate circumstances.

II. Quantitative Asymptotic Modelling of Nonlinear Wave - Kinetic Interactions in Reacting Gases

Here we present several simplified quantitative asymptotic models which incorporate nonlinear wave - kinetic interactions in the reacting shock wave regime. In particular, by following a paper of Rosales and the author ([3]), we will present a quantitative asymptotic derivation of the model equations in (1.5) of the previous chapter by taking a suitable distinguished asymptotic limit of the reacting compressible Navier Stokes equations. Then we will discuss an interesting asymptotic equation of Clarke and Gant ([15]) and Jackson and Kapila ([16]) which exhibits temperature amplification of low frequency waves in the induction zone. Finally we will derive the scalar asymptotic equation first given independently by Blythe ([31]) and Clarke ([32]) for nonlinear simple sound waves in the induction zone.

One of the principal goals of the quantitative asymptotic development of simplified asymptotic equations in the high Mach number regime is to gain insight into the substantial nonlinear amplification and interaction of waves which is observed experimentally in reacting shock wave instabilities, initiation of detonation, and transition to detonation. These phenomena are created through nonlinear wave - exothermic kinetic interactions and are poorly understood theoretically. The objective is to gain physical insight into the onset and development of such instabilities and wave amplifications by identifying the underlying mechanisms and phenomena in the simpler asymptotic limit equations. The work in this area is only just beginning.

Preliminaries:

The asymptotic tools used in this chapter combine the methods of weakly nonlinear geometric optics for nonlinear hyperbolic systems with high activation energy asymptotics in various regimes. The methods of weakly nonlinear geometric optics were introduced by Lighthill, Landau, and Whitham in the late 1940's and early 1950's; however, it is only recently in work of J. Keller, Hunter, Rosales, and the author that these methods were developed systematically for general hyperbolic systems (see the references in [17]). Futhermore, these methods are even better than the formal predictions of the theory and even apply after shock waves have formed - rigorous proofs of these facts have been presented for general homogeneous nonlinear hyperbolic systems in a single space variable by DiPerna and the author ([18]). We begin our discussion in this Chapter with a terse review of simple wave expansions for weakly nonlinear geometric optics for homogeneous nonlinear hyperbolic systems - we use general notation here and elsewhere to streamline the presentation although in this review, the reader should regard the abstract system presented below as the inviscid (non-reacting) compressible Euler equations. A much expanded discussion and review of the theoretical and systematic asymptotic developments of these methods are presented in the author's recent paper ([17]).

Nonlinear Geometric Optics for Inviscid (Non-Reacting) Compressible Fluid Flow

We consider the compressible Euler equations in 1-D as a special case of the general MxM hyperbolic system,

$$u_t + (F(u))_x = 0 , \qquad t > 0 \qquad -\infty < x < \infty \tag{2.1}$$

where $u = {}^t(u_1,\ldots,u_M)$ is an M-vector - for compressible flow u is a three-vector with components given by density, momentum, and total energy. We assume that $F(u)$ has the Taylor expansion about a constant state, u_o, given by

$$F(u_o + \varepsilon v) = F(u_o) + \varepsilon Av + \varepsilon^2/2\ B(v,v) + O(\varepsilon^3) \tag{2.2}$$

where A is the Jacobian matrix of $F(u)$ at u_o and $B(v,v)$ is the Hessian matrix of quadratic terms. Since the system in (2.1) is hyperbolic, we assume that A has m real eigenvalues $\{\lambda_j\}_{j=1}^M$, with corresponding left and right eigenvectors, $\{\ell_j\}_{j=1}^M$, $\{r_k\}_{k=1}^M$, satisfying

$$Ar_j = \lambda_j r_j , \ \ell_k A = \lambda_k \ell_k , \ \ell_k \cdot r_j = \delta_{kj} . \tag{2.3}$$

To develop simple wave expansions in the method of weakly nonlinear optics, we attempt to construct formal asymptotic solutions of (2.1) with the form,

$$u^\varepsilon = u_o + \varepsilon u_1(\tfrac{\phi}{\varepsilon},t) + \varepsilon^2\ u_2(\tfrac{\phi}{\varepsilon},t) + \tag{2.4}$$

which are valid high frequency asymptotic solutions of (2.1) for bounded time intervals. We note that u_o is a trivial solution of the equations in (2.1) and the solutions in (2.4) represent ε amplitude waves with a short wavelength proportional to ε which perturb this background state - the phase function, ϕ , needs to be determined. By substituting the ansatz from (2.4) into (2.1), we compute that the terms of order zero in ε vanish if

$$(\phi_t I + A\phi_x) \frac{\partial u_1}{\partial\theta} (\theta,t) = 0 \tag{2.5}$$

The only way to satisfy these equations is to choose ϕ to be a solution of the characteristic equation (eikonal equation)

$$\phi_t + \lambda_j \phi_x = 0 \quad \text{for some} \quad j,\ 1 \leq j \leq M\ . \tag{2.6a}$$

With the explicit choice, $\phi_j = x - \lambda_j t$, the equation in (2.5) for $u_1(\theta,t)$ is satisfied provided that

$$u_1(\theta,t) = \sigma_j(\theta,t) r_j \tag{2.6b}$$

where σ_j is an arbitrary function at this stage in the argument. By the method of multiple scales, the equation in (2.1) is satisfied to order $o(\varepsilon)$ provided that we can find $u_2(\theta,t)$ (with sublinear growth in θ) so that

$$(-\lambda_j I + A) \frac{\partial u_2}{\partial\theta} (\theta,t) = (\sigma_j)_t r_j + (\tfrac{1}{2}\sigma_j^2)_\theta\ B(r_j,r_j)\ . \tag{2.7}$$

The matrix $(-\lambda_j I + A)$ has a left null-vector, ℓ_j and by Fredholm's alternative, such a solution $u_2(\theta,t)$ can be found provided that σ_j solves the celebrated inviscid Burgers equation,

$$(\sigma_j)_t + b_j(\tfrac{1}{2}\sigma_j^2)_\theta = 0 \tag{2.8}$$

with $b_j = \ell_j \cdot B(r_j,r_j) - b_j \neq 0$ when the wave field is genuinely nonlinear at u_o . To summarize, we have constructed formal asymptotic solutions for the general hyperbolic system in

(2.1) by solving the much simpler scalar inviscid Burgers equation in (2.8) via

$$u^{\varepsilon} = u_o + \varepsilon\sigma_j\left(\frac{x-\lambda_j t}{\varepsilon}, t\right)r_j + O(\varepsilon^2) \; . \qquad (2.9)$$

Thus, solutions of the inviscid Burgers equation are not only a qualitative model for nonlinear sound waves in compressible flow but are a quantitative model as well. These arguments can be generalized to interacting waves in multi-D (see [17]). We have presented the details of the argument in (2.5)-(2.8) because we will use the ideas of this simple argument in the remainder of Chapter 2 and in a rather complex application in Chapter 3 as well. We remark that is we added small diffusion terms to the right hand side of (2.1), i.e., if we considered

$$u_t + (F(u))_x = \varepsilon^2(Du_x)_x \; , \qquad (2.10)$$

the same reasoning as above applies - the only change is that the scalar, $\sigma_j(\theta,t)$, satisfies the Burgers equation

$$(\sigma_j)_t + b_j(\tfrac{1}{2}\sigma_j^2)_\theta = d_j(\sigma_j)_{\theta\theta} \qquad (2.11)$$

with $d_j = \ell_j D r_j$.

The Nondimensional Compressible Reacting Euler Equations

It will be convenient to describe all of the asymptotic derivations in the chapter through one convenient non-dimensionalized version of the compressible reacting Navier-Stokes equations - for simplicity in exposition, we set all diffusion coefficients in (1.1) equal to zero since they can be trivially incorporated in all results presented below with similar modifications and assumptions as discussed above in (2.10) and (2.11).

With the above assumption the reacting compressible Euler equations for an ideal gas mixture can be written in the form,

$$\rho_t + (v\rho)_x = 0$$

$$(\rho v)_t + (\tfrac{1}{2}\rho v^2 + \frac{1}{\gamma} p)_x = 0 \qquad (2.12a)$$

$$(\rho E)_t + (\rho v E)_x + \frac{1}{\gamma}(pv)_x = \frac{\beta}{\gamma-1}\rho W$$

$$(\rho Z)_t + (\rho v Z)_x = -\rho W \qquad (2.12a)$$

(con't.)

where the equation of state is

$$p = \rho T \qquad (2.12b)$$

and the total specific energy, E , is given by

$$E = \tfrac{1}{2}v^2 + \frac{1}{\gamma-1} T \, . \qquad (2.12c)$$

The source terms, W , representing Arrhenius kinetics have the form

$$W = \frac{\varepsilon Z}{\beta} \exp\left(\frac{\tilde{\theta}}{T_0} - \frac{\tilde{\theta}}{T}\right) \qquad (2.13a)$$

with

$$\varepsilon = \frac{T_0^2}{\tilde{\theta}} \qquad (2.13b)$$

The variables T , ρ , p have been made dimensionless by reference to corresponding quantities in the cold unshocked gas; T_0 is a second higher reference temperature representing either an ignition temperature or a convenient higher reference temperature in the shocked gas (i.e., $T_0 \geq 1$). The velocity scale is the frozen sound speed in the cold gas while time is scaled by the homogeneous induction time at the second reference temperature T_o .

The quantity, $\tilde{\theta}$, is the dimensionless activation energy.

(2.14)

The quantity, β , is the heat release parameter.

This convenient nondimensionalization is the one used by Jackson and Kapila in their recent interesting paper ([16]).

IIa. Asymptotic Derivation of the Qualitative-Quantitative Model from Chapter I

We will derive the inviscid qualitative-quantitative model equations from (1.5), i.e.,

$$\sigma_t + (\frac{b}{2}\sigma^2 - q_o Z)_x = 0$$

$$Z_x = \phi(\sigma) Z \tag{2.15}$$

from the reacting compressible Euler equations in (2.12) using some of the ideas from weakly nonlinear asymptotics as described in (2.1)-(2.9) coupled with high activation energy asymptotics following the treatment by Rosales and the author ([3]). First, we will take $T_o = 1$ and expand about a reference state with $\rho_o = 1$, $p_o = 1$, $v_o = 0$. We will regard $T_o = 1$ as the critical reference temperature where substantial reaction occurs for $T > 1$ and negligible reaction occurs for $T < 1$ - the derivation we present here will apply to Arrhenius kinetics but the model obtained in (2.15) will suffer from the standard cold boundary difficulty unless we truncate the kinetics - the reader can consult [3] for a much more general derivation under different hypotheses than we present.

The crucial assumptions in our derivation are the following ones:

1. The non-dimensional activation energy, $\tilde{\theta} = \frac{1}{\varepsilon}$ is large.
2. The non-dimensional heat release, β is small and in fact $\beta = \hat{q}\,\varepsilon^2$. (2.16)
3. The perturbations of the reference state have amplitudes of order ε balanced with wave length of order ε.

Under the three assumptions in (2.16), we will derive the asymptotic equations in (2.15) as a uniformly valid leading order asymptotic approximation from the equations in (2.12) for non-dimensional times of order one, i.e., comparable to the induction time. We note that only a small amount of reactant is consumed during the induction time; on the other hand, under small heat release assumption in 2 of (2.16), there is a lean mixture for the reactant so substantial consumption of the reactant occurs on this time scale. Following (2.1), we denote the inviscid (nonreacting) compressible Euler equations by the notation,

$$u_t + (F(u)_x) = 0 .$$

Then, under the assumptions 1 and 2 from (2.16), the reacting compressible Euler equations from (2.12) have the nondimensional form,

$$\begin{aligned} u_t + F(u)_x &= \varepsilon W(u,u_o,\varepsilon) Z\vec{e}_o \\ Z_t + vZ_x &= -(\varepsilon\hat{q})^{-1} W(u,u_o,\varepsilon) Z \end{aligned} \tag{2.17a}$$

with $\vec{e}_o = {}^t(0,0,1)$, v the fluid velocity, and $W(u,u_o,\varepsilon)$ given by

$$W(u,u_o,\varepsilon) = \exp(\tfrac{1}{\varepsilon}(1 - \tfrac{1}{T})) .$$

Consistent with the assumptions in 3 of (2.16), we seek small amplitude, high frequency asymptotic solutions of (2.17) through the ansatz,

$$
\begin{aligned}
u &= u_o + \varepsilon u_1(\tfrac{\phi}{\varepsilon},t) + \varepsilon^2 u_2(\tfrac{\phi}{\varepsilon},t) ++ \\
Z &= Z_o(\tfrac{\phi}{\varepsilon},t) + \varepsilon Z_1 ++ \quad .
\end{aligned}
\tag{2.18}
$$

First, we substitute the ansatz for u from (2.18) into the first equation in (2.17). The order zero terms in ε are identical to those which we have already discussed in (2.5) and (2.6) leading to the form,

$$
\begin{aligned}
\phi_k &= x - \lambda_k t \\
u_1 &= \sigma_k(\theta,t) r_k
\end{aligned}
\tag{2.19}
$$

At this stage in the argument, σ_k is arbitrary and

> λ_k can be either of the two non-dimensional sound waves with speed ± 1 or the entropy wave with speed 0 . (2.20)

By expanding the smooth function, $\exp(1 - \frac{1}{T})$ around $T = 1$ in a power series, as in (2.7), the terms of order ε have the form,

$$(-\lambda_k + A)\ \frac{\partial}{\partial\theta} u_2(\theta,t) = \sigma_t r_k + (\tfrac{1}{2}\sigma^2)_\theta B(r_k,r_k) + e^{\alpha_k\sigma} Z_o(\theta,t)\vec{e}_3 \tag{2.21}$$

where we have dropped the subscript on σ and $\alpha_k > 0$ is a coefficient which we don't write here explicitly. Next, we apply the same solvability condition as we used below (2.7) to arrive at the equation,

$$\sigma_t + b_k(\tfrac{1}{2}\sigma^2)_\theta = c_k e^{\alpha_k \sigma} Z_o(\theta,t) \tag{2.22}$$

with $c_k \neq 0$ given by $\ell_k \cdot e_3 \neq 0$.

Next, we do the asymptotics in the mass fraction equation from (2.17a). Since the fluid is at rest at the constant state, u_o , it follows that the fluid velocity, v , satisfies $v = O(\varepsilon)$. With $\phi = x - \lambda_k t$, by substituting the ansatz for Z from (2.18) into (2.17), we see that the leading terms of order ε^{-1} vanish provided that

$$-\lambda_k (Z_o)_\theta = -(\hat{q})^{-1} e^{\alpha_k \sigma} Z_o \tag{2.23}$$

Since $\lambda_k = \pm 1$ for the sound waves (we only treat waves travelling to the right for simplicity so that $\lambda_k = +1$), we obtain the equation

$$(Z_o)_\theta = (\hat{q})^{-1} e^{\alpha_k \sigma} Z_o \ . \tag{2.24}$$

We remark that the expansion as presented is inconsistent if we considered the entropy wave with $\lambda_k = 0$ instead of the non-linear sound waves. Now, if we simply rescale the θ variable and define $q_o = c_k \hat{q}$, then with the specific function $\phi(\sigma) = \exp(\alpha_k \sigma)$, the equations in (2.22) and (2.24) are exactly the quantitative-qualitative model equations from Chapter 1 with x replaced by θ as we listed earlier in (2.15) of this chapter. For a more detailed derivation of the quantitative-qualitiative asymptotic model from (1.5) including the effects of diffusion which produces viscosity but not species diffusion in the asymptotic model, we refer the reader to [3].

The Asymptotic Model Equations in Multi-D

One advantage of the derivation which we have presented above is that it extends readily to dynamic problems in multi-dimensional with minor modifications involving the theory of weakly nonlinear geometric optics (see [3], [17]). Here we present a special case of the formulae in (1.02)-(1.06) of [3] and then we make a few comments about the behavior of these equations as compared with the behavior of the 1-D equations in (1.5). We copy the formulae from (1.02) and (1.06) of [3] with the special choices, $m = 1$, $\bar{\rho} \equiv \bar{T} \equiv \bar{u} \equiv 0$ and arrive at equations where the reaction front is defined by

$$\Psi(x) - t = 0$$

and Ψ solves the eikonal equation,

$$|\nabla\Psi|^2 = 1 \quad . \tag{2.25}$$

The coordinate θ is defined by $\theta = (\Psi(x)-t)/\varepsilon$, the obvious generalization of $(x-\lambda_k t)/\varepsilon$ from 1-D, and the time variable τ is proportional to arc-length integration along the characteristic rays of (2.25). Specifically, τ is defined by

$$\frac{dx}{d\tau} = \nabla\Psi(x)$$

with the initial condition $\tau = 0$ at $\Psi(x) = 0$, the position of the reaction front at time $t = 0$. In this case, we have

<u>The Multi-D Asymptotic Equations</u>

$$\begin{aligned} &\sigma_\tau + \tfrac{1}{2}b(\sigma^2)_\theta - (q_o Z)_\theta + \tfrac{1}{2}(\Delta\Psi)\sigma = \beta\sigma_{\theta\theta} \\ &Z_\theta = K\phi(\sigma)Z \end{aligned} \tag{2.27}$$

where $\Delta\Psi$ has the well-known explicit formulae,

$$\Delta\Psi = \sum_j K_j^o \, (1 + K_j^o\tau)^{-1} \tag{2.28}$$

with K_j^o, the principal curvatures of the reaction front at time, $t = 0$. <u>The equations in (2.27) are identical to those</u>

for 1-D presented in (1.5) except for the additional term from $\Delta\Psi$. This term acounts for the amplitude growth or decay according to the local geometric compression or expansion of areas in the wave front of geometric optics. In the inviscid case and for an initially unit outgoing spherical wave front in N-space dimensions, the equations in (2.27) reduce to

$$\begin{aligned} &\sigma + b(\tfrac{1}{2}\sigma^2)_\theta - (q_o Z)_\theta + \frac{N-1}{2}(1+\tau)^{-1}\sigma = 0 \\ &Z_\theta = k\phi(\sigma)Z \ . \end{aligned} \tag{2.29}$$

Obviously, the additional linear undifferential term in σ causes attenuation of the σ amplitude and this competes with the wave amplification produced by the term with $(q_o Z)_\theta$ which causes wave amplification through exothermic heat release. Thus, the multi-D quantitative model confirms the trends observed experimentaly that initiation of detonation or transition to detonation are much less likely to occur in expanding spherical geometries than in plane geometries due to wave attenuation from the expanding geometry. A more detailed investigation of these competing effets by Roytburd, Rosales, and the author is currently in progress.

Wave Amplification in the Asymptotic Model - An Example Involving Initiation of Detonation

Here we give two numerical examples illustrating the competition between the wave amplifying mechanisms of exothermic heat release and the wave attenuation caused by the nonlinear spreading of rarefaction waves. We study a caricature of the problem of initiation of detonation in the inviscid asymptotic model equations - we use Arrhenius kinetics with a large fixed activation energy for the kinetics structure function $\phi(u)$. The initial data consisted of a step of width, .5, with the two different heights, 1.5 and 1.4, for the u-profile at time zero while at $t = 0$, only reactant was present so that $Z=1$.

The front edge of the u-wave initially generates a shock which triggers combustion while the back of the u-wave generates a rarefaction. In Figures 2.1 and 2.2 we present the graphs of successive equally spaced snapshots of the waves that emerged from a fully resolved time dependent numerical calculations for the two initial data. In the first case with height, 1.4 , insufficient wave amplification is caused by the exothermic heat release, the spreading of the rarefaction attenuates the amplitude, and the solution ultimately decays to zero so that a detonation wave is not generated. In the second example with an initial height of u given by 1.5 , initiation of detonation occurs and beyond the time $t = 2$, a C-J detonation emerges followed by the expected rarefaction region. These calculations are taken from work in progress by Rosales, Roytburd, and the author. We mention to the reader that the spatial scales in the two graphs are different - these initial pulses have the same width and differ in height by only .1 (i.e., 7.5%) yet the higher pulse initiates detonation while the slightly lower pulse fails. Incidentally, the numerical pulsations in the second figure disappear under mesh refinement.

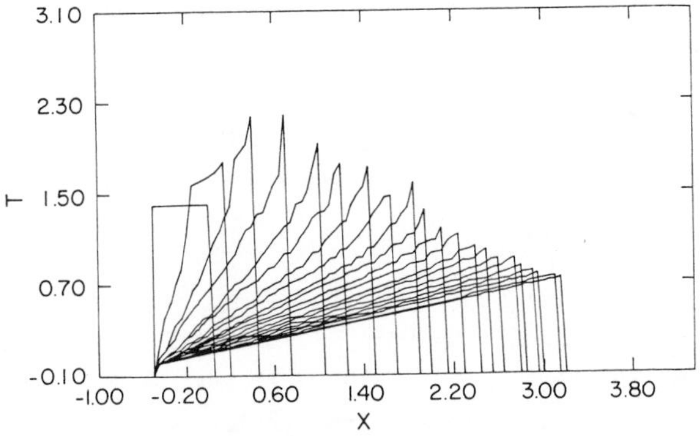

Figure 2.1: An Example of Failure to Initiate Detonation in the Quanlitative-Quantitative Model

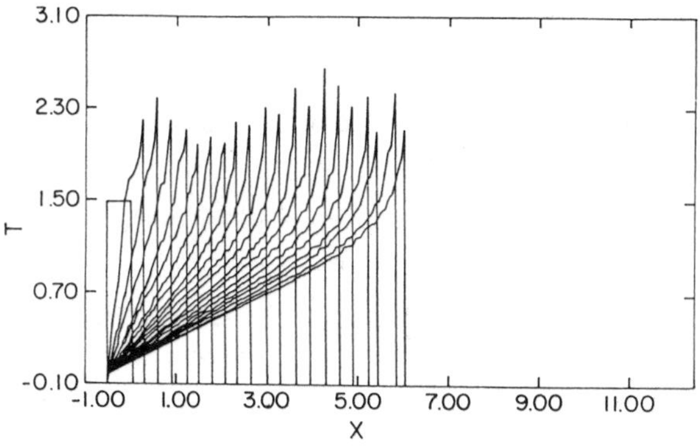

Figure 2.2: An Example of Initiation of Detonation in the Qualitative-Quantitative Model

IIb. Low Frequency Nonlinear Wave Amplification in the Induction Zone - The Equation of Clarke -Gant - Jackson -Kapila

We will study small amplitude low frequency waves on the induction time scale superimposed over a constant background state with reference temperature $T_o > 1$. We refer to the non-dimensional form of the equations discussed above (2.12) which is appropriate for this problem. We can construct low frequency asymptotic solutions for the compressible reacting Euler equations from (2.12) which are valid for times of order one (induction times) by the straightforward power series expansion

$$u = u_o + \varepsilon u_1(x,t) + \varepsilon^2 u_2(x,t) ++$$
$$Z = 1 + \varepsilon Z_1 + \varepsilon^2 Z_2 ++ \quad . \tag{2.30}$$

under the critical assumption that the non-dimensional activation energy has the order ε^{-1} with $\varepsilon << 1$. Thus, with the low frequency power series ansatz from (2.30), we anticipate that in the leading order asymptotics, the hydrodynamics is linear but the kinetics is nonlinear. As we have done earlier in (2.17), the reacting compressible Euler equations from (2.12) have the form,

$$u_t + F(u)_x = \varepsilon W(u,u_o,\varepsilon) Z\vec{e}_o$$
$$Z_t + vZ_x = -\frac{\varepsilon}{\beta} W(u,u_o,\varepsilon) Z \tag{2.31a}$$

with

$$W = \exp \frac{1}{\varepsilon} \left(\frac{1}{T_o} - \frac{1}{T}\right) . \tag{2.31b}$$

If we let T_1 denote the perturbed temperature and assume that $v_o = 0$, we substitute the ansatz from (2.30) into (2.31) and collect the terms of order ε to derive the equations,

$$\frac{\partial u_1}{\partial t} + A \frac{\partial u_1}{\partial x} = e^{T_1} \vec{e}_o$$
$$\frac{\partial Z}{\partial t} = - \beta^{-1} e^{T_1} . \tag{2.32}$$

These are a coupled system of semi-linear hyperbolic equations - the gas dynamics is linearized and the temperature changes in the kinetics are nonlinear. In [16], Jackson and Kapila have written the equations in (2.32) in characteristic form and have observed temperature amplification via numerical integration of these equations. It was observed independently in [15] and [16] that the equations in (2.32) can be written as the following single third order semilinear hyperbolic equations, the C-G-J-K Equations:

$$\left(\frac{\partial^2}{\partial t^2} - \frac{\partial^2}{\partial \xi^2}\right) \frac{\partial}{\partial t} T_1 = \left(\gamma \frac{\partial^2}{\partial t^2} - \frac{\partial^2}{\partial \xi^2}\right) e^{T_1} \tag{2.33}$$

with T_1 the perturbed temperature. The reader should consult the references ([15], [16]) for a more detailed discussion. Obviously, it is an interesting mathematical problem to determine initial data for either (2.32) or (2.33) so that the temperature increases rapidly and perhaps becomes infinite in finite time, i.e., thermal runaway occurs. The author believes that this is an accessible open mathematical question.

It is also natural to attempt to analyze the mutual resonant interaction of both high frequency and low frequency waves in the induction zone. A uniformly valid set of asymptotic equations to describe these resonant interactions with large heat release will appear in a forthcoming paper of Rosales and the author ([19]).

IIc. High Frequency Simple Sound Waves in the Induction Zone

Here we will apply similar ideas as we used in Section IIa) to derive the equations first given independently by Blythe ([31]) and Clarke ([32]) for nonlinear simple sound wave propagation in the induction zone. A variable coefficient Burgers equation with source terms is the final result of the derivation -- this equation also exhibits substantial sound wave amplification although we will not discuss this here (see [32] for some results).

With the non-dimensionalization discussed in (2.12), (2.13), the reacting compressible Euler equations from (2.12) have the form given in (2.31). The crucial assumptions needed in the asymptotic derivation are the following:

1) The activation energy is of order ε^{-1} with $\varepsilon \ll 1$.
2) The wave is a simple sound wave with amplitude of order ε and wavelength of order ε. (2.34)
3) The wave form decays rapidly in space (of order ε^N, N large) in a way to be made precise below.

Anticipating the same order zero equations as in (2.5) and (2.6), we attempt to construct asymptotic solutions of the equations in (2.31) with the ansatz,

$$u = u_o + \varepsilon(\bar{u}_1(t) + \sigma_k(\frac{\phi_k}{\varepsilon}, t)r_k) + \varepsilon^2 u_2(\frac{\phi_k}{\varepsilon}, t) ++$$

$$Z = 1 + \varepsilon(\bar{Z}_1(t)) + \varepsilon^2(Z_2(\frac{\phi_k}{\varepsilon}, t)) ++ . \tag{2.35}$$

First, we concentrate on the equation for the fluid dynamic variables, u. We choose $\phi_k = x - t$ as the solution of the eikonal equation, $\phi_t + \lambda_k \phi_x = 0$, with wave speed $\lambda_k = 1$ corresponding (non-dimensionally) to the right sound wave; then, the ansatz in (2.35) automatically guarantees that the order zero terms in powers of ε vanish. Our assumption in (2.35) is the requirement that

$$\frac{1}{2T}\int_{-T}^{T} \sigma(\theta,t)d\theta = O(T^{-N}) \tag{2.36}$$

for N sufficiently large. We will see below that this assumption is always satisfied trivially provided that

> the initial wave form, $\sigma_o(\theta)$, has compact support. (2.37)

The term, $\bar{u}_1(t)$, represents a correction to the mean field caused by homogeneous nontrivial chemical reaction. As in (2.7), the u-equation from (2.31) vanishes to order ε provided that we can choose $u_2(\theta,t)$ with sublinear growth in θ and satisfying,

$$(-\lambda_k I + A)\frac{\partial u_2}{\partial \theta}(\theta,t) = \{-(\sigma_k)_t r_k - (\tfrac{1}{2}\sigma_k^2)_\theta B(r_k,r_k)$$

$$- (\sigma_k)_\theta B(\bar{u}_1(t),r_k) + \vec{e}_o \exp(\vec{u}_1(t)\cdot\vec{w}_o)[e^{\alpha_k\sigma} - 1]\} +$$

$$\{-\frac{\partial \bar{u}_1(t)}{\partial t} + \vec{e}_o \exp(\vec{u}_1(t)\cdot\vec{w}_o)\} \equiv \{1\} + \{2\} \tag{2.38}$$

All of the terms in braces, $\{1\}$, have mean zero as a consequence of (2.36); on the other hand, the terms in braces $\{2\}$, are terms with individual non-zero means. The constant, $\alpha_k > 0$, and the vector, $\vec{w}_o$, we don't compute explicitly here. There are two solvability conditions to guarantee that $u_2(\theta,t)$ can be found with sublinear growth in θ. The first arises because the left hand side always has zero mean in θ; to avoid secularity, the right hand side must also have zero mean. Thus, the collection, $\{2\}$, must vanish, i.e., the mean field should satisfy the equation

$$\frac{\partial \bar{u}_1(t)}{\partial t} = \vec{e}_o \exp(\bar{u}_1(t) \cdot \vec{w}_o) . \tag{2.39a}$$

By explicitly computing this equation, it is not difficult to see that it reduces to the scalar O.D.E. for $T_1(t)$, the perturbed temperature,

$$\frac{\partial T_1(t)}{\partial t} = e^{T_1(t)} \tag{2.39b}$$

The equation in (2.39) is the famous high activation energy, leading order, induction time solution arising in the classical theory of homogeneous explosiions and is trivially integrated (with $T_1(o) = 0$, $T_1(t) = -\ell n(1-t)$ with $t \equiv 1$ the nondimensional time for homogeneous explosion). With term $\{2\}$ vanishing in (2.38), the solvability condition for $u_2(\theta,t)$, is exactly the same as the familiar solvability condition discussed below (2.7) and used earlier in section 2.A)--the inner product of $\{1\}$ with the left eigenvector, ℓ_k, necessarily must vanish. The result is the non-homogeneous Burger equation describing nonlinear simple sound waves in the induction zone,

$$\sigma_t + a_k(t)\sigma_\theta + b_k(\tfrac{1}{2}\sigma^2)_\theta = c_k \exp(T_1(t))[e^{\alpha_k \sigma} - 1] \tag{2.40a}$$

with $c_k = \ell_k \cdot \vec{e}_o \neq 0$, $\alpha_k > 0$, and $a_k(t)$ given by

$$a_k(t) = \ell_k \cdot B(\bar{u}_1(t), r_k) \ . \tag{2.40b}$$

We recall that $\bar{u}_1(t)$ is determined via the homogeneous induction time solution in (2.39). We leave to the reader, the straightforward exercise to check that the asymptotic expansion for Z as presented in (2.35) is self-consistent with the expansion from (2.35) which we have just derived for u.

The scalar equation for $\sigma(\theta,t)$, derived in (2.40), needs to be supplemented by initial conditions,

$$\sigma(\theta,t)_{t=0} = \sigma_o(\theta) \ .$$

Provided that this initial data, $\sigma_o(\theta)$, satisfies the requirement in (2.36), the solution of (2.40) continues to satisfy this condition for $t > 0$ as long as the amplitude remains finite; thus, the expansion is self-consistent. The equation in (2.40) is substantially simpler than the reacting compressible Euler equations and an initial study of the properties of solutions was begun by Clarke ([32])--much more remains to be done on this accessible open problem. As we have mentioned earlier, a uniformly valid set of asymptotic equations to describe mean field--high frequency sound wave, resonant wave amplification in the induction zone will be presented in a forthcoming paper of Rosales and the author.

III. Instabilities in Detonations and Complex Wave Bifurcations

The 1-D theory discussed in Chapter I with predictions of wave speed using the Chapman-Jouget theory gives excellent agreement (within a few percent) with the wave speeds occurring in reacting shock tube experiments (see the Chapters of [5] on detonations). Therefore, experimental observations from the late 1950's and 1960's by Oppenheim, Soloukhin, and others that the actual structure of the propagating reacting shock waves is enormously complex and multi-dimensional with embedded Mach stems propagating transverse to the main uni-directional front startled the theoretical combustion community, firmly committed, at that time, to the one-dimensional Chapman-Jouget and Z-N-D structure theories. Here is a schematic figure and schematic soot film experimental record of the complex bifurcating wave patterns actually observed.

Somewhat later, in the late 1960's and early 1970's, Toong and his co-workers ([20]) observed pulsating instabilities in experiments involving exothermic hypersonic flows about blunt bodies - the pictures on pages 540-541 of [20] document these striking facts. Unlike the typical instabilities mentioned in the first paragraph which are inherently multi-dimensional, these pulsations are purely one-dimensional instabilities but are capable of raising the local pressure spikes by as much as a factor of two when compared with the von Neumann spikes in the predictions of the Z-N-D theory.

These two striking and different experimental observations led to a vigorous amount of theoretical activity in the stability of detonation waves throughout the 1960's. The most significant contributions to this theory with the most complete stability theory are developed in a series of interesting papers by Erpenbeck (see the references in [4]; other interesting theories attempting a qualitative explanation of the multi-D instabilities observed in experiments were developed by Strehlow and his co-workers (see [4] for the references). The author recommends the review paper of Erpenbeck in the Twelfth Symposium on Combustion ([9], 1969) for an excellent summary and critique of that activity in detonation stability theory. Despite all of this work, a theory which might explain the mechanisms which yield multi-dimensional bifurcations to complex embedded Mach configurations remained inaccessible; in particular, an adequate explanation for the remarkable recurrence and regular spacing structure for the embedded Mach configurations has not been given. A lucid summary is given in the recent book by Fickett and Davis ([4]).

An explanation of all these complex multi-D phenomena represents an extremely challenging class of theoretical problems for applied mathematicians. In this chapter, we discuss a recently developed theory by Rosales and the author ([21], [22]) which provides mechanisms and a theory for the spontaneous formation of Mach stems in reacting shock fronts. We will only describe the asymptotic derivation and conclusions of that theory here - an independent confirmation using steady wave bifurcations (see [22]) as well as the numerical studies for the scalar integro-differential equation derived through asymptotics are discussed in the paper of Rosales at this meeting. One shortcoming of this theory is that the Chapman-Jouget structure is postulated at the outset and finite reaction zone effects are ignored. The authors used this simplification to yield a manageable asymptotic problem and did succeed in developing a theory which predicts the spontaneous formation of transverse propagating Mach stem. However, the instability bounds for such wave bifurcations to develop are somewhat unrealistic and this initial theory precludes any discussion of a theory for regular spacings. An important problem remaining is to include finite reaction zone effects both to develop more realistic stability bounds and also to predict the regular spacing.

1-D Instabilities: Pulsations and Galloping Detonations

The theoretical problem of explaining the purely one-dimensional pulsating instabilities is a challenging one but not nearly as difficult as the multi-D problem mentioned above. The theoretical reasons for this statement are clear -in the 1-D pulsating instabilities, there is no change in wave structure (the reacting shock remains a reacting shock) and the phenomena occur at essentially low frequencies; in the multi-d instabilities with complex Mach configurations, the wave structure changes dramatically so there must be a nonlinear transfer of energy to the high frequencies to trigger the discontinuous wave bifurcation.

There is no doubt that Toong and his coworkers (see [20], [23]) have developed an excellent qualitative understanding of the mechanisms responsible for this pulsating instability. The recent paper by Abouseif and Toong (see [23]) studies the development of linearized instabilities in a Z-N-D wave profile at sufficiently high activation energies by utilizing this large parameter in the calculations. With that paper as a starting point, it is clear to this author that a rational perturbation expansion at large activation energies can be developed to predict the nonlinear onset of the galloping detonation in 1-D.

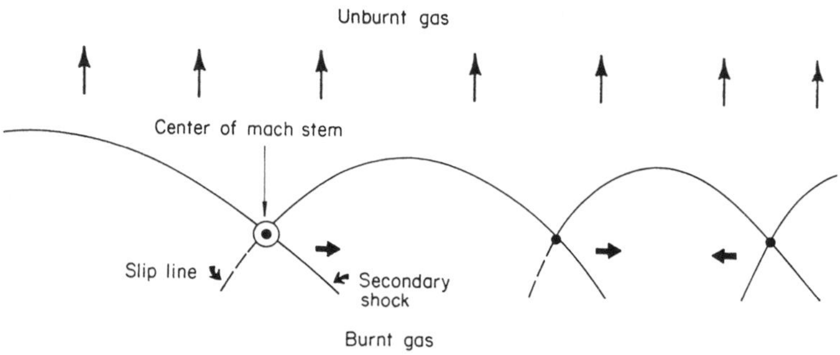

Figure 3.1: The overall direction of propagation of the radiating shock front is along the y-axis, but the embedded Mach stems have a transverse direction of propagation along the x-axis.

A Theory for Spontaneous Formation of Mach Stems in Reacting Shock Fronts

The theory which we will develop in this chapter will predict the development of complex transverse wave structures like those in Figure 3.1 as instabilities in a planar reacting shock front. The basic issue which we address here is the following:

Why do transverse Mach stems form in reacting gases? What are the mechanisms responsible for Mach stem formation? (3.1)

Any explanation for spontaneous formation of such transverse wave structures must be consistent with the following experimental fact:

Transverse Mach stems never form spontaneously in unidirectional shock fronts in inert polytropic gases. (3.2)

IIIa. The Argument for Spontaneous Mach Stem Formation

Step I. The Chapman-Jouget Structure is Postulated

We simplify the physical situation be making the Chapman-Jouget approximation for the physical description of the propagating detonation front. Thus, we assume

1. Reacting shock fronts are infinitely thin and are idealized as discontinuities.
2. The gas particles crossing the front instantaneously adjust from thermodynamic equilibrium in the unburnt gas ahead of the front to thermodynamic equilibrium in the burnt gas behind the front. (3.3)

In a standard fashion, as we discuss earlier, the assumptions in (1) and (2) above allow us to model the physics by using the Euler equations of compressible gas dynamics with a very general equation of state, $e(\tau,p)$, where e is the specific internal energy, $\tau = \frac{1}{\rho}$ is the specific volume with ρ the density, and p the pressure.

Step II: Detailed Linearized Stability for Multi-D Shock Fronts

Next, we study the detailed linearized stability of a planar over-driven detonation front to multi-dimensional perturbations. For definiteness we assume that this planar wave front is steady in a reference frame described by $y = 0$, and in the remainder of this section we use the shorthand notation, u_o^+, u_o^-, to describe the four vectors of basic unperturbed flow quantities, $(\tau_\pm, u_\pm, v_\pm, p_\pm)^T$ where $u_\pm$, $v_\pm$ are the unperturbed x and y components of fluid velocity, and $\tau_\pm$, $p_\pm$ are the unperturbed specific volume and pressure. The critical observation in our analysis is the existence of "radiating boundary waves" for certain parameter ranges in the equation of state. Except for the analysis by Majda in [24], such modes of propagation have been ignored in the literature on shock wave stability. These radiating boundary waves are outgoing modes of propagation with the following structure:

> There is a critical boundary velocity $c_* > 0$ so that the linearized shock front equations have special solutions in the burnt gas region, $y < 0$, with the form

$$\phi = \sigma(x - c_* t)$$

$$u_1^- = \sigma'(x - c_* t - \lambda_1 y)\hat{e}_1 + \sigma'(x - c_* t - \lambda_2 y)\hat{e}_2 , \tag{3.4}$$

where $\sigma(s)$ is an arbitrary smooth function and $\sigma'(s)=\partial\sigma/\partial s$. Here ϕ is the function describing the linearized perturbed shock front position and the numbers λ_1, λ_2 have the physical interpretation,

1. λ_1 is associated with <u>outgoing acoustic wave</u> radiation from the shock front.
2. λ_2 is associated with <u>outgoing convection along the particle paths</u> defined by c_* and the flow velocity. (3.5)

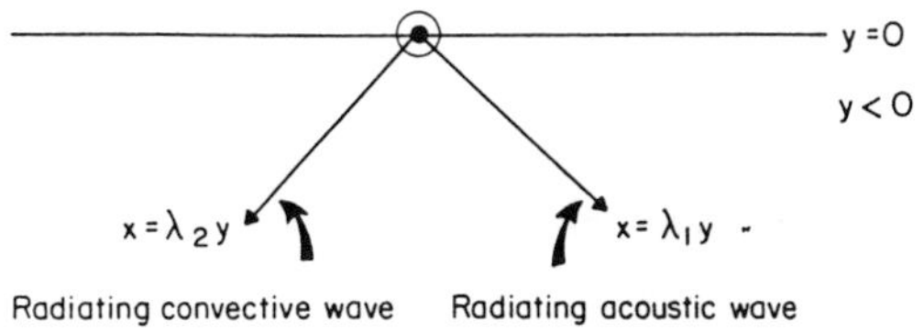

Figure 3.2: Spatial projection at a given time (t=0) of the characteristic surfaces emanating from (0,0) into $y < 0$ associated with a radiating boundary wave at a critical velocity c_*.

(see Figure 3.2). Here $\hat{e}_1$, $\hat{e}_2$ are four-vectors with a complicated algebraic form which we display in section 3. The outgoing radiation condition is determined by the natural requirement that a wave associated with the characteristic surfaces

$$x - c_* t - \lambda_i y = \text{constant}$$

is outgoing provided energy is transmitted from the shock boundary, $y = 0$, into the region, $y < 0$, as time increases. Thus, the bicharacteristic rays, which transport this energy, should move into the region $y < 0$ from the shock boundary as time increases. The y-component of such a ray is given by the formula

$$y = -\left(\frac{\partial \lambda_i}{\partial c_*}\right)^{-1} (t - t_o) ,$$

thus, the outgoing radiation condition is equivalent to the assertion that

$$\frac{\partial \lambda_i}{\partial c_*} > 0 , \quad i = 1,2 . \tag{3.6}$$

(We have been especially careful in this definition because the unperturbed flow is anisotropic in $y < 0$ as a consequence of the fact that the y-component of the fluid velocity, v_-, does not vanish.) We can characterize precisely when such boundary waves defined in (3.4) occur. We consider the three parameters,

Mach number, $M_1 = -\frac{v_-}{c_-}$, c_- the sound speed, $0 < M_1 < 1$

Compression ratio, $\mu = \frac{\rho_-}{\rho_+}$, $\mu > 1$

Gruneisen coefficient, $\Gamma = (\rho_- e_p(\tau_-, p_-))^{-1} > 0$.

<u>Radiating boundary waves exist precisely when these three parameters satisfy</u>

$$\frac{1}{\Gamma+1} < (\mu-1)M_1^2 < \frac{1+M_1}{\Gamma} . \tag{3.8}$$

The regime described by the inequality

$$\frac{1+M_1}{\Gamma} < (\mu-1)M_1^2 \tag{3.9}$$

is an equivalent form (see [4], pp. 254-255) of the well-known conditions for exponentially growing modes of instability for the linearized shock front first studied by Erpenbeck [4] for a general equation of state - very strong violent instability occurs in this regime. The regime described by the inequality

$$(\mu-1)M_1^2 < \frac{1}{\Gamma+1} \tag{3.10}$$

is also interesting. Majda [24] has recently given a rigorous proof of the full nonlinear structural stability of the wave front under these conditions -spontaneous Mach stem formation is not possible for the regime described by (3.10). Thus, the conditions in (3.8) for the existence of radiating boundary waves represent an intermediate regime between a regime of strong stability defined by (3.10) and a regime of strong instability defined by (3.9). Also, the existence of radiating boundary waves as a mechanism for Mach stem formation is consistent with the experimental observations in (3.2) since for shocks in ideal gases $\Gamma = \gamma - 1$ while $(\mu - 1)M_1^2 = \frac{1}{\gamma}(1 - \frac{p_+}{p_-})$ so that the inequality in (3.10) defining the region of strong stability is always satisfied.

<u>Step III: Weakly Nonlinear Asymptotics at the Shock Front</u>

We develop an appropriate weakly nonlinear perturbation theory based upon the radiating boundary waves discussed in Step II. Thus, for the regime where the inequalities in (3.8) are satisfied, we find small amplitude perturbed shock front solutions of the Euler equations about the basic unperturbed planar shock front. We introduce the slow time scale, $\tau = \varepsilon t$, and find asymptotic solutions with the form

$$\phi = \varepsilon\sigma(x - c_* r, \tau) + 0(\varepsilon^2)$$

$$u^+ \equiv u_0^+ \quad \text{for} \quad y > 0 \tag{3.11}$$

$$u^- = u_0^- + \varepsilon\sigma'(x - c_* t - \lambda_1 y, \tau)\hat{e}_1 + \varepsilon\sigma'(x - c_* t - \lambda_2 y, \tau)\hat{e}_2 + 0(\varepsilon^2) \quad \text{for } y < 0$$

where the perturbed shock front location is described by $y = \phi$. We use the notation $\hat{x} = x - c_* t$ and $\sigma'(\hat{x}, \tau) = \frac{\partial\sigma}{\sigma\hat{x}}(\hat{x}, \tau)$. The solvability requirements needed to continue the perturbation expansion to second order yield a nonlinear equation for the function $\sigma'(\hat{x}, \tau)$. In fact, $\sigma'(\hat{x}, \tau)$ satisfies the <u>integro-differential scalar conservation law</u>, for $\tau > 0$, $-\infty < \hat{x} < \infty$,

$$0 = \sigma' + a_1 (\tfrac{1}{2}(\sigma')^2)_{\hat{x}} + a_2 \left(\int_0^\infty \sigma'(\hat{x} + \beta s)\sigma'_{\hat{x}}(x + s)\,ds\right)_{\hat{x}}$$

$$\sigma'(\hat{x}, 0) = \sigma'_0(\hat{x}) \tag{3.12}$$

where β, a_1, a_2 are constants satisfying $\beta > 1$, $a_1 - a_2 \neq 0$ and have complicated algebraic expressions. Here $\sigma_0'(\hat{x})$ is a smooth initial shock front perturbation function vanishing rapidly as $|\hat{x}| \to \infty$. The reasons for the appearance of the integro-differential expression in (3.12) are clarified later, but we mention here that it arises from the following physical mechanism: The first order outgoing radiating waves produce nonlinear interactions in the second order which generate incoming waves which strike the front--the integro-differential term is a manifestation in the asymptotics of this nonlinear wave interaction.

Completing the Argument Given Steps I-III

Next, using Steps I-III, we complete the argument for spontaneous Mach stem formation. First, we remark that if $\beta \equiv 1$ in (3.12), then the integro-differential term in (3.12) can be integrated resulting in the inviscid Burgers' equation,

$$\sigma'_\tau + (a_1 - a_2)(\tfrac{1}{2}(\sigma')^2)_{\hat{x}} = 0 , \quad \tau > 0$$

$$\sigma'(\hat{x},0) = \sigma_0'(\hat{x}) .$$

Then it is well known ([17]) that for any smooth initial data, $\sigma_0'(\hat{x})$, vanishing for $|\hat{x}|$ large, there is a time τ_0, so that σ' develops a jump discontinuity for times $\tau \geq \tau_0$ at some shock location described by a curve, $\hat{x} = \hat{x}_0(\tau)$, $\tau \geq \tau_0$ (in general there might be several such curves). We anticipate a similar behavior for solutions of the equation in (3.12) where $\beta > 1$ and continue the argument. We recall from (3.11) that to first order, the perturbed shock front location is described by $\phi = \varepsilon\sigma(\hat{x},\tau)$ so that

$$\text{at the breaking points, } \hat{x}_0(\tau),\ \tau > \tau_0 \text{ for } \sigma', \text{ the main shock surface develops a corner.} \tag{3.13}$$

Also from (3.11), (3.5) we see that

$$\bar{u}=\bar{u}_o+\varepsilon(\sigma'(\hat{x}-\lambda_1 y,\tau)\hat{e}_1+\sigma'(\hat{x}-\lambda_2 y,\tau)\hat{e}_2)+0(\varepsilon^2) \qquad (3.14)$$

so that simultaneously, associated with the times and locations described in (3.13) where the surface develops this kink, the flow behind the wave front becomes discontinuous along the two space-time surfaces, with projections in (x,y) space given by

1. $x-c_* t-\lambda_1 y=\hat{x}_o(\tau)$, $\tau>\tau_o$, $y<0$,
   the radiating acoustic sound wave direction (3.15)
2. $x-c_* t-\lambda_2 y=\hat{x}_o(\tau)$, $\tau>\tau_o$, $y<0$,
   the radiating convective wave direction.

By putting together the facts in (3.13)-(3.15), we deduce that all the ingredients required to give the spontaneous formation of the triple-shock, slip-line, Mach stem configuration have been produced by the above perturbation analysis. In particular,

represents the breaking of the main front into two leading shock waves (3.13)

1. represents the simultaneous formation of the secondary shock wave emanating from this breaking location on the main front (3.15)
2. represents the simultaneous formation of the required vortex sheet emanating from the main front (3.15)

(see Figure 3.3). Furthermore, the directions in (3.15) 1 and 2 predict further details in the location of the secondary shock and slip line for the "small amplitude" Mach stem that has formed.

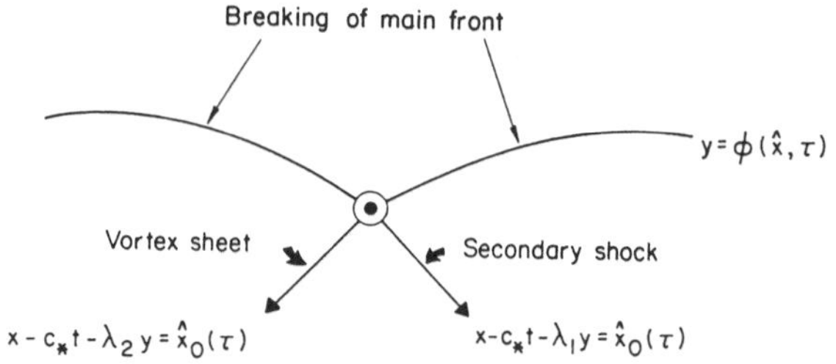

Figure 3.4. Spatial projection at a fixed time t of the spontaneous formation of the entire embedded Mach stem structure predicted through the weakly nonlinear approximation.

In Rosales lecture here, he will discuss the numerical experiments in [22] where we have shown that for a wide range of initial data $\phi_0(x)$ and for wide ranging values of the constants a_1, a_2, solutions of the integro-differential conservation law have breaking waves in the same fashion as the Burger's equation although the nonlocal term produces some prominent new effects in this breaking. If we take this fact for granted, then the argument for spontaneous Mach stem formation is complete. In fact, one week prior to this meeting, R. Gardner has sent the author a letter announcing a rigorous proof of the breakdown of solutions in the scalar integro-differential equation under some restrictions on the constants R, β and the initial data. The physical assumptions in Step I can be criticized as too simplified to yield a complete answer to the question of spontaneous Mach stem formation because of the competing 1-dimensional pulsating instability. Actual stability criteria should involve at least a combination of these two competing effects as well as others.

Below, to concentrate on the essential conceptual details, as we have done in Chapter 2, we will present the argument in an abstract set-up. The reader can see [21] for the tedious algebraic details for gas flow. Strictly speaking, the asymptotic expansion described in (3.11) is not uniformly valid for $|y| \to \infty$. If the reader remembers the weakly nonlinear expansions discussed in (2.1)-(2.9), the conceptual reason for this is clear. In the third equation from (3.11), we have ignored the cummulative effects of the nonlinear waves in the far field - for gas dynamics these are easily incorporated through the addition of other slow space-time scales, ϵx, ϵy, ϵt and is developed in gas dynamics in Section 4.2 of [3]. On the other hand, $|y| \to \infty$ is a comparatively uninteresting regime for the problem studied here - the derivation of (3.11) is uniformly valid where all the action occurs, the shock front, and we can concentrate then on the new asymptotic details.

We have already described the Chapman-Jouget theory in detail in Chapter 1 of these lectures. Thus, it is not necessary to describe part I in detail here. We only remind the reader that with the Chapman-Jouget theory in combustion, we are really regarding the detonation front as a shock wave for the compressible Euler equations with a general equation of state. Obviously, we also can choose a reference frame so that the basic unperturbed planar shock front is steady in that coordinate system.

IIIb. The Detailed Linearized Stability for Shock Fronts and Radiating Boundary Waves

A General Set-Up

With the vector notation $u = (\rho, \vec{m}, E)^T$, the compressible Euler equations can be written as a special case of the general system of conservation laws

$$\frac{\partial u}{\partial t} + \frac{\partial}{\partial x} F_1(u) + \frac{\partial}{\partial y} F_2(u) = 0 , \qquad (3.16)$$

where u is an m-vector, and $F_i(u)$ are nonlinear mappings from R^m to R^m with $A_i(u) = \partial F_i/\partial u$ the corresponding $m \times m$ Jacobian matrices. The basic unperturbed state u_0 is a steady shock located at $y = 0$ and has the form

$$u_o = \begin{cases} u_o^+ , & y > 0 \\ u_o^- , & y < 0 \end{cases}$$

where u_o^+, u_o^- are constant states satisfying the jump conditions

$$F_2(u_o^+) = F_2(u_o^-) \tag{3.17}$$

(The $+(-)$ notation always corresponds to $0(1)$ or unburnt(burnt) in the concrete case.) Without explicitly stating these conditions here, we assume Lax's κ-shock entropy inequalities are satisfied with $\kappa = m$; i.e., strong detonations are supersonic from the front and subsonic from behind.

In regions where solutions of (3.16) are smooth we use the simpler nonconservative form of (3.16)

$$\frac{\partial u}{\partial t} + A_1(u) \frac{\partial u}{\partial x} + A_2(u) \frac{\partial u}{\partial y} = 0 \tag{3.18}$$

Finally, we record the jump conditions for (3.16) for a discontinuity along a general curved surface moving with speed D along the unit normal $\hat{n}$ (pointing toward u^+),

$$-D[u] + [\hat{n}_1 F_1(u) + \hat{n}_2 F_2(u)] = 0 . \tag{3.19}$$

Here, for simplicity in exposition, we only treat κ-shocks with $\kappa = m$ so that only perturbations of u_o^- need to be considered. However, the constructions easily generalize to the case $\kappa \neq m$ where perturbations of both u_o^+ and u_o^- simultaneously need to be incorporated (see [33], where this is done)--this remark is useful for potential applications in other disciplines.

Perturbed Shock Fronts

A perturbed shock front solution of (3.16) is described by the perturbed shock front location, given by the equation

$$y = \phi^{\varepsilon}(x,t) \ , \quad \phi^{\varepsilon}\Big|_{\varepsilon=0} = 0$$

and perturbed states u_{ε}^{+} , u_{ε}^{-} defined in the regions $y > \phi^{\varepsilon}$, $y < \phi^{\varepsilon}$ with

$$u_{\varepsilon}^{+}\Big|_{\varepsilon=0} = u_{o}^{+} \ , \quad u_{\varepsilon}^{-}\Big|_{\varepsilon=0} = u_{o}^{-}$$

so that the equations in (3.18) and (3.19) are satisfied independent of ε where ε is a small parameter. The equations in (3.18) become

$$\frac{\partial u_{\varepsilon}^{\pm}}{\partial t} + A_1(u_{\varepsilon}^{\pm}) \frac{\partial u_{\varepsilon}^{\pm}}{\partial x} + A_2(u_{\varepsilon}^{\pm}) \frac{\partial u_{\varepsilon}^{\pm}}{\partial y} = 0 \qquad y \gtrless \phi^{\varepsilon}(x,t) \qquad (3.20)$$

while the equation in (3.19) reduces to

$$\phi_t^{\varepsilon}[u_{\varepsilon}] + \phi_x^{\varepsilon}[F_1(u_{\varepsilon})] - [F_2(u_{\varepsilon})] = 0 \qquad (3.21)$$

on the surface $y = \phi^{\varepsilon}$. In developing asymptotic expansions of the equations in (3.20) , (3.21), we follow a standard procedure and map the perturbed shock surface (which depends on ε) to the fixed surface $y = 0$. We set

$$\tilde{y} = y - \phi^{\varepsilon}(x,t) \ , \quad \tilde{x} = x \ , \quad \tilde{t} = t \ ,$$

and the equations in (3.20), (3.21) become

$$\frac{\partial u_{\varepsilon}^{\pm}}{\partial \tilde{t}} + A_1(u_{\varepsilon}^{\pm}) \frac{\partial u_{\varepsilon}^{\pm}}{\partial \tilde{x}} + (A_2(u_{\varepsilon}^{\pm}) - \phi_{\tilde{x}}^{\varepsilon} A_1^{\pm} - \phi_{\tilde{t}}^{\varepsilon}) \frac{\partial u_{\varepsilon}^{\pm}}{\partial \tilde{y}} = 0 \ \text{for} \ \tilde{y} \gtrless 0 \qquad (3.23)$$

and

$$\phi^{\varepsilon}_{\tilde{t}}[u_{\varepsilon}] + \phi^{\varepsilon}_{\tilde{x}}[F_1(u_{\varepsilon})] - [F_2(u_{\varepsilon})] = 0 \tag{3.24}$$

on $\tilde{y} = 0$.

Linearized Shock Front Equations

We assume we can construct formal solutions of (3.23) and (3.24) with the form

$$\begin{aligned} \phi^{\varepsilon} &= \varepsilon\phi_1(\tilde{x},\tilde{t}) + \varepsilon^2\phi_2(\tilde{x},\tilde{t}) + O(\varepsilon^3) \\ u^-_{\varepsilon} &= u^-_o + \varepsilon u^-_1(\tilde{x},\tilde{y},\tilde{t}) + \varepsilon^2 v^-_2 + O(\varepsilon^3) \\ u^+_{\varepsilon} &= u^+_o \ . \end{aligned} \tag{3.25}$$

By substituting (3.25) into (3.23), (3.24), we observe as a consequence of (3.16) that the terms of order zero vanish. By setting the terms of order one in equal to zero, we obtain the linearized shock front equations

$$\begin{aligned} &\text{a)} \quad \frac{\partial u^-_1}{\partial \tilde{t}} + A^-_1 \frac{\partial u^-_1}{\partial \tilde{x}} + A^-_2 \frac{\partial u^-_1}{\partial \tilde{y}} = 0 , \quad \tilde{y} < 0 \\ &\text{b)} \quad (\phi_1)_{\tilde{t}}[u_o] + (\phi_1)_{\tilde{x}}[F_1(u_o)] - A^-_2 u^-_1 = 0 , \quad \tilde{y} = 0 . \end{aligned} \tag{3.26}$$

Here and below we use the notation G^- as short-hand for $G(u^-_o)$. In particular, the nonlinear vector fluxes from (2.10) have the Taylor expansions,

$$F_i(u^-_o+\varepsilon v) = F^-_i + \varepsilon A^-_i v + \varepsilon^2 B^-_i(v,v) + O(\varepsilon^3) \tag{3.27}$$

for $i = 1,2$, where B^-_i is the vector-valued quadratic form in v given by the Taylor expansion terms of second order evaluated at u_0^- . We begin the next section by recording the equations in (3.26) in the case of gas dynamics.

<u>An Instructive Example for Linear Stability Theory with Radiating Waves</u>

We consider solutions of the wave equation

$$u_{tt}-u_{xx}-u_{yy} = 0 , \quad -\infty<x<\infty , \quad y>0 , \quad t>0 , \tag{3.28}$$

together with the boundary conditions

$$(u_y + \gamma u_t)\Big|_{y=0} = 0 , \quad t > 0 , \quad -\infty < x < \infty . \tag{3.29}$$

Here γ is a parameter which ranges from $-\infty$ to ∞ . We compute that the time derivative of the total energy of a solution u satisfying (3.28) and (3.29) is given by

$$\frac{\partial}{\partial t} \int_0^\infty \int_{-\infty}^\infty (\tfrac{1}{2}u_t^2+\tfrac{1}{2}u_x^2+\tfrac{1}{2}u_y^2)\,dxdy=\int_{y=0} u_t u_y \, dx=\gamma\int_{y=0} u_t^2 . \tag{3.30}$$

Thus,

$$\begin{aligned} &\text{For } \gamma < 0 \text{, the boundary conditions remove energy at } y = 0 . \\ &\text{For } \gamma > 0 \text{, the boundary conditions add energy at } y = 0 . \end{aligned} \tag{3.31}$$

Furthermore, <u>the boundary value problem in (3.28), (3.29) is certainly strongly stable for</u> $\gamma < 0$ since the energy is a decreasing function of time. Next we derive the classical stability analysis for the boundary value problem in (3.28), (3.29). For any s with $\text{Re}\, s > 0$ and ω real

$$v_+(s,\omega) = e^{st + i\omega x - \sqrt{s^2 + \omega^2}\, y} \tag{3.32}$$

is an explicit solution of the wave equation which remains bounded at $t = 0$ as $y \to \infty$ (because $\mathrm{Re}(\sqrt{s^2+\omega^2}) > 0$ for $\mathrm{Re}\, s > 0$). The boundary value problem in (3.28), (3.29) is strongly unstable provided there are values of s, ω with $\mathrm{Re}\, s > 0$ so that $v_+(s,\omega)$ satisfies the boundary conditions in (3.29), i.e.,

$$\left(\frac{\partial v_+}{\partial y} + \gamma \frac{\partial v_+}{\partial t}\right)\bigg|_{y=0} = 0 .$$

This condition is equivalent to requiring

$$-\sqrt{s^2 + \omega^2} + \gamma s = 0 \quad \text{for some } s,\ \omega,\ \mathrm{Re}\, s > 0 . \tag{3.33}$$

The equation in (3.10) has a required solution of this form if and only if $\gamma > 1$; therefore,

For $\gamma > 1$, the boundary value problem is strongly unstable. (3.34)

What happens in the regime $0 < \gamma < 1$ where energy is still pumped in at the boundary $y = 0$? We observe that there are special solutions of the wave equation satisfying the boundary conditions in (3.29) with the form

$$u = \sigma\left(x - \frac{1}{\sqrt{1-\gamma^2}}\, t + \frac{\gamma}{\sqrt{1-\gamma^2}}\, y\right) \tag{3.35}$$

for any γ with $|\gamma| < 1$ where σ is an arbitrary function. The bicharacteristic rays associated with the characteristic surfaces,

$$\sqrt{1 - \gamma^2}\, x - t + \gamma y = \text{constant}$$

have a y-component given by

$$y = \gamma(t - t_o)$$

so that the waves in (3.36) are outgoing radiating waves in the sense described above (3.6) only when $\gamma > 0$. For $\gamma < 0$ and $\gamma > -1$, these waves radiate into the region $y > 0$ as time decreases and do not effect the forward time stability of (3.28), (3.29) - in fact, we have already shown above that the kinetic energy decrease for $\gamma < 0$ as time increases and strong stability occurs. We summarize our observations regarding the model problem in (3.28), (3.29) by the following:

a) $\gamma < 0$ is a region of strong stability

b) $\gamma > 1$ is a region of strong instability (3.36)

c) $0 < \gamma < 1$ is a transition regime where outgoing radiating wave solutions of the form in (3.35) exist.

Thus, a similar transition regime with radiating waves as we anticipated in (3.8), (3.9), (3.10) of the introduction for the physical equations already occurs in a transparent fashion in the simpler model problem described above.

The Structure of Radiating Boundary Waves for Shock Fronts

We formulate the conditions for the existence of radiating boundary wave solutions for the linearized shock front equations in (3.26) a, b for the general planar shock front solution of (3.16).

First, we construct all outgoing radiating solutions of the equation

$$\frac{\partial u}{\partial t} + A_1^- \frac{\partial u}{\partial x} + A_2^- \frac{\partial u}{\partial y} = 0 \ , \quad y < 0 \ , \qquad (3.37)$$

which radiate from the boundary $y = 0$ into the interior $y < 0$ in the direction of increasing time for a given boundary wave velocity $c_* > 0$. We postulate the following:

For the given boundary wave velocity c_* all m eigenvalues $\lambda_1,\ldots,\lambda_m$ for the matrix $A_2^{-1}(-c_*I + A_1)$ are real with corresponding eigenvectors e_j, (3.38)

satisfying

$$(A_2^-)^{-1}(-c_*I + A_1^-)e_j = \lambda_j e_j \;, \quad 1 < j < m \;.$$

For future reference, we let e_K^* denote the corresponding left eigenvector with $e_K^* \cdot e_j = \delta_{K,j}$. Furthermore, we postulate that

$$\frac{\partial\lambda_j}{\partial c_*} > 0 \qquad \text{for} \quad j = 1,\ldots,m-1$$
$$\frac{\partial\lambda_m}{\partial c_*} < 0 \;. \tag{3.39}$$

Therefore, by looking back at the discussion above (3.6), we observe that

The λ_j, $1 < j < m-1$, describe the outgoing radiating modes of propagation for the region $y < 0$. (3.40)

With the postulated information in (3.36)-(3.38), it is a simple matter to write down all the outgoing radiating wave solutions of (3.14) with the boundary wave velocity c_*. These have the form

$$u = \sum_{j=1}^{m-1} \sigma_j'(x - c_*t - \lambda_j t)e_j \;, \tag{3.41}$$

where $\sigma_j'(s)$ is an arbitrary scalar function for $1 \leq j \leq m-1$.

Next, we concentrate on the structure of the boundary terms from (3.26b). We set $\vec{\ell} = (\ell_0, \ell_1, \ldots, \ell_{m-1})^T$ and define the mxm matrix associated with (3.26b) and the boundary velocity c_* by

$$M\vec{\ell} = \left(-c_*(u_0^+ - u_0^-) + F_1(u_0^+) - F_1(u_0^-)\right)\ell_0 + \sum_{j=1}^{m-1} \ell_j A_2^- e_j \; . \tag{3.42}$$

Our main postulate regarding the structure of (3.42) is the following one:

For some $c_* > 0$ satisfying (3.36)-(3.38), M has zero as a simple eigenvalue with eigenfunction $\vec{\ell}^T = (1, \ell_1, \ldots, \ell_{m-1})^T$ satisfying $M\vec{\ell} = 0$ and corresponding left eigenvector $\vec{\ell}^*$. (3.43)

With the postulate structure in (3.43), we can immediately write down explicit radiating boundary wave solutions for the linearized shock front problem in (3.26). We set

$$\phi = \sigma(x - c_* t) \tag{3.44}$$

$$u = \sum_{j=1}^{m-1} \ell_j \sigma'(x - c_* t - \lambda_j y) e_j \; , \tag{3.44}$$

then the above conditions in (3.36)-(3.41) guarantee that these are explicit radiating boundary wave solutions for (3.26).

IIIc. The Weakly Nonlinear Perturbation Expansion

In the general framework, the tacit assumption is that the linearized shock front equations have outgoing radiating boundary wave solutions with the form in (3.44) for a boundary wave velocity $c_* > 0$ with the structure discussed in (3.38)-(3.43). The basic principle applied in the asymptotic expansions developed in this section is the following causality principle:

> Additional incoming radiating waves are never generated at the perturbed shock front - only outgoing radiating waves are generated there. (3.45)

However, we remark that incoming waves can be generated through the interaction of outgoing waves in the interior region $y < 0$ without violating causality principles. In fact, as we shall see below, the nonlocal term in (3.12) is produced through such an effect.

Here our objective is to construct appropriate asymptotic solutions of the equations in (3.23) and (3.24) consistent with the causality principle in (3.45). We introduce the variables

$$\hat{x} = \tilde{x} - c_* \tilde{t} , \quad \hat{y} = \tilde{y} , \quad \tau = \varepsilon \tilde{t}$$

so that

$$\frac{\partial}{\partial \hat{x}} = \frac{\partial}{\partial \tilde{x}} , \quad \frac{\partial}{\partial \hat{y}} = \frac{\partial}{\partial \tilde{y}} , \quad -c_* \frac{\partial}{\partial \hat{x}} + \varepsilon \frac{\partial}{\partial \tau} = \frac{\partial}{\partial \tilde{t}}$$

and seek asymptotic solutions of the equations in (3.23), (3.24) with the form

$$\begin{aligned} &u^- = u_0^- + \varepsilon u_1^-(\hat{x},\hat{y},\tau) + \varepsilon^2 u_2^-(\hat{x},\hat{y},\tau) + 0(\varepsilon^3) , \quad \hat{y} < 0, \quad \tau \geq 0 \\ &u^+ \equiv u_0^+ \qquad\qquad\qquad\qquad\qquad\qquad\qquad\qquad\qquad\qquad (3.46) \\ &\phi = \varepsilon \phi_1(\hat{x},\tau) + \varepsilon^2 \phi_2(\hat{x},\tau) + 0(\varepsilon^3) . \end{aligned}$$

By equating the terms of order ε in (3.23), (3.24), we derive the linearized shock front equations

$$(A_2^-)^{-1}(A_1^- - c_* I)\frac{\partial u_1^-}{\partial \hat{x}} + \frac{\partial u_1^-}{\partial \hat{y}} = 0\ , \quad \hat{y} < 0$$

$$\left(-c_*(u_o^+ - u_o^-) + F_1(u_o^+) - F_1(u_o^-)\right)(\phi_1)_{\hat{x}} + A_2^- u_1^- = 0\ , \quad \hat{y} = 0\ . \tag{3.47}$$

The general solutions of (3.47) consistent with the causality principle in (3.45) are the general outgoing radiating boundary wave solutions constructed in (3.44); namely,

$$\phi_1(\hat{x},\tau) = \sigma(\hat{x},\tau)$$

$$u_1^-(\hat{x},\hat{y},\tau) = \sum_{j=1}^{m-1} \ell_j \sigma'(\hat{x}-\lambda_j \hat{y},\tau) e_j \tag{3.48}$$

where $\sigma'(s,\tau) = \frac{\partial \sigma}{\partial s}$ and the remaining quantities in (3.48) have already been defined. At this stage in the argument, $\sigma(\hat{x},\tau)$ is a completely arbitrary function. Below, we will only require that $\sigma(\hat{x},\tau)$ decays sufficiently rapidly as $|x| \to \infty$; this condition will automatically be satisfied provided that the initial disturbance

$$\sigma(\hat{x},\tau)\big|_{\tau=0} = \sigma_o(\hat{x})$$

satisfies these same conditions as we shall see below.

A tedious calculation of the terms of order ε^2 in (3.24) at the boundary $\hat{y} = 0$ yields the equation

$$(-c_*(u_o^+ - u_o^-) + F_1^+ - F_1^-)(\phi_2)_{\hat{x}} + A_2^- u_2^- =$$

$$-((\phi_1)\ (u_o^+ - u_o^-) + (\phi_1)_{\hat{x}}(c_* I - A_1^-)u_1^- + B_2^-(u_1^-, u_1^-)) \tag{3.50}$$

where the notation from (3.27) has been used. Similarly, the terms of order ε^2 from (3.23) yield the following equation in the interior region $\hat{y} < 0$,

$$(A_2^-)^{-1}(-c_*I+A_1^-)\frac{\partial u_2^-}{\partial \hat{x}} + \frac{\partial u_2^-}{\partial \hat{y}} = -(A_2^-)^{-1}\{\frac{\partial u_1^-}{\partial \tau} + \frac{\partial}{\partial \hat{x}}(B_1^-(u_1^-,u_1^-))$$

$$+ \frac{\partial}{\partial \hat{y}}(B_2^-(u_1^-,u_1^-)) - (\phi_1)_{\hat{x}}(A_1^- - c_*I)(\frac{\partial u_1}{\partial y})\} \equiv F \,. \quad (3.51)$$

From (3.38) we observe that the general solution of the second-order equations in (3.51) is given by

$$u_2^- = \sum_{j=1}^{m} \ell_{2,j} e_j$$

where $\ell_{2,j}$ satisfies the scalar equation

$$\lambda_j \frac{\partial}{\partial \hat{x}} \ell_{2,j} + \frac{\partial}{\partial \hat{y}} \ell_{2,j} = F_j \equiv e_j^* F \,, \quad j = 1,\dots,m \,. \quad (3.52)$$

In particular, provided $F_{m\hat{}}$ can be integrated, one special solution determining $\ell_{2,j}(x,y,\tau)$ is given by

$$\ell_{2,m}(\hat{x},\hat{y},\tau) = \int_{-\infty}^{y} F_m(\hat{x} - \hat{\lambda}_m(y-s),s,\tau)ds \,, \quad (3.53)$$

and the general solution to (3.52) when $j = m$ has the form

$$\ell_{2,m}(\hat{x},\hat{y},\tau) + \tilde{\ell}(\hat{x} - \ell_m\hat{y},\tau)$$

where $\tilde{\ell}$ is an arbitrary function. However, if we apply the causality condition in (3.45) to the terms of second order, we deduce that necessarily

$$\tilde{\ell}(\hat{x},\tau) \equiv 0 \,;$$

otherwise, incoming waves would be generated at the shock front

$\hat{y} = 0$. From the detailed form of F_m computed below, it will be clear that F_m can always be integrated under the condition in (3.49) and is generated through the nonlinear wave interactions of outgoing radiating waves. The other outgoing radiating wave equations from (3.52) have the general solutions for $\hat{y} < 0$ given by

$$\ell_{2,j} = \tilde{\ell}_{2,j}(\hat{x}-\lambda_j\hat{y}) - \int_{\hat{y}}^{0} F_j(\hat{x}-\lambda_j(\hat{y}-s), s, \tau)\,ds, \quad j=1,\ldots m-1. \quad (3.55)$$

By substituting the form of u_1^- from (3.48) into the right-hand side of (3.51), we get

$$F_j = \Big(\sum_{\ell=1}^{m=1} p_{j,\ell}\frac{\partial\sigma'}{\partial\tau}(\hat{x}-\lambda_\ell\hat{y},\tau)\Big) + \Big(-\sigma'(\hat{x},\tau)d_j\lambda_j^2\frac{\partial\sigma'}{\partial\hat{x}}(\hat{x}-\lambda_j\hat{y},\tau)\Big)$$

$$+\Big\{\sum_{\ell,s=1}^{m-1} r_{j\ell s}\sigma'(\hat{x}-\lambda_\ell\hat{y},\tau)\frac{\partial\sigma'}{\partial x}(\hat{x}-\lambda_s\hat{y},\tau)\Big\}, \quad (3.55)$$

where $d_j = \ell_j$, $1 \leq j \leq m-1$, $d_m = 0$, and the constants $p_{j,\ell}$, $r_{j\ell s}$ are determined from (3.51) (but the explicit form of these constants is not important here). In particular, by using (3.55) in (3.53) and evaluating at $\hat{y} = 0$, we obtain

$$\ell_{2,m}(\hat{x},\hat{y},\tau)\Big|_{\hat{y}=0} = \sum_{\ell=1}^{m-1}\frac{p_{m,\ell}}{\lambda_m-\lambda_\ell}\frac{\partial\sigma}{\partial\tau} + \sum_{\ell=1}^{m-1}\frac{r_{m\ell\ell}}{\lambda_m-\lambda_\ell}\tfrac{1}{2}(\sigma_{\hat{x}})^2$$

$$+ \sum_{\substack{\ell\neq q\\ \ell,q\neq m}} r_{m\ell q}\int_{-\infty}^{0}\sigma'(\hat{x}-(\lambda_\ell-\lambda)s,\tau)\,\sigma'_{\hat{x}}(\hat{x}-(\lambda_q-\lambda_m)s,\tau)\,ds. \quad (3.56)$$

Since generally $\lambda_\ell \neq \lambda_q$ when $q \neq \ell$, the last terms in (3.56) are not the integral of a perfect derivative and cannot be simplified. As mentioned previously, we observe that the nonlocal terms arise from nonlinear wave interactions of outgoing modes; also, provided $\sigma(\hat{x},\tau)$ vanishes sufficiently rapidly as $|\hat{x}| \to \infty$, it follows from (3.55), (3.56) that the

<u>contributions to</u> $\ell_{2,m}(\hat{x},\hat{y},\tau)$ <u>are never resonant</u> and can always be integrated.

In order to satisfy the perturbed shock front equations from (3.50) at the second order on $\hat{y} = 0$ subject to the causality principle from (3.45) (as expressed in (3.54), with the notation for M from (3.42), we need to choose $(\phi_2)_{\hat{x}}$, $\tilde{\ell}_{2,j}$, $1 \leq j \leq m-1$, so that

$$M \begin{bmatrix} (\phi_2)_x \\ \tilde{\ell}_{2,1}(\hat{x},\tau) \\ \vdots \\ \tilde{\ell}_{2,m-1}(\hat{x},\tau) \end{bmatrix} + A_2^- e_m \ell_{2,m}(\hat{x},0,\tau) + \sigma_\tau (u_o^+ - u_o^-) + (\sigma_{\hat{x}})^2 \vec{w} = 0 ,$$

$$-\infty < \hat{x} < \infty \qquad \tau > 0 ,$$

where $\vec{w}$ is an m-vector which we don't compute explicitly. Recall by the assumption in (3.43), M has zero as a simple eigenvalue with a left eigenvector $\vec{\ell}^*$; thus, the solvability of the perturbation expansion to second order requires that

$$(\vec{\ell}^* \cdot A_2^- e_m)\ell_{2,m}(\hat{x},0,\tau) + \vec{\ell}^* \cdot (u_o^+ - u_o^-)\sigma_\tau + \vec{\ell}^* \cdot \vec{w}\sigma_{\hat{x}}^2 = 0 . \tag{3.58}$$

By plugging in the form of $\ell_{2,m}$ from (3.56) and assuming that the coefficient of $\partial/\partial\tau$ is nonzero, we see that (3.58) is equivalent to the integro-differential Hamilton-Jacobi equation

$$0 = \frac{\partial\sigma}{\partial\tau} + \tfrac{1}{2} a_1 (\sigma_{\hat{x}})^2 + \sum_{m-1 \geq q > \ell \geq 1} d_{q,\ell} \int_{\infty}^{0} \sigma_{\hat{x}}(\hat{x} + (\lambda_m - \lambda_\ell)s,\tau)\, \sigma_{\hat{x}\hat{x}}(\hat{x} + (\lambda_m - \lambda_q)s,\tau)\, ds ,$$

$$\sigma(\hat{x},0) = \sigma_o(\hat{x}) , \tag{3.59}$$

or equivalently, by differentiating and using $\sigma' = \sigma_{\hat{x}}$, the integro-differential conservation law

$$0=\frac{\partial\sigma'}{\partial\tau}+a_1\frac{\partial}{\partial\hat{x}}(\tfrac{1}{2}(\sigma')^2)+\sum_{m-1\geq q>\ell\geq 1} d_{q,\ell}\int_{-\infty}^{0}\sigma'(\ )\sigma'_{\hat{x}}(\ )ds$$

$$\sigma'(\hat{x},0)=\sigma'_0(\hat{x}) \qquad (3.60)$$

where the arguments of integration are the same as in (3.59). This completes Step III in the abstract framework.

IV. The Transition to Detonation - The Experimental and Numerical Evidence

Transition to detonation is the remarkable phenomenon observed where initially highly subsonic flames (speeds ≅ 5m/sec.) develop into fully developed detonation waves (speeds ≅ 2000m/sec.) in times on the order of one hundred microseconds under the appropriate conditions. To identify these appropriate conditions and the mechanisms responsible for transition to detonation is the most outstanding theoretical unsolved problem for applied mathematicians to attack in the high Mach number regime. This chapter is very short because the author is unaware of any significant theoretical progress in explaining the complex mechanisms responsible for the observed sequence of events which correlates with the experimental record. Next, we describe the result of a series of famous experiments which indicate the remarkable variety of the different modes for transition to detonation and the corresponding complexity of the fluid mechanic - chemical-kinetic processes involved.

IVa. The Experimental Record for Transition to Detonation in Gaseous Mixtures

This section is written as if the reader has a copy of the famous paper by P. Urtiew and A.K. Oppenheim, "Experimental observations of the transition to detonation in an explosive gas," Proc. Royal Society, A, 295, 13-28 (1966).

Those authors document four distinct ways in which transition to detonation occurs in hydrogen-oxygen mixtures.

Case 1: Formation of an exothermic hot spot in the vicinity of the flame front near the wall. The origin of this hot spot can be associated with weak transverse waves in the region between the percursor shock and flame front. This exothermic hot spot generates a reacting blast wave which interacts with the percursor shock producing a self sustained detonation wave. Thus, transition occurs as result of weak transverse waves generating an exothermic hot spot (see Figure 6 of [25]).

Case 2: Transition occuring as result of a shock wave formed at the tip of a turbulent flame.

This shock wave heats the gas and leads to formation of second reacting blast wave emanating from an exothermic hot spot at the tip of the turbulent flame -this blast wave triggers the transition to detonation (see Figure 7 of [25]).

Case 3: Transition to detonation occurring at the shock front.

Through multi-D turbulence and pressure wave interactions, the flame propagates extremely fast along the boundary layer. The flame tip ignites the top of the shock wave and leads to an exothermic hot spot. This hot spot generates the reacting blast wave which engulfs the shock front and triggers the transition to detonation (see Figure 8 of [25]).

Case 4: Two precursor shock waves interact ahead of the flame front forming a contact discontinuity and a region of higher local temperature than before. After an induction time, the medium of the contact discontinuity becomes ignited leading to the transition (see Figure 9 of [25]).

Conclusions:

The various modes of transition to detonation depend on particular patterns of shock waves which have been created by the accelerating flame as it forms a turbulent combustion front. In Cases 1-3, the effects of turbulence are prominent. Once localized exothermic hot spots are generated, the transition process has an essentially universal character. The hot spot forms a reacting blast wave. If the strength of this blast wave is sufficiently large, a transition to detonation occurs. In general, purely one-dimensional gas dynamic processes are completely inadequate for explaining the transition (see the discussion in [26] and the review article in [27], where among other facts it is documented that turbulent mixing without shock waves can trigger the transition to detonation).

In the experiments which we discussed eariler, the final stages of the transition to detonation are described. It is also interesting and much more involved to describe the turbulence - pressure wave resonances which have already accelerated the turbulent combustion fronts to speeds which are a significant fraction of the sound speed; these events occur prior to the stage of transition described by the schlieren photographic record in [25]. The author recommends the discussion in the review paper by Lee and Moen ([27]) as well as the references given there for a clear presentation of the evidence.

IVb. Numerical Experiments on the Transition

First, we mention the papers, [28], [29]. In that work simplified 1-D models with temperature dependent flames speed are used to discuss the possibility of transition to detonation. When laminar flame laws are used, transition does not occur; on the other hand, when turbulent flame laws are used, a transition to detonation can happen. This provides very convincing supporting numerical evidence that the transition to detonation is not a purely one dimensional effect for gaseous mixtures.

The paper by Kailasanath and Oran (Comb. Sci. Tech. 31, 345-362, 1983), represents the only attempt (in the author's knowledge) to study transition to detonation in gases through purely 1-D effects by direct numerical integration of the reacting compressible Euler equations for the hydrogen-oxygen system. Here we make several comments about this paper:

1. This paper also indicates the role of the development of exothermic hot spots as reaction centers which trigger the transition to detonation providing numerical evidence in agreement qualitatively with experiments.

2. In these calculations, the exothermic hot spots develop after approximately 600 μs. This time is ten times larger than the time of development of the explosive reaction centers in the experiments discussed earlier. Thus, the wave amplification process from purely 1-D laminar considerations proceeds on a slower time scale than that in the (Multi-D) experiments described earlier where turbulence is important. A careful comparison with the experimental record in the papers in the bibliography of [30] should clarify this point.

3. The following quote from page 357 of [30] and describing the final stage of the transition process is a controversial one: "At 609 μs the forward moving reaction wave is supersonic. The velocity of the fluid with respect to the wave decreases across the wave but is supersonic on either side. The pressure rise across the wave is just over a factor of two. The reaction wave behaves like a weak detonation wave at this time."

   The wave being discussed is the reacting blast wave which triggers the transition detonation. Can such a wave be a weak detonation in the inviscid compressible Euler equations with finite reaction rates? The author believes that the answer to this question is no!! He believes that this wave is a numerical artifact. Here is the reasoning. The numerical methods used are operator splitting and flux-corrected-transport, (F-C-T), for the hydrodynamics. As a numerical algorithm, this method has properties very similar to the higher Godunov algorithms analyzed by Colella, Roytburd, and the author in [7]. For these schemes, numerical artifact weak detonations can be generated by the difference method (see [7] and Roytburd's lecture at this meeting where such waves are even constructed explicitly). The mesh in [30] is highly nonuniform and perhaps generates the weak detonation with a varying speed as a computational artifact. The author is most interested in hearing other opinions and explanations regarding the above sentences quoted from [30].

Bibliography

1. Fickett, W., Amer. J. Physics, 47, 1050-1059 (1979).
2. Majda, A., SIAM J. Appl. Math., 41, 70-93 (1981).
3. Rosales, R., and Majda, A., SIAM J. Appl. Math., 43, 1086-1118 (1983).
4. Fickett, W., and Davis, W., Detonation, University of California Press, Berkeley (1979).
5. Williams, F.A., Combustion Theory, Addison-Wesley, Reading, MA (1965).
6. Courant, R. and Friedrichs, K.O., Supersonic Flow and Shock Waves, Wiley-Interscience, New York, (1948).
7. Colella, P., Majda, A., and Roytburd, V., to appear in SIAM J. Sci. Stat. Computing.
8. Erpenbeck, J., Ninth Symposium on Combustion, 442-453 (1963).
9. Erpenbeck, J., Twelfth Symposium on Combustion, 711-721 (1969).
10. Gardner, R., Trans. Amer. Math. Soc., 277, 431-468 (1983).
11. Lu, G.C. and Ludford, G., SIAM J. Appl. Math., 42, 625-635 (1982).
12. Holmes, P. and Stewart, C.S., Studies Appl. Math. (1983).
13. Larrouturou, B., to appear in Nonlinear Analysis (1985).
14. Oppenheim, A.K., and Rosciszewski, J., Ninth Symposium on Combustion, 424-434 (1963).
15. Clarke, J., and Gant, R., to appear in Progress in Aeronautics and Astronautics (1985).

16. Jackson, T.L. and Kapila, A.K., to appear in SIAM J. Appl. Math. (1985).
17. Majda, A., to appear in Proceedings of Workshop on Oscillations, IMA, Univ. of Minnesota (April 1985).
18. DiPerna, R., and Majda, A., Comm. Math. Physics (April 1985).
19. Majda, A., and Rosales, R., in preparation.
20. Alpert, R.L., and Toong, T.Y., Astronaut. Acta, 17, 539-560 (1972).
21. Majda, A., and Rosales, R., SIAM J. Appl. Math., 43, 1310-1334 (1983).
22. Majda, A., and Rosales, R., Studies Appl. Math., 117-148 (1984).
23. Abouseif, G.E. and Toong, T.Y., Comb. and Flame, 45, 67-94 (1982).
24. Majda, A., Memoirs AMS, 41, #275 (1983), and #281 (1983).
25. Urtiew, P. and Oppenheim, A.K., Proc. Royal Soc., A, 295, 13-28 (1966).
26. Urtiew, P., and Oppenheim, A.K., Eleventh Symposium on Combustion, 665-670 (1967).
27. Lee, J.H., and Moen, I.O., Prog. Energy Combust. Sci., 6, 359-389 (1980).
28. Kurylo, J., Dwyer, H.A., and Oppenheim, A.K., Paper 79-0290, AIAA (1979).
29. Deng, Chorin, A.J., and Liu, T.P., SIAM J. Appl. Math. (1983).
30. Kailasanath, K. and Oran, E.S., Comb. Sci. Tech., 34, 345-362 (1983).
31. Blythe, P.A., 17th International Symposium on Combustion, 909-916 (1978).
32. Clarke, J.F., J.F.M. 89, 343-379, (1978) and J.F.M. 94, 195-208, (1979).
33. Majda, A., Memoir of American Mathematical Society, 41, No. 275, (1983).

Princeton University
Department of Mathematics
Washington Road, Fine Hall
Princeton, NJ 08544

Lectures in Applied Mathematics
Volume 24, 1986

NUMERICAL CALCULATION OF THE INTERACTION OF PRESSURE WAVES AND FLAMES*

H. A. Dwyer, A. Lutz* and R. J. Kee*
University of California, Davis
*Sandia National Laboratories, Livermore

INTRODUCTION

The goal of this paper is to describe and discuss the problems and progress which have developed for the numerical solution of the important physical problem of the interaction of a flame with pressure waves[1,2,3]. From a mathematical point of view this problem is extremely challenging since it involves a strong interaction between a hyperbolic or wave system with a parabolic or flame system. This challenge is a direct result of the incompatible properties of the two types of partial differential equation systems, which will be shown to act in a way to destroy the numerical accuracy of the total system solution.

In the present paper an implicit numerical method and an adaptive numerical technique will be combined to achieve a solution procedure which shows good promise. It will be shown that the present method limits the excessive interaction of a shock wave with a flame and leads to a good prediction of both the flame and pressure wave speeds. Also, there will be given detailed discussions of the conflicting properties of these two

* work supported by U.S. Department of Energy Office of Basic Energy Sciences

systems which both satisfy the Rankine-Hugoniet relationships[4] in steady state, but which have drastically different overall properties. Finally, some calculations will be presented for the rapid ignition of a premixed H_2/Air gas mixture in a closed vessel by the use of rapid energy deposition.

DESCRIPTION OF THE PROBLEM

The physical problem which will be addressed is the one dimensional ignition of a combustible mixture by the rapid deposition of energy which could be caused by a laser beam. The rapid heating of the gas causes both the temperature and pressure to increase in a small volume and leads to the formation of both a flame and a shock wave. The flame travels at a speed at least two orders of magnitude slower than shock wave and has a thickness which is two orders of magnitude greater. However, the flame is very thin in itself and is usually two orders of magnitude smaller than the typical combustion chamber length.

One partially simplifying aspect of the problem is that the correct shock speed and amplitude can be obtained without resolving the internal structure of the shock wave[5], and therefore artificial thickening (artificial viscosity) can be employed to reduce oscillatory solutions. The challenge presented by the shock/flame interaction is that this artificial thickening and transport must not interact to change the physical transport properties in the flame. In order to correctly determine the correct structure and speed of a flame the internal transport processes must be determined accurately and significant artificial transport will lead to incorrect results.

The approach which will be taken in this paper is to employ an adaptive grid technique to resolve both the flame and the shock. The flame will be resolved to its correct physical size, but the shock will be resolved only to a scale significantly smaller than the flame thickness. The reason that the shock is not resolved to the correct physical size is becasue it is extremely small and this small size would cause excessively small time steps (explicit numerical methods) or large Courant Numbers (implicit) numerical methods.

In order to stop the basic system equations from resolving the shock two numerical features were added, and these were a minimum step size and artificial numerical viscosity. The basic scientific problem is then to prove that these two numerical tools do not change the properties of the flame system. Up until the present time the method has been implemented but not fully tested for accuracy and the present result therefore represent work in progress.

BASIC EQUATIONS AND METHOD OF SOLUTION

The system of equations used to describe the problem are the one-dimensional conservation equations for mass, momentum, energy and chemical species in planar, cylindrical, or spherical coordinates and which are:

$$\frac{\partial \rho}{\partial t} + \frac{1}{r^i}\frac{\partial}{\partial r}(r^i m) = 0$$

$$\frac{\partial m}{\partial t} + \frac{1}{r^i}\frac{\partial}{\partial r}(r^i \rho u^2) + \frac{\partial p}{\partial r} + \frac{1}{r^i}\frac{\partial}{\partial r}(r^i \tau_{rr}) - \frac{(\tau_{\theta\theta} + \tau_{\phi\phi})}{r} = 0$$

$$\frac{\partial e}{\partial t} + \frac{1}{r^i}\frac{\partial}{\partial r}(r^i u\,(e + p)) + \frac{1}{r^i}\frac{\partial}{\partial r}(r^i q) - \frac{1}{r^i}\frac{\partial}{\partial r}(r^i u\tau_{rr}) - S(t,r) = 0$$

$$\frac{\partial \rho_k}{\partial t} + \frac{1}{r^i}\frac{\partial}{\partial r}(r^i u\rho_k) + \frac{1}{r^i}\frac{\partial}{\partial r}(r^i \rho Y_k V_k) - \dot{\omega}_k W_k = 0 \quad k = 1, \ldots, K$$

$$p = \frac{\rho R_u T}{\overline{W}}$$

where the energy flux is

$$q = -\lambda \frac{\partial T}{\partial r} + \sum_{k=1}^{K} \rho Y_k V_k h_k ,$$

and the stress components are given as

$$\tau_{rr} = -\mu \left[2 \frac{\partial u}{\partial r} - \frac{2}{3} \frac{1}{r^i} \frac{\partial}{\partial r} (r^i u)\right]$$

$$\tau_{\theta\theta} = \begin{cases} 0 & \text{if } i = 0 \\ -\mu \left[\frac{2u}{r} - \frac{2}{3} \frac{1}{r^i} \frac{\partial}{\partial r} (r^i u)\right] , & \text{if } i = 1,2 \end{cases}$$

$$\tau_{\phi\phi} = \begin{cases} 0 & \text{if } i = 0,\ 1 \\ \tau_{\theta\theta} , & \text{if } i = 2 \end{cases}$$

The index i is equal to 0, 1 or 2 for planar, cylindrical, or spherical coordinates respectively; ρ is the pressure; τ_{rr}, $\tau_{\theta\theta}$, and $\tau_{\phi\phi}$ are the normal stress components; λ is the mixture thermal conductivity; T is the temperature; V_k are the species diffusion velocities; T_k are the species mass fractions; h_k are the species enthalpies; $\dot{\omega}_k$ are the species chemical production rates; W_k are the species molecular weights; R_u is the universal gas constant; K is the total number of species; $\bar{W}$ is the mean molecular weight of the mixture; and S(t,r) is the energy source to be defined.

For the chemical kinetics model and the transport coefficients extensive use has been made of the software programs that have been developed at Sandia Laboratories[8,9], and the details can be found in Reference [10]. The shortened version of the chemical kinetics model we have employed is shown in Table (I).

TABLE I.

Reaction Mechanism Rate Coefficients In Form $k_f = AT^{\beta} \exp(-E_0/RT)$.

Units are moles, cubic centimeters, seconds, Kelvins and calories/mole.

REACTION	A	β	E
1. $H_2 + O_2 \rightleftarrows 2OH$	1.70E13	0.000	47780
2. $H_2 + OH \rightleftarrows H_2O + H$	5.20E13	0.000	6500
3. $H + O_2 \rightleftarrows OH + O$	5.13E16	−0.816	16507
4. $O + H_2 \rightleftarrows OH + H$	1.80E10	1.000	8826
5. $2OH \rightleftarrows O + H_2O$	6.00E08	1.300	0
6. $H_2 + M \rightleftarrows H + H + M$[a]	2.23E12	0.500	92600
7. $H + OH + M \rightleftarrows H_2O + M$[b]	7.50E23	−2.600	0

[a] Third body efficiencies: $k_6(H_2O) = 6k_6(Ar)$, $k_6(H) = 20k_6(Ar)$, $k_6(H_2) = 3k_6(Ar)$.

[b] Third body efficiencies: $k_7(H_2O) = 19k_7(Ar)$.

In order to proceed further both the adaptive grid method and a numerical solution procedure must be developed. The first step in employing the adaptive grid method, needed to resolve the time-dependent shock and flame, is to transform the basic equations from an arbitrary physical space distribution of nodes to a uniform computational space. This is done by transforming the derivatives as

$$\frac{\partial(\ )}{\partial t} = \frac{\partial(\ )}{\partial \tau} + \frac{\partial(\ )}{\partial n}\frac{\partial n}{\partial t}$$

$$\frac{\partial(\ )}{\partial r} = \frac{\partial(\ )}{\partial n}\frac{\partial r}{\partial n}$$

where the functional transformations

$$\tau = f(r,t)$$

$$\text{and} \quad n = g(r,t)$$

are to be determined by time dependent properties of the numerical solution itself. A general relationship for the relationship from the physical coordinates to the transformed coordinates is given by

$$\frac{dr}{dn} \; w(r, u) = \text{Constant}$$

where $w(r,u)$ is a positive weight function, which depends on the other components of the dependent variable vector, denoted by $U = (\rho, \rho u, e, \text{ and } \rho k)$ for species $k=1,\ldots, K$. The specific form used in the present calculations is

$$w(r, v) = 1 + b_1 \left|\frac{\partial T}{\partial r}\right| + b_2 \left|\frac{\partial p}{\partial r}\right| + b_3 \left|\frac{\partial u}{\partial r}\right|$$

where the dynamic constants b_1, b_2, and b_3 were found from specifying the ratios

$$R_1 = \frac{b_1 \int_0^L \left/ \partial T \right/ \partial r \left/ \partial r\right.}{\int_0^L w(r,u) \; \partial r}$$

$$R_2 = \frac{b_2 \int_0^L \left/ \partial p \right/ \partial r \left/ \partial r\right.}{\int_0^L w(r,u) \; \partial r}$$

and

$$R_3 = \frac{b_3 \int_0^L \left/ \partial u \right/ \partial r \left/ \partial r\right.}{\int_0^L w(r,u) \; \partial r}$$

Also, the entire transformation had to be renormalized in a slight but simple way when the minimum step size criterion was invoked.

The basic numerical method used to solve the transport equations was to employ central differences for all n-derivatives and a Crank Nicolson method in time. For the non-linear terms Newton's method was employed and the overall method is similar to that of Beam and Warming[11] except for the adaptive gridding.

In the present and previous works the transformation was lagged one time step behind the solution, and this is usually satisfactory except at wall reflections. In general it seems that a two step method will have to be employed in the future, since the shock usually moves more than one grid point on the small grid points. The first step of this new method will be to predict the transformation while the second will involve the solution itself. The full implementation and testing for accuracy is currently being carried out.

The final problem that has to be resolved is the oscillations caused by waves when central differences are employed with the Navier-Stokes equations. The approaches taken in the present work is to employ both second order[5] and fourth order smoothing[11], and at the present time the best use of either of these two artifacts has not been fully optimized. The ideal situation would be to add enough smoothing to eliminate oscillations on the minimum grid size. Since the minimum grid size is an order of magnitude smaller than the flame grid size, this smoothing would not interfer with the flame transport processes. Another practical consideration is that the shock is moving two orders of magnitude faster than the flame and quickly passes through the flame structure. Therefore it may be possible to increase the shock thickness in a practical sense without degrading significantly the overall accuracy of the flame propagation. For the results presented in the next section only second order smoothing[5] applied to the pressure term has been employed, and is given by

$$\hat{p} = p + q$$

$$\text{where } q = -\rho a^2 \Delta r^2 \Big/ \frac{\partial u}{\partial r} \Big/ \frac{\partial u}{\partial r}$$

In this manner the oscillations on the minimum step size were controlled.

SAMPLE CALCULATION

A sample calculation in planar coordinates is presented for a 25 percent hydrogen in air mixture. In the planar geometry the boundary at x=1 centimeter is a solid wall and the boundary at x=0 is a plane of symmetry. The premixed gas is initially at 1 atmosphere and 300 K. The source term used here has the form

$$S(r) = \begin{cases} Ae^{-\omega r^2}, & \text{if } t < \tau \\ 0, & \text{if } t \geq \tau \end{cases}$$

where $\omega = -\ln(0.1)/r_s^2$ and r_s is the radius where the Gaussian profile is ten percent of the peak amplitude A. In this example, $A = 2 \times 10^5$ joules per cubic centimeter per second and $\tau = 5$ microseconds. This amounts to approximately 100 millijoules per unit area added over the time interval. Figure 1 shows the calculation at 200 microseconds, at which time the flame has ignited and the chemical heat release has caused the centerline temperature to increase to 2600 K. At this point pressure waves have traversed the domain a number of times and fluid motion is driven by a combination of the flame expansion and pressure redistribution processes.

The mesh adaptation allows us to follow thin fronts on a large spatial domain with relatively few gridpoints. The squares in Figure 1 indicate the positions of the individual gridpoints. Figure 2a shows the relation of the physical coordinate to the transformed coordinate. The reader can see by comparison to Figure 1 that the slope of this curve is driven by the gradients in the physical variables. Figure 2b shows the variation in the physical mesh spacing across the domain. This calculation uses 51 gridpoints with a minimum grid spacing of 0.005 centimeters and 80 percent of the points are dedicated to regions of high gradients.

Figure 3 displays the temperature profiles at the time when the source turned off and at time intervals of 50 microseconds. Notice that the initial flame propagation appears to be quite unsteady, but becomes more steady at later times when the pressure waves have dissipated. Figure 5 shows that the flame position oscillates in time as might be expected. The flame position is considered to be the point where the temperature reaches 2000 K. Note that the amplitude of the oscillation is as large as one millimeter, which is on the order of the flame thickness.

SUMMARY

The present computational model shows good potential to study the problem of ignition of premixed gas mixtures. The initial calculations show that the pressure waves influence ignition and have a significant influence on the position of the flame. At the present time the influence of artificial smoothing on the damping and dispersion characteristics of the system has not been fully studied.

The adaptive-gridding techniques developed seem to offer great economies in the solution of problems of this type, and to properly capture the physics of the problem. Considerably more research and verification is needed until the problem is completely solved or before the techniques can be employed accurately in multiple space dimensions.

ACKNOWLEDGEMENT

The authors would like to thank Dr. Nagi Monsour for many helpful discussions regarding the conservative transformations of the Navier-Stokes equations.

REFERENCES

1. D. E. Kooker, "Numerical Study of a Confined Premixed Laminar Flame: Oscillatory Propagation and Wall Quenching," Combustion and Flame, Vol. 49, pp. 141-149 (1983).

2. S. Wiriyawit and E. K. Dabora, "Modeling the Chemical Effects of Plasma Ignition in One-Dimensional Chamber," XXth International Symposium on Combustion, (1984).

3. K. Kailasanath, E. Oran, and J. Boris, "A Theoretical Study of the Ignition of Premixed Gases," Combustion and Flame, Vol. 47, pp. 173-190, (1982).

4. L. D. Landau, and E. M. Lifshitz, Fluid Mechanics, Addison-Wesley Publishing Company, 1959.

5. R. D. Richtmyer and K. W. Morton, "Difference Methods for Initial-Value Problems," Wiley, New York, 1967.

6. H. A. Dwyer, R. J. Kee, B. R. Sanders, "Adaptive Grid Methods for Problems in Fluid Mechanics and Heat Transfer," AIAA Journal, Vol. 18, No. 10, pp. 1205-1212 (1979).

7. H. A. Dwyer, "Grid Adaption for Problems with Separation, Cell Reynolds Number, Shock-Boundary Layer Interaction, and Accuracy," AIAA 21st Aerospace Sciences Meeting, AIAA-83-0449 (1983).

8. R. J. Kee, J. A. Miller, and T. H. Jefferson, "CHEMKIN: A General-Purpose, Problem-Independent, Transportable, Fortran Chemical Kinetics Package," Sandia National Laboratories Report SAND80-8003, March 1980.

9. R. J. Kee, J. Warnatz, J. A. Miller, "A FORTRAN Computer Code Package for the Evaluation of Gas-Phase Viscosities, Conductivities, and Diffusion Coefficients," Sandia National Laboratories Report SAND83-8209, March 1983.

10. A. E. Lutz, R. J. Kee, H. A. Dwyer, "Ignition Modeling with Grid Adaption", 10th International Colloquium Explosions and Reactive Systems, Berkeley, Calif., August 1985, to be published.

11. R. F. Warming and R. M. Beam, "On the Construction and Application of Implicit Factored Schemes for Conservation Laws," SIAM-AMS Proceedings of the Symposium on Computational Fluid Dynamics, New York, April 1977.

H. A. Dwyer, University of California, Davis

A. Lutz and R. J. Kee, Sandia National Laboratories, Livermore

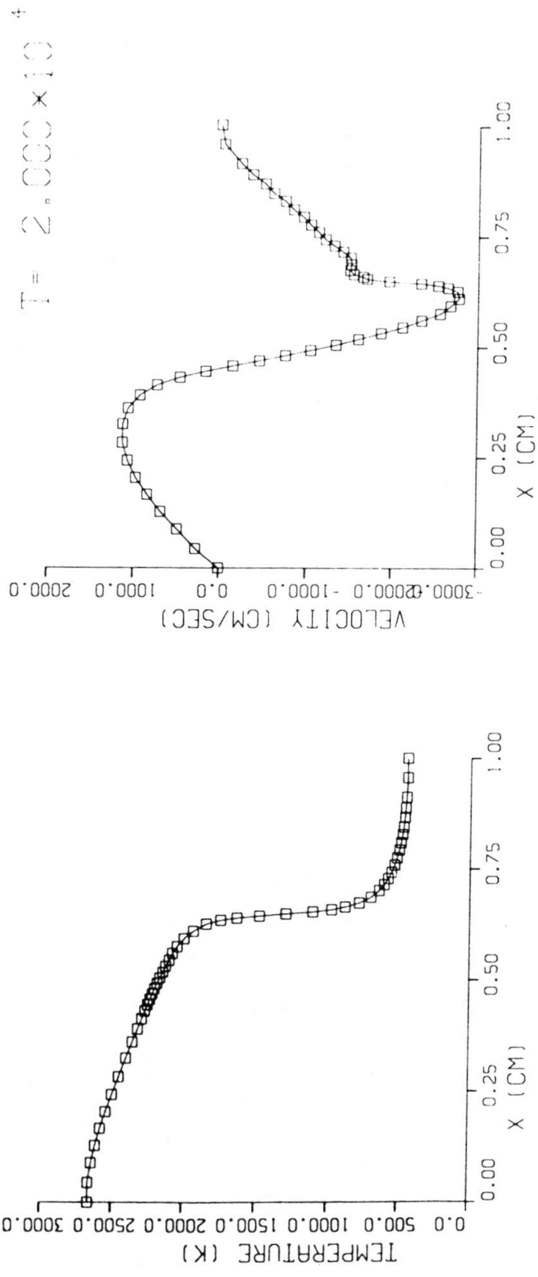

Figure 1a. Temperature and velocity solutions for 25% hydrogen-air mixture.

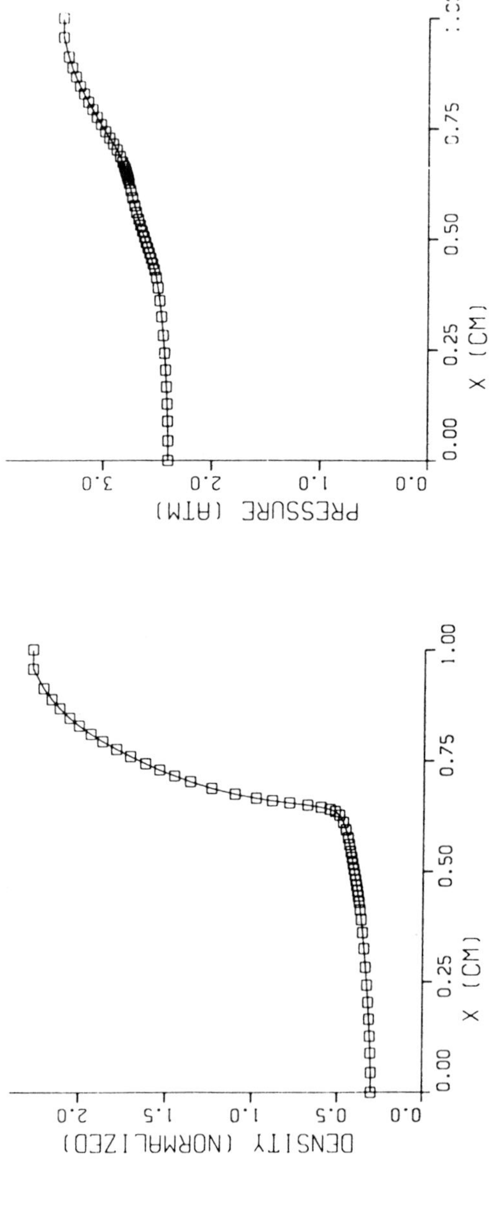

Figure 1b. Density and pressure solutions for 25% hydrogen-air mixture.

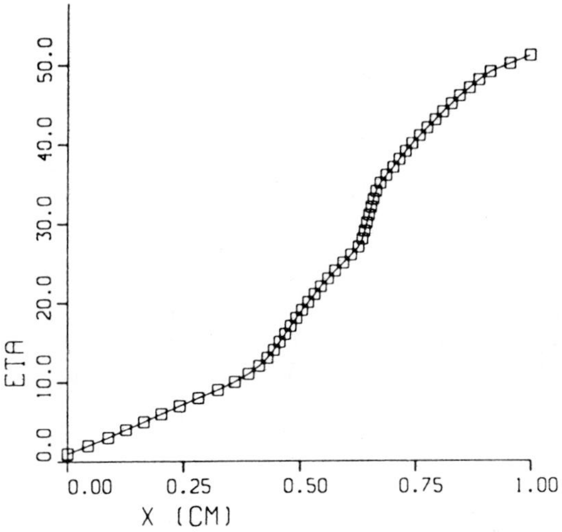

Figure 2a. Relation of physical coordinate (x) to transformed coordinate (η) for the solution shown in Figure 2.

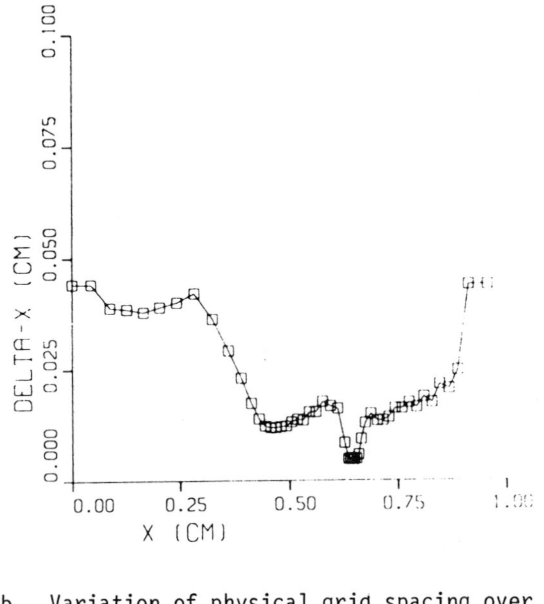

Figure 2b. Variation of physical grid spacing over the physical domain.

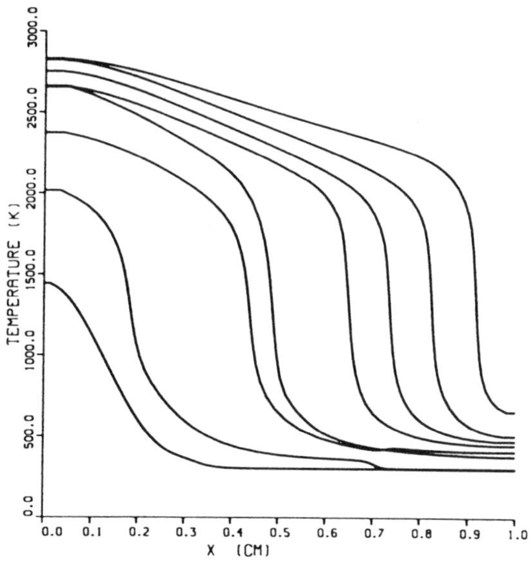

Figure 3. Temperature profiles at times 5, 50, 100, 150, 200, 250, 300, and 350 microseconds.

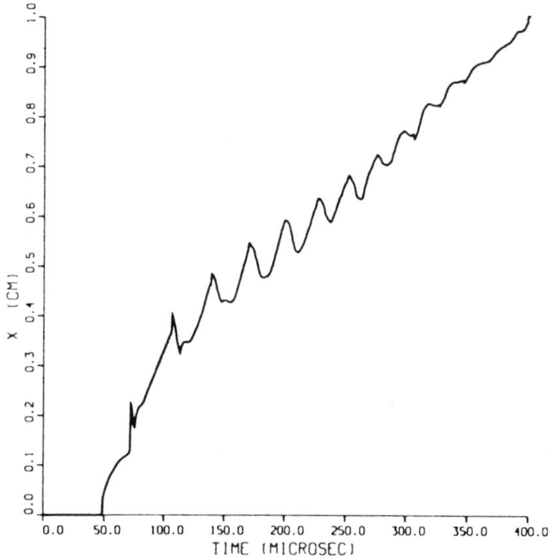

Figure 4. Flame position as measured by the point where the temperature is 2000 K versus time in microseconds. The oscillatory behavior damps out as the pressure waves are dissipated.

Lectures in Applied Mathematics
Volume 24, 1986

COMPUTATIONAL METHODS IN TURBULENT REACTING FLOW

Ahmed F. Ghoniem

ABSTRACT. Computations of the dynamics of reacting flows are presented. Numerical simulation of the Navier-Stokes equations using vortex methods is discussed in detail. Results are presented for shear flow and recirculating flow, portraying the time-dependent flow structure at a wide range of Reynolds number. Propagation of flames in premixed gases at conditions of low Mach number and fast chemistry is analyzed in terms of an active burning interface. Combustion-hydrodynamics interactions inside confined circular chambers and within recirculation zones are discussed.

1. INTRODUCTION

Flow fields associated with combustion are multi-component, compressible, multi-dimensional and time-dependent. In most cases, they combine regions of laminar and turbulent flows. They are characterized by three distinctive features: (1) confinement, which emphasizes the importance of boundary layer effects; (2) shear mixing, which occurs between material streams that move at two different velocities and leads to their homogenization; and (3) recirculation, which is often induced to stabilize combustion at high velocity streams by establishing stationary hot zones. Fluid mechanical phenomena which commonly occur under these circumstances are strongly unstable, i.e., they promote the growth of perturbations well beyond their linear range, leading

---

1980 Mathematics Subject Classification. 22-02, 73K12.

Supported by the National Sciences Foundation Grant CPE-840481, and the Air Force Office of Scientific Research Grant AFOSR 84-0356.

to the formation of finite structures that dominate the dynamics of the field. Moreover, the dynamic interaction between the flow and the burning process in terms of the expansion across the flame and the generation of vorticity along its surface results in new forms of instabilities that complicate the phenomenon further.

The accurate simulation of the flow field in which one allows for proper account of various interactions without extraneous numerical effects that may obscure the underlying physical phenomena is not an easy task. It requires a careful examination of the properties of the model equations and the computational schemes utilized in their solution. In this article, recent development in a class of numerical approximations of the Navier-Stokes equations which have been used successfully to study phenomena associated with combustion is reviewed, focusing on Lagrangian vortex methods and their applications to problems in shear flow. Moreover, a model that describes the propagation of thin flames at low Mach number is analyzed. Recent progress on the analysis of flame front instability mechanisms is summarized in connection with this model. Results obtained for turbulent burning rates and turbulent flame stabilization, both governed by flame-hydrodynamic interactions, are discussed.

The use of vorticity in numerical simulations of Navier-Stokes equations is motivated in part by its fundamental role in fluid dynamics. Knowledge of the vorticity field allows the calculation of the velocity field which, in turn, is used to transport the vorticity. In that regard, the laws of transport of vorticity can be considered as the kinematic interpretation of the conservation of momentum in which the balance of forces is replaced by the description of motion. Moreover, Lagrangian vortex methods offer some desirable computational properties: (1) they avoid the convective non-linearity of the governing equations; (2) they provide time-dependent, grid-free, self-adaptive schemes that focus the computational effort around areas of high concentration of vorticity; and (3) they reduce the numerical diffusion that acts to damp growth of perturbations at high Reynolds numbers.

Flame sheet models have been suggested to study turbulent combustion at conditions of high Damkohler numbers where the chemical reaction is fast compared to the flow speed. In the case of premixed gas reaction, the transition between reactants and products occurs across an essentially thin surface acting as an interface. If the turbulent field is formed of eddies with scales larger than the thickness of the laminar flame, its surface will wrinkle and its area will increase accordingly. This approximation simplifies the analysis without sacrificing the intrinsic mechanisms in the phenomenon. Turbulent eddies provide the stirring mechanism between reactants and products, while burning establishes a density jump across the flame associated with a large volumetric expansion that affects the turbulent field. The tight interaction between hydrodynamics and combustion complicates the theory of turbulent combustion and encumbers the progress in the numerical treatment of its governing equations.

This article is divided into four major sections. In section 2, vortex approximation of the two-dimensional Navier-Stokes equations is reviewed. Applications to two important classes of fluid flow problems: shear layers and recirculating flows are presented in section 3. In section 4, the equations describing the proposed combustion model are derived, followed by a brief discussion of the associated problems in flame stability. Finally, results for flame propagation in premixed gases, under conditions of uniform turbulence and in recirculating flows, are discussed in section 5.

2. VORTEX SIMULATION OF FLUID FLOW

Vortex methods have been applied successfully in accurate simulations of flow fields associated with combustion systems. The reason for this success is described in the following. Combustion processes which occur in flow fields where the flow velocity is higher than the laminar burning speed are controlled by the mixing between the two streams of cold reactants and hot products in premixed flames, or between the two streams of fuel and oxidizer in diffusion flames. The rate of mixing is governed by the cross-stream currents which accompany the formation of large vortex eddies within the shear layer established between the two streams. The field of a shear layer is characterized by the formation of highly localized, time-dependent regions of vorticity with sharp transitions between "rotational fluid" and "irrotational fluid." To resolve this field, it is necessary to use computational elements that can identify the rotational fluid and trace its motion accurately. Vortex methods are constructed to meet this requirement.

In this section, we show that vortex methods are designed to implement the actual dynamics of vorticity. Vorticity generation is governed by the no-slip condition on solid boundaries, vorticity convection is implemented by the Lagrangian displacement of finite vortex elements using the velocity field, and vorticity diffusion is simulated by the dispersion of the vortex elements according to stochastic solutions of the diffusion equation. The form of the computational elements is designed to provide improved resolution, and is derived by analyzing the error of the discretization in such a way that high order approximations can be constructed.

This method is then compared to other numerical schemes used in numerical solution of fluid flow equations; all regarded as different means of approximating the vorticity field using integral interpolation. Next, the satisfaction of the potential boundary condition for complicated geometries, which require special treatment if the grid-free nature of the scheme is to be preserved, is described. The implementation of the inlet and outlet boundary conditions, and the integration of the ordinary differential equations that govern the motion of vortex elements are discussed.

1.1. FORMULATION

Vortex simulation of viscous, two-dimensional incompressible flow starts by casting the governing equations in a vorticity transport form. The dynamics of fluid motion is governed by the Navier-Stokes equations, expressing the conservation of mass and momentum (Landau and Lifshitz [46]):

$$\nabla \cdot u = 0 \tag{2.1}$$

$$\frac{\partial u}{\partial t} + u \cdot \nabla u = -\frac{1}{\rho} \nabla p + \nu \nabla^2 u \tag{2.2}$$

where $u=(u_1,u_2)$ is the two-dimensional velocity vector, $x=(x_1,x_2)$, x_1 and x_2 are the streamwise and cross-stream directions, respectively, t is time, ρ is density, p is pressure, ν is kinematic viscosity and

$$\nabla = \left(\frac{\partial}{\partial x_1}, \frac{\partial}{\partial x_2}\right) \quad \text{and} \quad \nabla^2 = \frac{\partial^2}{\partial x_1^2} + \frac{\partial^2}{\partial x_2^2}$$

Vorticity is defined as the curl of the velocity vector, $\omega=\nabla\times u$. In two dimensions, ω is a scalar:

$$\omega = \frac{\partial u_2}{\partial x_1} - \frac{\partial u_1}{\partial x_2} \tag{2.3}$$

The vorticity transport equation is derived by taking the curl of Eq. (2.2). Using the identity $\nabla\times\nabla p=0$, one obtains the following equation:

$$\frac{\partial \omega}{\partial t} + u \cdot \nabla \omega = \nu \nabla^2 \omega \tag{2.4}$$

expressing the transport of vorticity by convection and diffusion. Eqs. (2.1) and (2.3), supplemented with the appropriate boundary conditions, are sufficient to evaluate the velocity if the vorticity field is known. Given the velocity field, the vorticity can be updated using Eq. (2.4), and the solution can be advanced further in time. The pressure is recovered from the velocity field by integrating the following equation:

$$\nabla^2 p = 2\rho \left(\frac{\partial u_1}{\partial x_1}\frac{\partial u_2}{\partial x_2} - \frac{\partial u_1}{\partial x_2}\frac{\partial u_2}{\partial x_1} \right) \tag{2.5}$$

which is derived by taking the divergence of Eq. (2.2).

Eq. (2.1) is satisfied by using a stream function ψ such that:

$$u_1 = \frac{\partial\psi}{\partial x_2} \quad \text{and} \quad u_2 = -\frac{\partial\psi}{\partial x_1} \tag{2.6}$$

Substituting into Eq. (2.3), the following Poisson equation is obtained:

$$\nabla^2\psi = -\omega(x) \tag{2.7}$$

The solution of this equation in a domain without boundaries is given by:

$$\psi(x) = \int G(x-x')\,\omega(x')\,dx' \tag{2.8}$$

where $G(x) = -\frac{1}{2\pi}\ln r$

$r^2 = (x_1^2+x_2^2)$, $dx=dx_1dx_2$ and the integration is performed over the area where $|\omega| > 0$. The velocity distribution is recovered by substituting Eq. (2.8) into Eq. (2.6):

$$u(x) = \int K(x-x')\,\omega(x')\,dx' \tag{2.9}$$

where
$$K(x) = -\frac{1}{2\pi}\frac{(x_2,-x_1)}{r^2} \tag{2.10}$$

K is the integral kernel of the Poisson equation.

In vortex methods, Eq. (2.4) is solved in two fractional steps (Beale and Majda [9]):

$$\frac{\partial\omega}{\partial t} + u\cdot\nabla\omega = 0 \tag{2.11}$$

$$\frac{\partial\omega}{\partial t} = \nu\nabla^2\omega \tag{2.12}$$

In the first step, the transport of vorticity due to convection is derived from the solution of Eq. (2.11) in Lagrangian coordinates. If $\chi(X,t)$ is the trajectory of a particle that starts at X, and ω° is the vorticity distribution at t=0, then Eq. (2.11) is equivalent to (Courant and Hilbert [23]) :

$$\omega(\chi(X,t)) = \omega^{\circ}(X) \tag{2.13}$$

while χ is the solution of the differential equation:

$$\frac{d\chi}{dt} = u(\chi(X,t)) \quad \text{and} \quad \chi(X,0) = X \tag{2.14}$$

Using the Lagrangian formulation of the vorticity transport equation circumvents some of the difficulties associated with the numerical treatment of the non-linearity of the inertial terms in the Navier-Stokes equations. Thus, in principle, diffusive or dispersive errors, aliasing errors or unstable finite-differencing of derivatives are avoided in the numerical treatment of the equations of motion (for definitions, see Peyret and Taylor [61]). It also allows the construction of efficient grid-free schemes in which the computational points follow the vorticity field.

Equation (2.14) is a set of uncountable ordinary differential equations which, supplemented with Eqs. (2.9), provides a solution to the inviscid part of the Navier-Stokes equations. In order to reduce this set to a finite number of equations, $\omega^{\circ}(X)$ is discretized among small area elements to form particles of vorticity. The two possible approximation schemes that have been proposed are described in the following sections. (We will return to Eq. (2.12), representing the diffusive transport, in Section 2.5.)

2.2. POINT VORTEX APPROXIMATION

In this scheme, a continuous vorticity field is replaced by a set of point vortices. The area A where $|\omega|>0$ is divided into a number N of adjoined area elements δA_i and the circulation assigned to each point vortex is calculated as:

$$\Gamma_i = \int_{\delta A_i} \omega^\circ \, dX \tag{2.15}$$

The center of this area element, X_i, is chosen as the coordinate of the vortex point at $t=0$. At later times, the approximate vorticity distribution is:

$$\omega(x) = \sum_{i=1}^{N} \Gamma_i \, \delta(x-\chi_i) \tag{2.16}$$

where δ is the Dirac delta function, $\int\delta(x)dx=1$; $\delta(x)=0$ for $x\neq 0$ and $\chi_i=\chi(X_i,t)$. The corresponding velocity distribution is obtained by substituting Eq. (2.16) into Eq. (2.9):

$$u(x) = \sum_{i=1}^{N} \Gamma_i \, K(x-\chi_i) \tag{2.17}$$

where $K(x)$ is defined by Eq. (2.10), and represents the velocity induced by a potential point vortex. Both the vorticity field and the velocity field possess a number of singularities equal to the number of computational elements, hence difficulty should be expected in the numerical calculations. The discrete analogue of Eq. (2.14), using Eq. (2.17) is:

$$\frac{d\chi_j}{dt} = \sum_{\substack{i=1 \\ j\neq i}}^{N} \Gamma_i \, K(\chi_j-\chi_i) \qquad j=1,2.....,N \tag{2.18}$$

where $\chi(X_i,0)=X_i$ and $i=j$ is excluded since a point vortex does not induce a velocity on itself in an open space. Eqs. (2.18) form a Hamiltonian system, its solution constitutes a weak or singular solution of the Euler equations (Aref [3]). This approximation proved to be of little utility in numerical calculations due to the strong singularities in the velocity field.

The consistency of the point vortex approximation of a thin vortex sheet--the simplest possible model of a shear layer--was established by Moore [56], using the van der Vooren correction

[69] of the discretization of Eq. (2.17). However, it was concluded that the growth of perturbations due to arithmetic truncation prevents the solution from describing the long term behavior of the sheet. Since the dynamics of the physical process is absolutely unstable, perturbations with the shortest wave length grow the fastest, and noise generated by the amplification of the error associated with finite-digit arithmetics supersedes the growth of long wavelength instability modes. Krasney [44] observed, by conducting careful numerical experiments, that the growth of this error causes the early onset of irregular motion of the small scales after the roll-up of the sheet. Employing a spectral filtering technique, he was able to remove the high frequency noise, hence maintaining the accuracy of the computations for longer times. Chorin and Bernard [16] showed that by removing the singularity at the center of point vortices, which amounts to filtering out the small scale perturbations, one can accurately follow the successive folding of the sheet.

In the case of a smooth flow, Beale and Majda [11] found that the numerical simulation of the flow field of a vortex disk with an initially smooth vorticity distribution fails to describe the behavior of the solution. The error in the calculation of the velocity at points in between the vortex points grows unpredictably at short times, and continues to grow at longer times. Thus, this scheme must be abandoned for a less singular approximation.

2.3. VORTEX BLOB APPROXIMATION

In order to overcome the difficulty associated with the singularities of the simple point vortex decomposition of a continuous vorticity field, point vortices are replaced by vortex elements with a finite core radius δ (Chorin [17], and Kuwahara and Takami [42]). Inside the core radius, vorticity is smooth, or almost smooth, guaranteeing a finite velocity at the center of the vortex element. In the vortex blob approximation, the vorticity distribution takes the form:

$$\omega(x) = \sum_{i=1}^{N} \Gamma_i \; f_\delta(x-\chi_i) \tag{2.19}$$

where f_δ is the core function, employed here as an approximation of the δ-function in Eq. (2.16). f_δ is radially symmetric and $\int f_\delta dx=1$. δ is a characteristic fall-off length such that f_δ is small for $r>\delta$. In order to approximate the initial vorticity distribution accurately, 2δ is often chosen to be larger than the initial distance between vortex centers h, such that a margin of overlap is allowed between vortex blobs. In a normalized form, f_δ can be written as:

$$f_\delta(x) = \frac{1}{\delta^2} f(\frac{r}{\delta}) \tag{2.20}$$

The velocity produced by a distribution of vortex blobs is obtained by substituting Eq. (2.19) into Eq. (2.9):

$$u(x) = \sum_{i=1}^{N} \Gamma_i \; K_\delta(x-\chi_i) \tag{2.21}$$

where $$K_\delta(x) = -\frac{1}{2\pi} \frac{(x_2, -x_1)}{r^2} \kappa(\frac{r}{\delta}) \tag{2.22}$$

and $$\kappa(r) = 2\pi \int_0^r r \; f(r) \; dr \tag{2.23}$$

κ represents the fractional circulation of the element within radius r. A choice of f that has been used extensively in applications is the Chorin's core function for which

$$f(r) = \begin{cases} 1/(2\pi r) \\ 0 \end{cases} \quad \text{and} \quad \kappa(r) = \begin{cases} r \\ 1 \end{cases} \quad \text{for} \quad r = \begin{cases} \leq 1 \\ \geq 1 \end{cases} \tag{2.24}$$

This is the simplest choice. While it possesses a singularity in the vorticity distribution and a discontinuity in the velocity field, it has been used successfully in numerous studies because of its compatibility with the vorticity generation algorithm along the walls suggested by Chorin [17]. A slight modification of this core is the Rankine vortex core used by Milinazzo and Saffman [54] for which:

$$f(r) = \begin{cases} 1/\pi \\ 0 \end{cases} \quad \text{and} \quad \kappa(r) = \begin{cases} r^2 \\ 1 \end{cases} \quad \text{for } r = \begin{cases} \leq 1 \\ \geq 1 \end{cases} \tag{2.25}$$

Hald [35] showed that the solution of a vortex method that uses Chorin's core converges to the solution of the Euler equations. He proved that the rate of convergence of the vorticity approximation is $O(\delta^2)$. He also showed that smooth and continuous forms of f, which possess both positive and negative values of vorticity, can be used to achieve higher order accuracy. Leonard [49] proved that if f has a Gaussian form, the vorticity distribution converges with second order accuracy in h, where h is the initial spacing between the vortex-blob centers, provided that the flow is smooth. However, that requires the flow to be smooth at all times. Beale and Majda [10] showed the convergence of a class of vortex schemes with arbitrary high order accuracy if f is smooth, rapidly decaying and with an increasing number of vanishing moments. One choice is a generalized Gaussian which was expressed elegantly by Monaghan [55] in the form:

$$f(r) = \frac{1}{\pi} \exp(-\frac{\nabla^2}{4})\, e^{-r^2} \tag{2.26}$$

where successively higher order core functions can be obtained by expanding the derivative operator and retaining more terms. The second-order Gaussian core used in Beale and Majda [11] and Nakamura et al. [57] is the first-order term in the expansion of the above expression:

$$f(r) = \frac{1}{\pi} e^{-r^2} \quad \text{and} \quad \kappa(r) = 1 - e^{-r^2} \tag{2.27}$$

Other choices for the core function have been constructed in Hald [35].

By requiring f to have an infinite number of vanishing moments, Leonard [49] showed that the spectral core, expressed in terms of the Bessel functions J_o and J_1, can be used as a basis of an infinite order scheme. Its characteristic functions are:

$$f(r) = \frac{1}{2\pi r} J_1(2r) \; ; \quad \text{and} \quad \kappa(r) = 1 - J_o(2r) \qquad (2.28)$$

The family of high order core functions that can be used to obtain very accurate schemes was extended recently by Hald [37] to a large class of Bessel functions. The actual rate of convergence depends on the smoothness of the underlying flow field which, in most cases, becomes very convoluted in a finite period of time. The deterioration of the rate of convergence of high order methods as the flow field evolves with time was observed computationally by Beale and Majda [11], and Anderson [2], who made some recommendations concerning how to reverse this trend by redistributing the computational elements. On the other hand, Nakamura et al. [57] showed that for highly accurate schemes, the time integration of Eqs. (2.18) may still limit the improvement in the overall accuracy of the results. In practice, the issue is complicated more by the error introduced in the simulation of the effect of diffusion and the vorticity generation along the boundaries.

The motion of vortex blobs is governed by a set of ordinary differential equations in the form of Eq. (2.18) with K replaced by K_δ. However, since the vorticity of each element is distributed over a finite area, it is not clear whether the velocity of a blob should be a vorticity-weighted average or the velocity at its center. The first choice is compatible with the definition of Γ in Eq. (2.15). Leonard [49] derived the appropriate formula for the velocity by averaging over the vorticity of a blob. However, Nakamura et al. [57] showed computationally that the accuracy of this scheme is inferior to the accuracy of the scheme which uses the velocity at the center of a vortex element. This is consistent with the analysis of Hald [37], who indicated that using pointwise vorticity at the center of the blob to define its circulation, i.e.,

$$\Gamma_i = \omega^\circ(X_i)\ \delta A_i \qquad (2.29)$$

is necessary to construct high order schemes. Here, we will only use this last choice, for which:

$$\frac{d\chi_j}{dt} = \sum_{\substack{i=1 \\ i\neq j}}^{N} \Gamma_i\ K_\delta(\chi_j - \chi_i) \qquad j=1,2,\ldots\ldots,N \qquad (2.30)$$

Finally, it should be noted that in actual computation, the vorticity field changes with time due to the generation of vorticity and the results of the theory may have to be modified accordingly. In most cases, extra vorticity is generated every time step to satisfy either the no-slip condition along solid boundaries or the Kutta condition at points of separation (sections 2.6 and 2.7).

2.4. NATURE OF APPROXIMATION

Equation (2.19) provides a non-diffusive approximation of the vorticity distribution (Leonard [49]). Heuristically, since the vorticity is discretized only at the initial conditions, and if the time integration of Eqs. (2.30) is error-free, the distribution of vorticity remains unsmoothed by the solution scheme. Moreover, solutions obtained using vortex methods do not depend on the spatial derivatives of ω. Thus, these methods are most appropriate to employ when the solution is expected to develop sharp gradients due to high concentrations of vorticity, such as shear layers or turbulent fields. Schemes that depend on the representation of functions on fixed grids may fail in such cases because of the excessive numerical diffusion associated with discretization of high-order gradients, or become unstable when the gradients are too sharp.

Equation (2.19) can be viewed as a discrete form of a core function, or kernel, approximation of the vorticity distribution (Monaghan [55]). This kernel form is derived from the expression:

$$\omega_\delta(x) = \int f_\delta(x-x')\ \omega(x')\ dx' \qquad (2.31)$$

where f_δ is an approximation of the δ-function, $\int f_\delta dx=1$, and $\omega_\delta \to \omega$ as $\delta \to 0$. δ is the radius of the circle, centered at x', within which the kernel f_δ provides the major contribution to ω_δ (core function or kernel have been used in the literature of the theory of distributions to refer to the same object). This representation helps in determining the error in the approximation of the vorticity field. Expanding $\omega(x')$ in Eq. (2.31) in a Taylor series, and assuming a one-dimensional flow for simplicity (more algebra is required for two dimensions), one finds that in the neighborhood of x:

$$\omega_\delta(x) = \omega(x) + \sum_{i=1}^{\infty} \frac{\omega^{(i)}(x)}{i!} \int (x-x')^i \, f_\delta(x-x') \, dx' \qquad (2.32)$$

where $\omega^{(i)}$ is the i-th order derivative of ω (the flow is assumed smooth and without boundaries). In order for the successive terms in the error to vanish, higher-order moments of f_δ should be identically zero. If f is radially symmetric, i.e. function of r only, the integration is zero for all the odd values of i and the vortex approximation is at least of second order in δ (showing, in essence, why a scheme using Eq. (2.24), Eq. (2.25) or Eq. (2.27) should have the same accuracy (Ng and Ghoniem [58])). Moreover, if we require that:

$$\int r^i \, f(r) \, dr = 0 \qquad \text{for } i=2,4\ldots \quad p-2 \qquad (2.33)$$

then the approximation is $O(\delta^p)$, where p is an even number. This important result shows that vortex approximations can be constructed with an arbitrary high order. The second contribution to the error, $|\omega_\delta(x)-\omega(x)|$, results from approximating the integral using a summation of finite terms, and was evaluated by Anderson and Greengard [1] using the analysis of the trapezoidal rule. More rigorous analysis in two and three dimensions is given by Hald [35,37] and Beale and Majda [10].

In terms of conventional schemes for the numerical approximation of fluid flow equations, and depending of the choice of f_δ in Eq. (2.19), vortex blob representation can be viewed as a combination of the two basic integral methods: finite element methods and spectral methods. In both methods, the function of interest is expanded in terms of a set of base functions $F_i(x)$ such that:

$$\omega(x,t) = \sum_{i=1}^{N} \omega_i(t)\, F_i(x) \tag{2.34}$$

In a finite element approximation, F_i exists only within the element i and ω_i are the nodal values of ω, while in a pseudo-spectral approximation, F_i exists on the whole space and ω_i are the values of ω in the spectral domain.

If the core functions f_δ are not allowed to overlap, i.e. the separation between the centers of vortex elements is larger than 2δ, vortex decomposition approaches a finite element representation of the vorticity field and f_δ can be interpreted as the shape function within each element. On the other hand, if f_δ extends over the whole domain and contributions from different elements overlap to approximate the vorticity field, Eq. (2.19) resembles a spectral decomposition of the vorticity field and f_δ becomes the mapping function. Indeed, Bessel core functions provide high order approximations in much the same way as conventional spectral methods. It should be mentioned that even with low order core functions, the corresponding integral kernels K_δ in the velocity distribution in Eq. (2.22) are highly overlapping, hence resulting in smooth velocity fields.

Vortex methods are different from integral methods in two fundamental aspects: (1) in vortex methods, the values of ω_i, or Γ_i, are known from either the initial conditions (for inviscid flow) or the boundary conditions (the no-slip condition in viscous flow) and are constant for each vortex element with time. In integral methods, however, the values of ω_i are defined at grid points and are sought in the solution that changes with time; and (2) the core functions f_δ for vortex methods are invariant with time and their position changes according to Eqs. (2.30), while the grid points in integral methods are fixed in space (except for a small class of adaptive grid methods). Therefore, vortex methods can be regarded as grid-free, Lagrangian finite-element or spectral methods in which the function is interpolated among a set of moving points and the solution is obtained in a vorticity-velocity form.

As mentioned earlier, computational vortex elements are distributed only in areas where $|\omega|>0$. In most physically interesting problems, such as boundary layers, shear layer and recirculating flows at high Reynolds numbers, the area of non-zero vorticity represents a small portion of the whole domain. Thus, the computational effort is concentrated within regions of large gradients to provide high resolution. Moreover, if viscous effects are neglected, the total area of $|\omega|>0$, while getting more distorted with time, remains constant. Thus, the computational effort remains almost constant while the resolution is maintained since the computational elements redistribute themselves to follow the regions of high vorticity concentrations.

2.5. DIFFUSIVE TRANSPORT

Eq. (2.12) expresses the contribution of diffusion to the transport of vorticity. In vortex methods, the effect of diffusion has been interpreted as either the dispersion of a finite number of vortex elements with finite and constant vorticity, or the expansion of the core of individual vortices without changing their total circulation. The first interpretation is due to Chorin [17] (with earlier applications to the diffusion of heat by Courant, Friedrichs and Levy [22] and by Doob [25]) and is summarized in the following. The Green function of the one-dimensional form of Eq. (2.11):

$$G(y,t) = \sqrt{\frac{1}{4\pi\nu t}}\, \exp(-\frac{1}{4\nu t}\, y^2) \qquad (2.35)$$

is identical to the probability density function of a Gaussian random variable η with a zero mean and a standard deviation σ

$$P(\eta,t) = \sqrt{\frac{1}{2\pi\sigma^2}} \exp(-\frac{1}{2\sigma^2}\eta^2) \tag{2.36}$$

if $\sigma=\sqrt{2\nu t}$. Thus, the distribution of $\omega(y,t)$ can be approximated by a number of particles N located at η_i, $i=1,2,...,N$, where $\{\eta\}$ is chosen from a Gaussian population with zero mean and standard deviation σ. At $t=0$, particles are placed at $y=0$ with each carrying vorticity $1/N$. If the solution is obtained by marching in time using time steps Δt, then the same particles are displaced every time step using $\{\eta\}$ with $\sigma=\sqrt{2\nu\Delta t}$ (exploiting the linearity of the diffusion equation). For an arbitrary initial condition $\omega(y,0)$, particles are distributed over the space at $t=0$ such that at each point $N=N(y)=\omega(y,0)/\omega_m$, ω_m is the maximum vorticity assigned to a particle.

Boundary conditions are satisfied by reflecting the particles that leave the boundaries back into the domain. For Dirichlet boundary conditions, the particle vorticity remains the same, while for Neumann boundary conditions, the sign of the particle vorticity is changed upon reflection (Chandraskehar [14]). The scheme is readily extendable to two dimensions by displacing the particles in two perpendicular directions using two independent sets of η (for more detail, see Ghoniem and Sherman [32]).

In terms of the velocity, which is evaluated by summing over the contributions of all vortex elements, Hald [35] showed that the expected value of this stochastic solution of the diffusion equation in one-dimension is the same as the exact solution. Moreover, the variance of the solution decreases linearly as the number of elements increases. A more important result is that the error is independent of ν. The last result was supported computationally by Ghoniem and Sethian [31] in their two-dimensional calculations for a flow behind a step. Thus, the scheme possesses the following important property: it does not introduce extra numerical diffusion as the inertial effect becomes more important. This property is particularly useful at high Reynolds number when the dynamics is controlled by the delicate balance between convection and diffusion.

In the random vortex method, the diffusion transport of vortex elements is simulated stochastically by adding to their convective motion an extra displacement drawn from a Gaussian population with a zero mean and a standard deviation σ. The total transport of vortex elements is obtained by adding the two fractional displacements:

$$\chi_i(X_i, t+\Delta t) = \chi_i(X_i, t) + \sum_k u_\delta(\chi_{ik})\,\Delta t + \eta_i \qquad (2.37)$$

where $\sum_k$ is a kth order time integration scheme and $\eta_i = (\eta_1, \eta_2)_i$ is a two-dimensional Gaussian random number.

The random walk algorithm is compatible with vortex schemes because of its grid-free Lagrangian form. It can also be applied near solid boundaries to move vortex elements which are generated to satisfy the no-slip condition without loss of resolution since it does not depend on the resolution of a grid. Ghoniem and Sherman [32] discuss in detail the stochastic solutions of the diffusion equation with different boundary conditions, its application to the reaction-diffusion equation governing laminar flame propagation (Ghoniem and Oppenheim [29]), and its application to the combined heat and momentum diffusion, characteristic of problems in natural convection.

The second mechanism of including the effect of diffusion in the vortex solution, used by Kuwahara and Takami [42] and Ashurst [5], is based on the analytical solution for the decay of an isolated vortex (see Batchelor [8]). The modified potential vortices in Eq. (2.18) are replaced by viscous vortices whose cores grow with time such that the core radius expands at a rate proportional to $\sqrt{4\nu t}$. This model produces reasonable results if one replaces ν by a "turbulent viscosity." However, this is against the notion of numerical simulations of turbulent flow in which ad hoc modeling assumptions are abandoned. Moreover, Greengard [34] proved that this scheme does not converge to the solution of Navier-Stokes equations.

2.6. BOUNDARY CONDITIONS

Navier-Stokes equations, when used to describe a viscous incompressible flow, contain a mix of hyperbolic and elliptic terms that necessitates the satisfaction of a set of conditions on the boundaries of the solution domain. On solid walls, the fluid adheres to the solid, acquiring the velocity of the boundary. Inlet velocity profiles are mostly well specified, albeit downstream activities may perturb the inlet section in a time-dependent flow. The situation is more subtle for outflow conditions since they are not known in advance and their variation with time can be unpredictably complicated.

The potential boundary condition is imposed in the direction normal to the boundaries. The total velocity produced by a vorticity distribution in a semi-confined space can be decomposed into two components:

$$v = u_\delta + u_p \tag{2.38}$$

where u_δ is the field induced by the vorticity within the domain, while u_p is an irrotational component added to satisfy the boundary conditions by not allowing the flow to leave through the walls (this component can be regarded as a result of image vortices which are located outside the domain.) In this case, u_δ in Eq. (2.37) should be replaced by v. Since u_p is irrotational, it can be represented in terms of a potential ϕ where $u_p = \nabla\phi$ where ϕ is governed by:

$$\nabla^2\phi = 0 \tag{2.39}$$

ϕ is uniquely determined if its value or its normal derivative $\nabla\phi.n$ is specified on a closed contour (in the second case, ϕ is determined up to an arbitrary constant). On a solid wall:

$$\nabla\phi \cdot n = -u_\delta \cdot n \tag{2.40}$$

where n is the local normal to the wall. In vortex schemes, the solution of the Laplace equation with Neumann conditions in an arbitrary domain has been obtained using panel methods (Chorin [17]), fast Poisson solvers on a grid (Ashurst [6] and Sethian

[65], Hsiao et al. [39]) and by conformal mapping employing image vortices (Peskin and Wolfe [60] and Ghoniem et al. [28]).

The computational efficiency of fast Poisson solvers is superior since they require $O(M \log M)$ number of operations, where M is the number of grid points used to represent ϕ. Therefore, they should be employed when available for the geometry of interest, which in general limits their applicability. On the other hand, conformal mapping methods produce exact solutions for the potential flow by defining the image system of the distribution of vortex elements. Moreover, they are grid-free and can provide higher resolution close to the boundaries and at points of separation where it is essential to determine accurately the flux of vorticity into the interior of the domain.

For a given distribution of N vortex elements in the physical z-plane, where $z=x_1+ix_2$ and $i=\sqrt{-1}$, the total velocity at the center of the jth element is

$$w(z_j) = \{ \sum_{\substack{i=1 \\ i\neq j}}^{N} \frac{-i\Gamma_i}{2\pi} \frac{1}{\Delta\zeta} \kappa(\frac{|\Delta\zeta|}{\delta_i}) + \sum_{i=1}^{N} \frac{i\Gamma_i}{2\pi} \frac{1}{\Delta\tilde{\zeta}} \kappa(\frac{|\Delta\tilde{\zeta}|}{\delta_i})\} F(\zeta_j)$$

$$- \frac{i\Gamma_j}{4\pi} (\frac{dF}{d\zeta})_j$$

where $w= v_1-iv_2$ is the complex velocity, ζ is the complex coordinate in the transform plane, $\Delta\zeta=\zeta_j-\zeta_i$, $\Delta\tilde{\zeta}=\zeta_j-\tilde{\zeta}_i$ and $\tilde{\zeta}$ is the complex conjugate, $\delta_i=\delta\ F(\zeta_i)$. $F=(d\zeta)/(dz)$ is the transformation function that maps the physical z-plane to the upper half of the ζ-plane.

The viscous boundary condition should also be imposed along solid walls. The action of viscosity stops the fluid along solid walls and annihilates the tangential velocity component:

$$v_s \cdot s = 0 \tag{2.41}$$

where v_s is the total slip velocity on the wall and s is the unit vector in the tangential direction. This equation defines the only mechanism of vorticity generation in incompressible isothermal flow. The total circulation produced by satisfying the no-slip condition is:

$$\Gamma = \int v_s \cdot ds \qquad (2.42)$$

This scheme of vorticity generation was first suggested by Lighthill [48], to construct a numerical solution of boundary layer flow. Chorin [17] was first to implement it in the random vortex method. In this method, vorticity is generated along the wall using Eq. (2.42) and discretized into elements spaced by a distance h. To improve the resolution, the element at each point is subdivided into several elements such that $\Gamma<\Gamma_{max}$. These elements leave the wall by diffusion, simulated by random walk, to become part of the interior vorticity field at later times. Boundary layer scaling can be used to reform the shape of vortex elements near the boundaries and insure high resolution in the direction normal to the wall. In this case, vortex sheet representation of new vorticity is used within a distance $\sqrt{\nu}$ normal to the wall (Chorin [18,19]). A smooth transition between vortex sheets and vortex blobs can be accomplished by introducing elliptical vortices with a variable geometry which start as sheets on the wall and become circles at distances larger than $\sqrt{\nu}$.

The vorticity generation algorithm satisfies the viscous boundary at each time step, while it introduces perturbations in the transport of vorticity from the walls to the interior. These perturbations will grow at conditions of high Reynolds number, amplified by the destabilizing effect of convection. On the other hand, at low Reynolds numbers, the scheme is capable of simulating viscous processes accurately, without noticeable effect for the noisy simulation of diffusion. Ghoniem and Sethian [31] showed that the recirculating flow behind a rearward-facing step, where the accurate satisfaction of the no-slip condition is essential, can be simulated successfully at Reynolds numbers as low as 50.

A complete convergence proof for the random vortex scheme with vorticity generation is not available yet, and one should rely on careful numerical experiments to judge the credibility of the method (except for the work of Marchioro and Pulvirenti [52], which showed that the random vortex method in two dimensions converges weakly to a weak solution of the Navier-Stokes equation without boundaries.) However, a convergence proof was obtained by Hald [36] for the one-dimensional form of the algorithm, even when complicated by the generation of vorticity in the interior of the field. In this case, diffusion is in one dimension only, but vorticity is generated by both baroclinic effects produced by a temperature gradient in a gravity field as well as by the condition of no-slip on the wall. This application of the random walk algorithm to the coupled momentum and energy equations for natural convection was suggested by Ghoniem and Sherman [32].

2.7. INFLOW AND OUTFLOW BOUNDARIES

The inflow and outflow boundary conditions, specified in terms of the values of u_p and ω or their normal derivatives, are applied at sections of the boundaries where there are no walls. At the inlet section, the flow may carry an amount of vorticity which was created due to an earlier separation or a developing boundary layer. If the point of separation is well defined, such as a trailing edge or a sharp corner, the rate of creation of vorticity can be evaluated by applying the Kutta condition, even if the flow is unsteady. For the development of a mixing layer and the flow behind a step, Ashurst [5,6] used this algorithm. A more precise representation of the flux of vorticity around points of separation may be obtained from the calculations of the development of the boundary layer upstream of this point. Ghoniem et al. [28] used the vortex sheet algorithm to simulate the effect of the inlet boundary layer on the recirculation zone behind a rearward-facing step.

Outlet flow conditions on ω are complicated by the fact that most shear flows of interest are time-dependent, and require a long distance downstream to reach a fully-developed state. In most cases, one is forced to delete vortex elements at a specified section x_{max} to limit the computational effort, implying that the opposite flux of vorticity at this section is negligibly small. At x_{max}, if the flow is not strongly elliptic, deleting vorticity should not have a strong influence on the flow upstream. In recirculating flow, x_{max} must be larger than X_R, where X_R is the recirculation zone length, to ensure the appropriate accumulation of vorticity within the recirculating flow. (In this case, it was found by Gagnon and Ghoniem [27] that if $x_{max} \geq 2X_R$, the solution becomes insensitive to x_{max}.) While in parabolic flows, such as shear layers, the situation is less stringent and the work of Ng and Ghoniem [58] shows that if x_{max} is chosen downstream of the section where eddies grow to the size of the channel height, the harmful effect of deleting vorticity does not propagate more than one channel height upstream.

The value of u_p at the inlet section should be adjusted to match the incoming flow, i.e., $u_p=V-u_\delta$, where V is the incoming flow velocity. At the exit section, several assumptions have been used. Sethian [65] assumed that if the channel is long enough, v achieves a fully developed profile. However, in a time-dependent flow, this assumption besides requiring a long computational domain may be too stringent. Hsiao et al. [39] used ϕ=constant at the exit section, which is equivalent to $u_{p2}=0$, hence the potential component of the flow is assumed fully developed. Their results showed some deviation from the experimental data around the exit section. Ghoniem et al.[28] and Ng and Ghoniem [58] employed conformal transformation to solve for the potential component. In this case, u_p is evaluated by constructing the images of the existing vortices, assuming infinitely long channel in the flow direction. Thus, an explicit boundary condition on u_p at the exit section is not required.

2.8. TIME INTEGRATION

In most applications a first order integration scheme was used to advance the vortices in time in Eq. (2.37). Vortex methods are unconditionally stable by virtue of the fact that each vortex element affects the velocity of all other elements in the field via the Biot-Savart Law. In that regard, vortex methods can be viewed as implicit schemes, in which each computational element has an infinite domain of influence, and relatively large time steps can be used. However, the accuracy and stability of the time integration scheme may have a strong effect on the overall accuracy. If viscous effects are not considered, a multi-step scheme may be used, otherwise, a single step method must be employed. For the first case, Spalart and Leonard [68] showed that the stability characteristics of the Adam-Bashforth second order method are most desirable. For the calculations with viscous effects, Sethian [64] found that Heun's second order scheme is sufficient to obtain accurate results. In both cases, a core function that results in a spatially second order accurate scheme was used.

In some studies, such as Kuwahara and Takami [43] and Inoue [40], the time integration error associated with a diffusive first order scheme was interpreted as turbulent diffusion. However, this is misleading, and more rigorous analysis of Anderson and Greengard [1], along with the careful computational experiments of Nakamura et al. [57], show that this error is proportional to the second derivative of vorticity with a large effective diffusivity. Thus, it behaves like molecular viscosity form and causes the effective Reynolds number of the calculations to be much lower than anticipated. This led Aref and Siggia [4] to conclude that Ashurst [5] results for a free shear layer bear more resemblances to laminar dynamics than turbulent behavior. The large effective viscosity associated with low order time integration schemes was also observed in the study of Ng and Ghoniem [58]. In this case, the eddy structures were smeared out and diffused together, while the turbulent statistics were reduced by a factor of 30%.

2.9. CLOSURE

This review concentrated on a sub-class of vortex methods that uses direct summation of the velocity produced by vortex elements, governed by the Biot-Savart law, to evaluate the induced field of vorticity. Although computationally inefficient, this form presents the most reliable way of calculating the velocity field and has proven to be the most accurate in numerical studies. Vortex-in-cell, particle-mesh and cloud-in-cell algorithms use a similar Lagrangian representation of vorticity with an Eulerian computation of velocity to economize the procedure when the number of vortex elements is very large. However, Christiansen [21] showed that they introduce a directional bias in the velocity calculations. Aref and Siggia [4] used the vortex-in-cell algorithm to follow the development of a temporally-growing shear layer, but observed that the scheme does not conserve the entropy. The articles of Leonard [49], and Saffman and Baker [63] should be consulted for a more comprehensive review of these methods.

3. RESULTS: SHEAR LAYER AND RECIRCULATION

The applications of computational vortex methods have been shown to produce results with reasonable accuracy at a wide range of values of ν, or the Reynolds number $R=1/\nu$. The true power of these methods can be appreciated from the details they reveal about the structure and dynamics of the flow. Computer-generated movies--regarded as numerical flow visualization experiments--have been used to examine the resemblance between the dynamics of the large scale structure in shear flows as seen experimentally and calculated numerically (Ghoniem et al. [30]).

In section 2, vortex methods were derived as rigorous numerical approximations of the Navier-Stokes equations. Heuristically, they can be viewed as a description of the dynamics of vorticity on the small scale level. It is conceivable that small scale vortex interactions follow a set of universal rules independent of the flow configuration and, if properly incorporated in a simulation, it should predict the dynamics of the large scales in accordance with the boundary conditions and the range of the dynamic parameters.

Here, we present numerical solutions for two generic shear flow problems: a confined shear layer, and a confined recirculating flow. The importance of the first problem stems from the fact that this flow has been used extensively to study the deterministic dynamics of turbulent flow. The formation of large scale vortical structures downstream the point of separation, and their growth and interaction, have been shown experimentally to follow a well-organized pattern even at high values of the Reynolds numbers (Cantwell [13]). Numerically, this problem is a critical test for the accuracy of the scheme since the structures tend to develop sharp boundaries which separate the vorticity from the irrotational flow. In the separating flow behind a rearward-facing step, the interaction between the shear layer and the boundary layers forming on the two side walls of the step causes the formation of the recirculation zone. Contrary to the previous case, the flow field is strongly dependent on the Reynolds number, and the structure of the recirculation zone changes with varying the relative effect of viscous forces.

In figures 1 and 2, the shear layer is formed between two streams flowing at velocities 1 and .333, with $\nu=0$, and $\nu=.00025$, respectively. The vorticity field is depicted by all the computational vortex elements, with their velocity shown by line vectors. Upstream of the end of the splitter plate, the two velocities are uniform. A vortex sheet is formed at the point where the two streams meet, and vorticity is generated at a rate proportional to the difference between the two velocities. The following stages of development of the vorticity structure are recognized from the figures.

A vortex sheet is linearly unstable to perturbations of wavelengths longer than its thickness via the Kelvin-Helmholtz instability mechanism. Perturbations grow exponentially downstream at a rate inversely proportional to their wavelength. This growth leads to the roll-up of the sheet into small vortex eddies of almost circular cross section that flow at the average speed of the two streams.

Beyond this stage, eddies merge by successive pairings. During this interaction two or three eddies wrap around each other by their mutual velocity field, and deform by their concomitant strain field, till their separate identities are lost when their common axis is about 90° of the primary flow direction. This process leads to the formation of larger eddies with high concentration of vorticity. Entrainment, a process during which irrotational fluid is pulled inside the vortex eddies, is observed only for large eddies during their motion between pairings. Ingestion of irrotational flow into a large eddy increases its size and allows molecular diffusion to homogenize the two initially unmixed fluids.

Figures 1 and 2 show that the free shear layer flow is independent of the Reynolds number, except where the roll-up starts. Average streamwise velocity profiles and average streamwise velocity fluctuations, computed at stations 1,2,3,4, and 5 channel heights downstream the point of separation, are shown in figures 3 and 4, respectively. The comparison with experimental data shows good agreement except for the spread of the fluctuations profiles due to the lack of the third dimension in the computations. The agreement between the computational results and the experimental data indicate that the method can capture some of the properties of turbulence, limited by the restrictions to two dimensions, without invoking modeling assumptions concerning turbulent transport mechanisms. For more details, see Ng and Ghoniem [58], Ashurst [5], and Inoue [40].

Contrary to the free shear flow problem, the growth of boundary layers along solid walls and their interactions with the separating shear layer leads to the establishment of the recirculation zone in the flow over a rearward-facing step. The accurate satisfaction of the boundary conditions on solid walls is, hence, of particular importance in this case, and predictions of the reattachment length can serve to validate the numerical scheme. Computations for a flow behind a rearward-facing step reveals the effect of the Reynolds number on the structure of the recirculation zone. Figures 5, 6, 7 and 8 show the instantaneous streamlines for the flow at Reynolds numbers of 50, 125, 500 and 5000, respectively. At very low Reynolds number, R=50, the recirculation zone is small and stagnant, and the flow is stable and steady with very little convective exchange between the main stream and the recirculating flow. The recirculation zone forms an ellipse of vorticity whose major axis is about three step heights. Statistical perturbations associated with the random walk simulation of diffusion decay without threatening the stability of the computations.

At low Reynolds numbers, R=125, the recirculation zone grows to a critical size where it breaks up into two eddies, one is confined to the step and one flows downstream. The confined eddy grows by entraining more vorticity which is being shed from the step, while the other eddy decays by the action of diffusion as it flows downstream. The process repeats itself regularly yet the flow is still stabilized by viscous effects and can be described as laminar. The reattachment length is six step heights, in accordance with experimental data. At moderate Reynolds numbers, R=500, the oscillations of the recirculation zone become more severe, indicating that the amplification of perturbations overcome the smoothing effect of diffusion, which leads to a state of transition. The separating shear layer breaks down into small eddies which grow by pairing, reminiscent of the simple shear layer case. The reattachment length grows to a maximum at this value of the Reynolds number, as has been shown experimentally, then decreases in the turbulent range.

At high Reynolds numbers, a wide spectrum of eddy scales is formed downstream of the step with a counter-rotating eddy trapped at the lower corner of the step. Small eddies are formed along the bottom wall of the channel, resembling a turbulent boundary layer, and strong separation appear on the opposite wall of the step, resulting in the formation of a separation eddy (Ghoniem and Sethian [31]). The reattachment length oscillates around seven step heights. The flexibility and robustness of the numerical scheme are confirmed by these calculations. The results predict the change in the flow structure as conditions change from diffusion-dominated to inertia-dominated dynamics without loss of stability. No modeling assumptions or adjustable constants are employed to fit a particular range of Reynolds number and the same set of numerical parameters, i.e. Δt, h, and Γ_i, were used for all the Reynolds numbers.

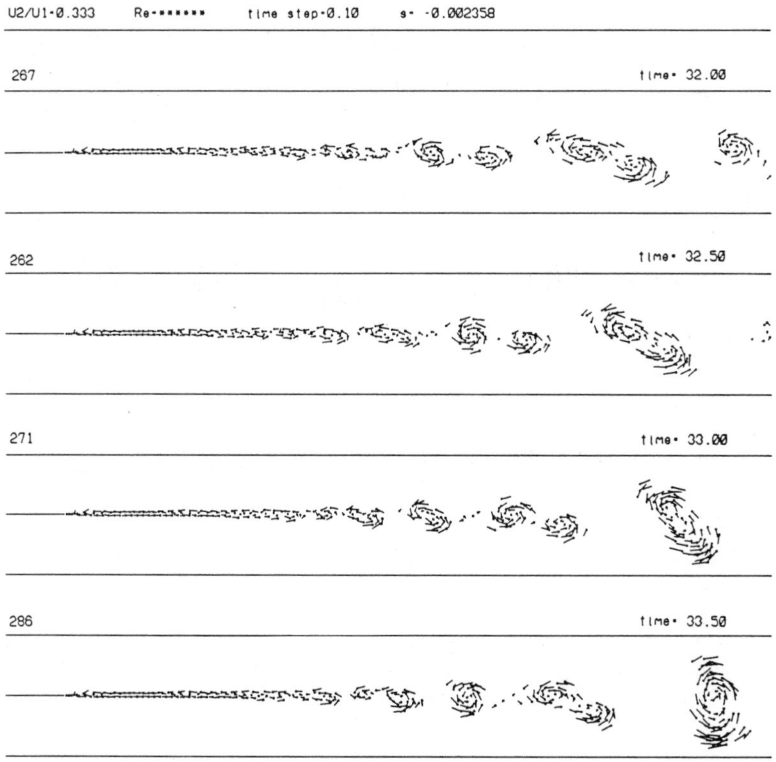

Figure 1. The vorticity field of a shear layer forming between two streams flowing at velocities 1 and 0.33 in an inviscid flow, showing the formation of large eddies and their pairings.

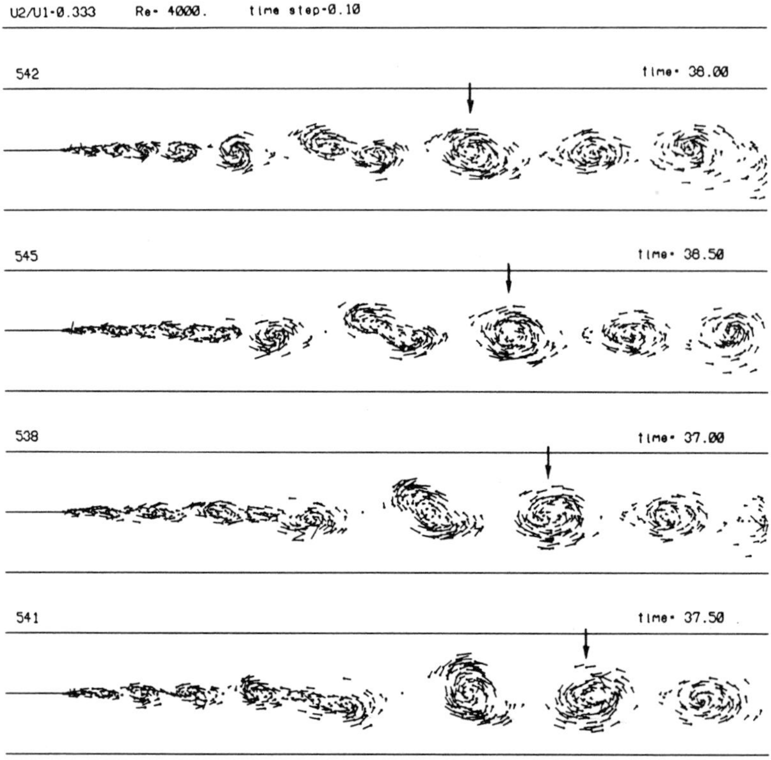

Figure 2. The vorticity field of a shear layer forming between two streams flowing at velocities 1 and 0.33, R=4000. Most of the entrainment of irrotational fluid occurs between pairings.

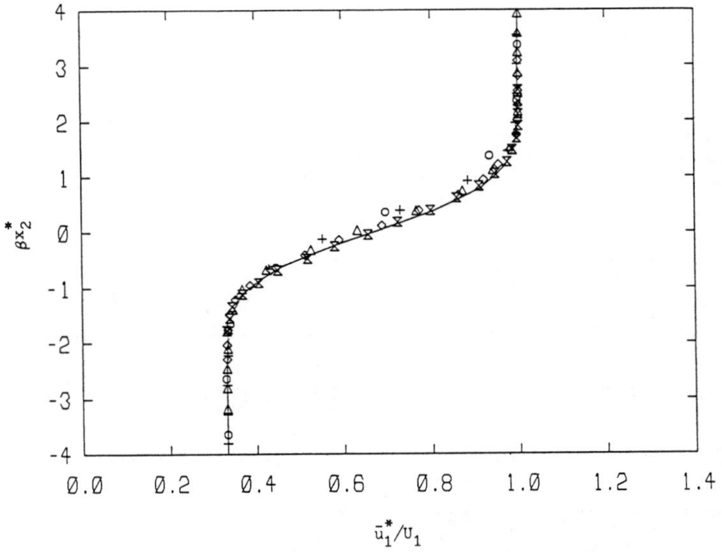

Figure 3. Average streamwise velocity profiles for the shear layer in figure 2, computed at 5 sections downstream (o: $x_1=1$; +: $x_1=2$; Δ: $x_1=3$; ◇: $x_1=4$ and ⧗: $x_1=5$), compared with experimental data.

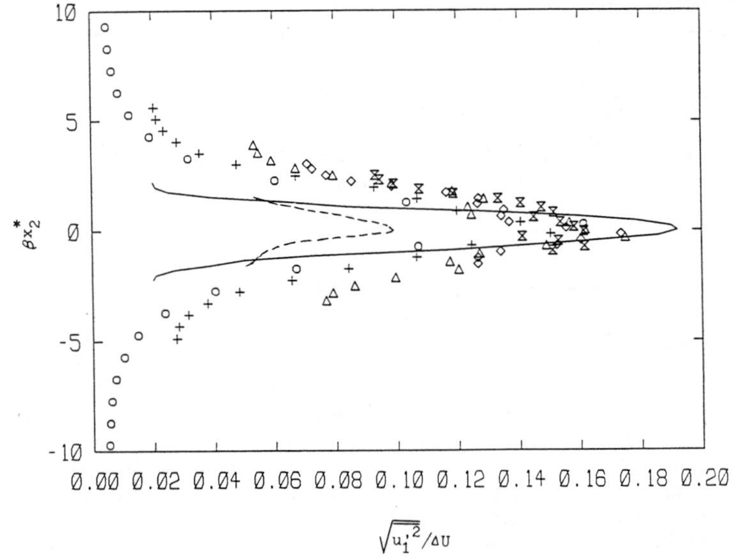

Figure 4. Average streamwise fluctuating velocity component for the shear layer of figure 2, compared with experimental data at high Reynolds number ——— and low Reynolds number - - - . (Symbols are as in figure 3.)

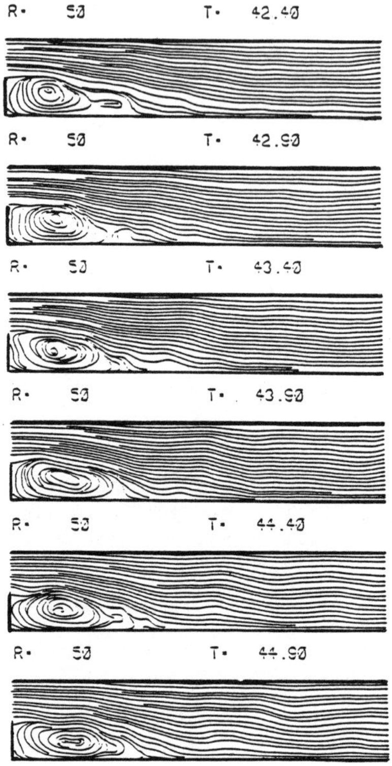

Figure 5

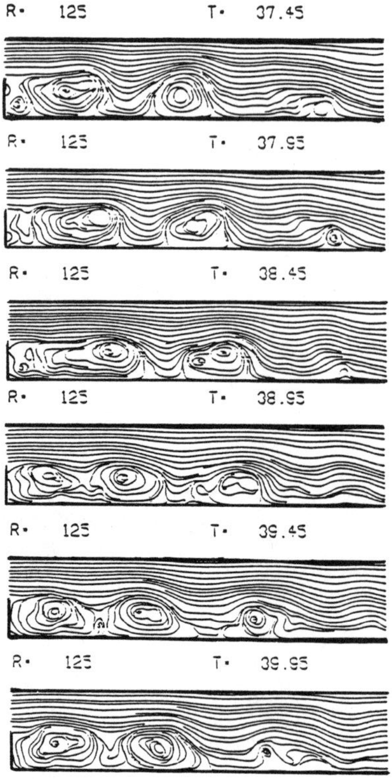

Figure 6

Figure 5. Streamline plots for a viscous flow behind a sudden expansion in a channel at R=50, plotted every 10 time steps of $\Delta t=0.1$, starting at t=84.8. The recirculation zone is formed of one stagnant eddy.

Figure 6. The break-up of the recirculation eddy into a two-eddy zone at R=125, signifying the establishment of laminar conditions. The vortex in the first frame expands by entrainment to a critical size, shown in the third frame, which disintegrates into two separate eddies. The first frame is plotted at t=74.9 with t=1 between each two frames.

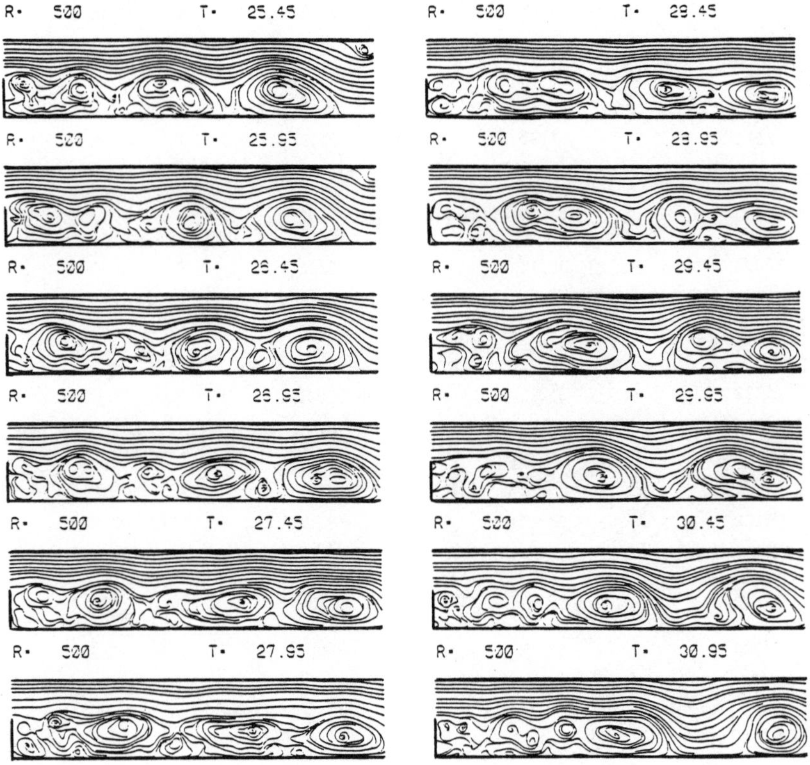

Figure 7. Streamlines for a flow at R=500, plotted every ten computational time steps, starting at t=50.9, depicting the formation of a wide spectrum of scales at the range of transition to turbulence. The pairing of large-scale eddies is clearly observed.

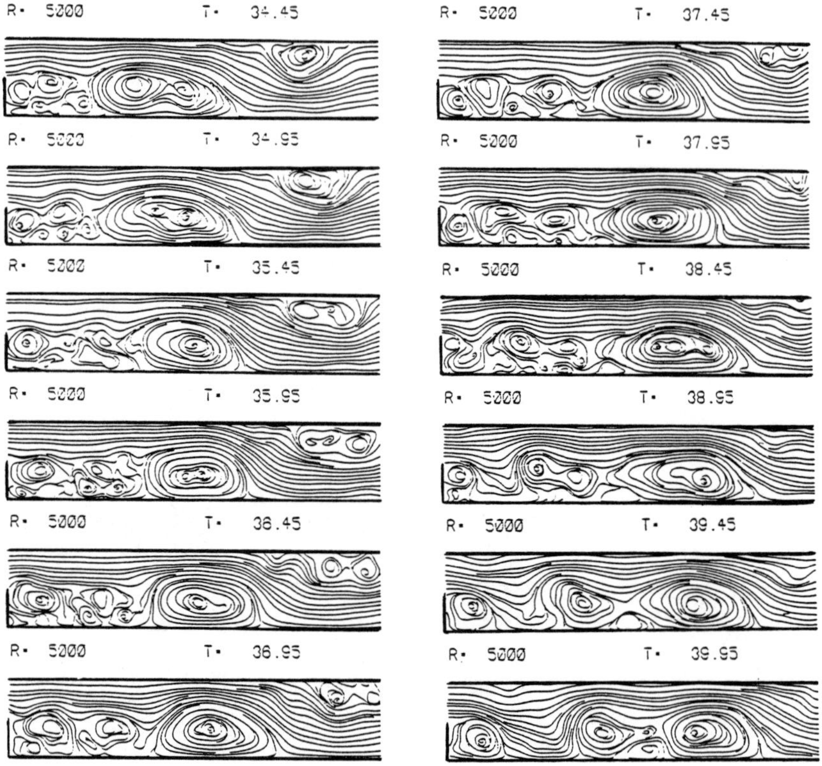

Figure 8. Recirculation zone structure at early stages of turbulence, R=5000. Two zones of eddy configuration are identified: small-scale formations near the step that pair to form larger scales downstream. A secondary separation appears on the opposite wall in synchronization with the large recirculating eddy. The first frame is plotted at t=68.9 with 10 computational time steps between each two frames.

4. THIN FLAME PROPAGATION AT SMALL MACH NUMBERS

In this section, the analysis of combustion at low Mach numbers, when acoustic waves propagate faster than a typical flow velocity so that the thermodynamic pressure remains spatially uniform, is presented. Combustion-hydrodynamics interactions are limited to the two modes of volumetric expansion and vorticity generation. When the flame structure is thin compared to a typical flow length scale, combustion is modeled as a reaction sheet or a wrinkled laminar flame (Damkohler [24] and Bray [12]). In this case, the flame is regarded as a hydrodynamic discontinuity that propagates through the field at an assigned normal burning speed.

The equations governing the propagation of a thin flame at low Mach number are derived. These equations are not amenable to analytical solution for any practical case, and the application of conventional numerical schemes may fail to illustrate some of the most interesting phenomena of turbulent combustion. A form that allows the application of numerical schemes which preserve the properties of the physical model is derived by expressing the equations in terms of Lagrangian coordinates. In particular, equations describing the evolution of the flame surface, the pressure field, the expansion field and the vorticity field are obtained. The question of the existence of a thin front is addressed in light of the recent studies of non-linear stability of laminar flames.

Numerical schemes used in the solution of the equations governing flame front motion are reviewed, and the validity of the solution algorithm for combustion in a closed volume is investigated by comparing the numerical solution to analytical solution of some special cases. Solutions pertaining to the turbulent flame speed and to flame stabilization in a high velocity stream are discussed as applications for the combustion model.

4.1. FORMULATION

The following formulation resembles the approach of Sivashinsky [66], with the three modifications: (1) an eikonal equation, suggested by Ghoniem et al. [28], is used to describe the motion of the flame front; (2) Helmholtz decomposition is applied to split the velocity field into two components, one is attributed to the expansion across the flame and the concomitant pressure rise in closed volumes, and the second is due to the vorticity field; and (3) the pressure equation, developed by Majda and Sethian [50], is implemented explicitly. The equations are derived here to show the original combustion-hydrodynamics interactions, and to provide a general framework for numerical solutions.

The reacting mixture is formed of two components, reactants R and products P, and the chemical reaction is described by a single step Arrhenius-rate reaction. The two components are assumed to behave as perfect gases with the same molecular weight and specific heat ratio. The formulation is restricted to two-dimensional Cartesian coordinates. The non-dimensional form of the conservation equations that govern this flow is given in table I. Eqs. (4.2-6) form a set of five equations in five gas dynamic variables, namely ρ, C, u, p, T, that should be supplemented with the appropriate initial and boundary conditions to determine the solution uniquely. Non-dimensional variables are listed in table II. A tilde denotes a dimensional variable, subscript "i" is for initial conditions, subscript "u" is for unburnt or reactants and subscript "b" is for burnt or products. Variables are normalized with respect to their respective values at the initial conditions. Non-dimensional groups are defined in table III.

TABLE I. GOVERNING EQUATIONS

$$R \rightarrow P \qquad (4.1)$$

$$\frac{D\rho}{Dt} + \rho\, \nabla \cdot u = 0 \qquad (4.2)$$

$$\rho \frac{DC}{Dt} = - \frac{\ell_T}{L_e} \nabla \cdot g - \frac{1}{\ell_T} \dot{W} \qquad (4.3)$$

$$\gamma M^2 \rho \frac{Du}{Dt} = -\nabla p + \frac{\gamma M^2}{R_e} \nabla \cdot \tau \qquad (4.4)$$

$$\rho \frac{DT}{Dt} - \frac{\gamma - 1}{\gamma} \frac{Dp}{Dt} = - \ell_T \nabla \cdot q + \frac{(\gamma - 1) M^2}{R_e} \Phi + \frac{Q}{\ell_T} \dot{W} \qquad (4.5)$$

$$p = \rho T \qquad (4.6)$$

where

$$\dot{W} = A \rho C\, e^{-T_a/T} \qquad (4.7)$$

$$\tau = \mu\left(\frac{\partial u_i}{\partial x_j} + \frac{\partial u_j}{\partial x_i} - \frac{2}{3} \delta_{ij} \nabla \cdot u\right)$$

$$\Phi = \mu\left[\, 2\left\{\left(\frac{\partial u_1}{\partial x_1}\right)^2 + \left(\frac{\partial u_2}{\partial x_2}\right)^2\right\} + \left(\frac{\partial u_1}{\partial x_2} + \frac{\partial u_2}{\partial x_1}\right)^2 - \frac{2}{3}(\nabla \cdot u)^2\right]$$

$$g = -d\, \nabla C$$

$$q = -k\, \nabla T$$

$$\frac{D}{Dt} = \frac{\partial}{\partial t} + u \cdot \nabla \qquad \text{and} \qquad \nabla = \left(\frac{\partial}{\partial x_1}, \frac{\partial}{\partial x_2}\right)$$

TABLE II. NOTATION AND NON-DIMENSIONALIZATION

A	is the normalized frequency factor; $A=\tilde{A}\tilde{k}/(\tilde{\rho}_b\tilde{c}_p\tilde{S}_b^2)$
$\tilde{c}_p$	is the specific heat at constant pressure, $\tilde{c}_p=\tilde{R}\gamma/(\gamma-1)$
C	is the mass fraction (concentration) of reactants R
d	is the diffusivity, normalized with respect to d_i
$\tilde{E}$	is the activation energy
g	is the diffusive mass flux
k	is the conductivity, normalized with respect to k_i
$\tilde{\ell}$	is the thermal flame thickness, $\tilde{\ell}=\tilde{k}/(\tilde{c}_p\tilde{S}_u\tilde{\rho}_b)$
$\tilde{L}$	is a characteristic dimension of the field
p	is pressure, $p=\tilde{p}/\tilde{p}_i$
q	is the conductive heat flux
Q	is the enthalpy of reaction at (T,p), $Q=\tilde{Q}/(\tilde{c}_p\tilde{T}_b)$
$\tilde{R}$	is the gas constant
S_b	is the normal burning speed at constant pressure; $S_b=\tilde{S}_b/\tilde{S}_{bi}$
t	is time, $t=\tilde{t}/(L/\tilde{S}_b)$
T	is temperature, $T=\tilde{T}/\tilde{T}_b(\tilde{T}_i,\tilde{p}_i)$
T_a	is the non-dimensional activation energy, $T_a=\tilde{E}/\tilde{R}\tilde{T}_b)$
$\tilde{T}_b$	is the adiabatic flame temperature; $\tilde{T}_b=\tilde{T}_{ui}+\tilde{Q}_i/\tilde{c}_p$
u	(u_1,u_2) is the velocity, $u=\tilde{u}/\tilde{S}_b$
$\dot{W}$	is the rate of production of reactants per unit volume
x	(x_1,x_2) is the coordinate, $x=\tilde{x}/\tilde{L}$
ρ	is density, $\rho=\tilde{\rho}/\tilde{\rho}_b(\tilde{T}_i,\tilde{p}_i)$
γ	is the specific heat ratio
τ	is the shear stress tensor
Φ	is the dissipation function

μ is the dynamic viscosity, normalized with respect to μ_i

δ_{ij} is the Kronecker delta

TABLE III. NON-DIMENSIONAL GROUPS

L_e :	Lewis number,	$L_e = \dfrac{k}{\rho_b c_p d_i}$
ℓ_T :	Flame thickness parameter,	$\ell_T = \ell/L = \dfrac{k}{c_p S_b \rho_b L}$
M :	Mach number,	$M = \dfrac{S_b}{\sqrt{\gamma p_i/\rho_b}}$
R :	Reynolds number,	$R = \dfrac{\rho_b S_b L}{\mu}$
Q :	Enthalpy of reaction	$Q = \dfrac{Q(T_i, p_i)}{c_p T_b}$
T_a :	Activation energy	$T_a = \dfrac{E}{RT_b}$

All the dimensionless groups are evaluated at the initial state $(\tilde{p}_i, \tilde{T}_i)$. Q and T_a are the Frank-Kamenetski first and second numbers. For combustion in open-flow systems, U is used instead of S_b, where U is the free stream velocity. The above represent the minimum set of groups which can be used to characterize a combustion system. Several alternatives can be used to replace one of the above groups, e.g., the Prandtl number $P_r = 1/(\ell_T R_e)$, and the Peclet number $P_e = (L_e/\ell_T)$.

4.2. LOW MACH NUMBER LIMIT

If the Mach number is small, and since it appears in the equations only as $\varepsilon=\gamma M^2$, one should be able to expand all the gas dynamic variables in terms of ε. Let ζ denote any gas dynamics variable $(\rho,u,p,T,\ldots)$, and:

$$\zeta(x,t) = \zeta_o(x,t) + \varepsilon\zeta_1(x,t) + \varepsilon^2\zeta_2(x,t) + \ldots. \tag{4.8}$$

Substituting into Eq. (4.4) and gathering terms that are independent of M, one finds that $\nabla p_o = 0$, which shows immediately that:

$$p_o = p_o(t) \tag{4.9}$$

This is the main result of the low Mach number approximation. The largest component of the pressure is constant throughout the field and changes only with time due to the heat release associated with combustion, if the flow is confined. p_o is called the thermodynamic pressure. This result agrees readily with the assumption that $M \ll 1$, since it implies that acoustic waves travel so fast with respect to the motion of the fluid that the pressure is equalized instantaneously. The second component of the pressure appears in the ε-component of the expansion of the momentum equation:

$$\rho_o \frac{Du_o}{Dt} = - \nabla p_1 + \frac{1}{R_e} \nabla\cdot\tau \tag{4.10}$$

p_1 is called the hydrodynamic pressure, and is generated to balance the changes in momentum within the flow field. Its contribution to the total pressure is restricted to ε. Applying the same procedure to the energy equation, Eq. (4.5) yields:

$$\rho_o \frac{DT_o}{Dt} - \frac{\gamma-1}{\gamma}\frac{dp_o}{dt} = - \ell_T\, \nabla\cdot q_o + \frac{Q}{\ell_T}\, \dot{W} \tag{4.11}$$

Thus, the energy balance is independent of the hydrodynamic pressure and hydrodynamic dissipation is negligibly small.

The continuity and species equations and the equation of state, Eqs. (4.2, 3 and 6) respectively, retain their forms to the zeroth order in ε. However, as shown by Majda and Sethian [50], a more convenient form of the continuity equation can be obtained by differentiating Eq. (4.6), and using Eq. (4.2) for $D\rho_o/Dt$ and Eq. (4.11) for DT_o/Dt to obtain :

$$\nabla\cdot u_o = -\frac{1}{\gamma p_o}\frac{dp_o}{dt} - \frac{\ell_T}{p_o}\nabla\cdot q_o + \frac{Q}{\ell_T p_o}\dot{W} \tag{4.12}$$

Eq. (4.12) defines combustion-hydrodynamics interactions in terms of the volumetric expansion produced by heat release. This effect can be visualized more readily by applying the condition of confinement for combustion in closed volumes, i.e., $V(t)$=constant where V is the total volume, to evaluate the total divergence of the velocity field.
This yields: $\int\nabla\cdot u_o dV=0$. (Here we are assuming that the field is confined, i.e., V is finite. However, the analysis produces the correct limit for an open-flow system by setting $V\to\infty$. An independent analysis, based on the overall continuity of mass instead of the overall conservation of volume, can be constructed for the latter case.) Since $\int\nabla\cdot u_o dV=\int u\cdot dA$, and integrating Eq. (4.12):

$$\frac{dp_o}{dt} = -\frac{\gamma\ell_T}{V}\,{}_A\!\int q_o\cdot dA + \frac{\gamma Q}{\ell_T V}\,{}_V\!\int \dot{W}\,dV \tag{4.13}$$

where A is the surface area corresponding to V. Eq. (4.13) shows that the temporal change of the thermodynamic pressure p_o is governed by the total rate of heat release due to combustion and the heat loss across the boundary of the system. In an open flow system, $V\to\infty$ and p_o remain constant since the extra volume produced by expansion is ventilated across the inlet and outlet sections of the reaction chamber.

The two modes of combustion-hydrodynamics interactions--volumetric expansion and vorticity production--can be analyzed separately by using the Helmholtz decomposition of the velocity field:

$$u_o = v + w \tag{4.14}$$

where v is an irrotational field; $\nabla\times v=0$, and w is a solenoidal field for which $\nabla\cdot w=0$. To ensure uniqueness for both fields, we require that: $v\cdot n(x)=0$ and $w\cdot n(x)=0$ where n(x) is the normal vector to solid boundaries. Since v is irrotational, it can be expressed in terms of a potential ϕ, where $v = \nabla\phi$, and the equation governing ϕ is obtained by substituting Eq. (4.14) into Eq. (4.12):

$$\nabla^2\phi = \frac{1}{p_o}\left[-\frac{1}{\gamma}\frac{dp_o}{dt} - \ell_T\ \nabla\cdot q_o + \frac{Q}{\ell_T}\dot{W}\right] \tag{4.15}$$

with $\nabla\phi\cdot n(x)=0$. Eq. (4.15) defines the first dynamic role of combustion as manifested by the volumetric expansion associated with heat release in terms of an acceleration field produced by the expansion across the flame and a deceleration field produced by the change in the thermodynamic pressure throughout the field.

Moreover, taking the curl of Eq. (4.10) and noting that $\nabla\times\nabla\phi=0$ and $\nabla\times\nabla p_1=0$, $\nabla\cdot w=0$ and $\omega=\nabla\times u=\nabla\times w$, and using Eq. (4.2), we obtain the vorticity transport equation in a variable density field:

$$\frac{D}{Dt}\left(\frac{\omega}{\rho_o}\right) = \frac{1}{\rho_o R_e}\nabla^2\omega + \frac{1}{\rho_o^3}\nabla\rho_o\times\nabla p_1 \tag{4.16}$$

Eq. (4.16) defines the second part of the dynamic role of combustion. Volumetric expansion establishes a variable density field that causes a baroclinic generation of vorticity within the reaction zone, while it forces the adjustment of the existing vorticity as it crosses the flame zone. The appropriate boundary condition on the vorticity equation is $u\cdot s(x)=0$ where s is the tangential unit vector to solid boundaries. This condition is used to generate extra vorticity along the solid boundaries, as explained in the description of the vortex method in section 2.

The governing equations of combustion at low Mach number, after removing all subscript o and rewriting p_o as P and p_1 as p, are listed in table IV. According to the non-dimensionalization in table II, $P(0)=1$. If the vorticity ω is used as a primary variable, then the velocity w can be evaluated from the solution of the equation $\omega=\nabla\times w$. Since ω is solenoidal, i.e., $\nabla\cdot\omega=0$, then using a stream function ψ such that $u_1=\partial_2\psi$ and $u_2=-\partial_1\psi$, we find $\nabla^2\psi=-\omega$. Eqs. (4.17-23) are seven equations in the new seven decoupled gas dynamic variables $(P,\phi,\psi,\omega,C,T,\rho)$. If the domain is open, $V\to\infty$ and Eq. (4.17) are reduced to the effect of heat transfer only.

TABLE IV. GOVERNING EQUATIONS FOR COMBUSTION AT LOW MACH NUMBER

$$\frac{dP}{dt} = -\frac{\gamma \ell_T}{V} \, {}_A\!\int q \cdot dA + \frac{\gamma Q}{V \ell_T} \, {}_V\!\int \dot{W} \, dV \tag{4.17}$$

$$\nabla^2 \phi = \frac{1}{P}\left[-\frac{1}{\gamma}\frac{dP}{dt} - \ell_T \nabla \cdot q + \frac{Q}{\ell_T}\dot{W}\right] \tag{4.18}$$

$$\nabla^2 \psi = -\omega(x,t) \tag{4.19}$$

$$u = \nabla\phi + (\partial_2 \psi, -\partial_1 \psi)$$

$$\frac{D}{Dt}\left(\frac{\omega}{\rho}\right) = \frac{1}{\rho R_e}\nabla^2\omega + \frac{1}{\rho^3}\nabla\rho \times \nabla p \tag{4.20}$$

$$\rho\frac{DC}{Dt} = -\frac{\ell_T}{L_e}\nabla \cdot g + \frac{1}{\ell_T}\dot{W} \tag{4.21}$$

$$\rho\frac{DT}{Dt} - \frac{\gamma-1}{\gamma}\frac{dP}{dt} = -\ell_T \nabla \cdot q + \frac{Q}{\ell_T}\dot{W} \tag{4.22}$$

$$P = \rho T \tag{4.23}$$

4.3. THIN FLAME LIMIT

Eqs. (4.17-23) can be solved numerically for a deflagration (slow combustion wave) propagating according to a specified reaction rate $\dot{W}(C,\rho,T)$. The solution of Eqs. (4.21, 22 and 23) provides the flame structure in terms of the distributions of C, T and ρ within the reaction zone. Similar equations were used by and Rehm and Baum [62] to describe fire spread in chambers and by McMurtry et al. [53] to study flame propagation through a shear layer. An interesting limit arises when the flame thickness is small compared to a characteristic geometry parameter and when the activation energy is large, i.e. the reaction is fast and the diffusivity is small. At that limit, the flame acts as a hydrodynamic discontinuity between reactants and products. Recent experimental observations confirm the existence of flame-reaction sheets, or wrinkled thin flame fronts at moderate levels of turbulence.

Sivashinsky [66] showed that a formal limit of the equations can be obtained for $\ell_T \to 0$ and $T_a \to \infty$ simultaneously. Since $(\rho S)=\rho_u S_u=\rho_b S_b$, we can write:

$$\frac{\dot{W}}{\ell_T} = (\rho S)\ \delta(x-\chi_f)$$

$$\text{and} \qquad \frac{1}{\ell_T}\ {}_V\!\int \dot{W}\ dV = {}_A\!\int\ (\rho S)\ n \cdot dA \tag{4.24}$$

where δ is the Dirac delta function and χ_f is the coordinate of the flame surface, $\chi_f=\chi(X,t)$, and X is the Lagrangian coordinate of a point on the flame surface (as the surface area of the flame expands, extra points are used in the numerical solution to improve the resolution). If the set of points $\{\chi_f\}$ is used to construct a surface function F(x,t) such that:

$$F(x,t) = 0 \qquad \text{for the flame front } x = \chi_f \tag{4.25}$$

while $F(x,t)<0$ for reactants, and $F(x,t)>0$ for products, then $F(x,t)$ can be used to describe the motion of the flame. An equation that governs the evolution of F is derived by satisfying the conservation of mass across the flame surface, i.e.

$$\rho_u(u_f - u_u \cdot n_f) = \rho_b(u_f - u_b \cdot n_f) = (\rho S) \qquad (4.26)$$

where u_f is the flame surface velocity, and u_u and u_b are the fluid velocity on the unburnt side and the burnt side, respectively, and n_f is the normal vector to the flame surface. Since $u_f=(\partial F/\partial t)/|\nabla F|$, and $n_f=-\nabla F/|\nabla F|$, then

$$\frac{\partial F}{\partial t} + u_u \cdot \nabla F = S_u |\nabla F| \qquad (4.27)$$

Equation (4.27) indicates that the flame surface is advected with the flow field at a velocity u_u while it propagates normal to itself at a speed S_u.

Using Eqs. (4.24), the equations in table IV can be reduced to:

$$\frac{dP}{dt} = \frac{\gamma Q}{V} \int_{A_f} (\rho S)\, dA \qquad (4.28)$$

$$\nabla^2 \phi = -\frac{1}{\gamma} \frac{d(\ln)P}{dt} + \frac{QS_u}{T_u} \delta(x-x_F) \qquad (4.29)$$

$$\frac{DC}{Dt} = S_u \delta(x-\chi_f) \qquad (4.30)$$

$$\rho\frac{DT}{Dt} - \frac{\gamma-1}{\gamma}\frac{dP}{dt} = \rho_u Q S_u \delta(x-\chi_f) \qquad (4.31)$$

$$\frac{D}{Dt}\left(\frac{\omega}{\rho}\right) = \frac{\omega_f}{\rho_u} \delta(x-\chi_f) \qquad (4.32)$$

where ω_f is the local baroclinic generation of vorticity across the flame front, resulting from the pressure-density interactions across its surface. At this asymptotic limit, combustion occurs across a sharp discontinuity in an inviscid, non-conducting, non-diffusive, rotational field.

4.4. PARTIAL INTEGRATION

The solution of Eq. (4.30) is $C(x,t)=H(x-\chi_f)$, where χ_f is defined by Eqs. (4.25 and 27), and H is the Heavyside function. Eq. (4.31) can be integrated and combined with the equation of state, Eq. (4.23), to obtain:

$$T(\chi) = s(\chi)\, P^{\left(\frac{\gamma-1}{\gamma}\right)} \qquad (4.33)$$

where s is a non-dimensional entropy and $\chi=\chi(X,t)$ is a particle trajectory. On the unburnt side, $s=s_i$, and the unburnt gas experiences isentropic compression (note that P is uniform). On the burnt side, $s(\chi)$ is determined by the jump conditions across the flame at the moment when particle X is ignited. This can be calculated from the thermodynamic pressure P and the temperature of the products T_b, defined by:

$$T_b(\chi_f,t) = T_u(t) + Q(T_u,P) \qquad (4.34)$$

If the reactants are assumed to have frozen composition, then Q is constant. Combining Eqs. (4.33, and 34)

$$T(\chi)\, P(\chi)^{\left(\frac{1-\gamma}{\gamma}\right)} = T(\chi_f)^{\left(\frac{1-\gamma}{\gamma}\right)} \qquad (4.35)$$

where $T(\chi_f)$ and P are the temperature and pressure behind the flame at the moment χ_f crossed its surface. Thus, the flow behind the flame surface is isentropic along particle paths.

For uniform initial conditions, and if the normal burning speed S_u is independent of χ_f, the right hand side of Eq. (4.28) can be integrated to obtain Eq. (4.37) (note that $S_u A_f=\partial V_b/\partial t$, where V_b is the total volume of the burnt gas. Both S_u and Q are, in general, functions of the conditions ahead of the flame in terms of T_u and P. The solution of Eqs. (4.28-32) requires the solution of Eq. (4.27) in terms of F(x,t)=0. For numerical convenience, the function F=0 is expressed in terms of a set of points $\{\chi_f\}$, where $\chi_f=\chi(X,t)$, representing the coordinates of the flame front. The corresponding equation, Eq. (4.36), is obtained by combining Eqs. (4.26 and 27).

4.5. SOLUTION ALGORITHM

The system of equations which describes the propagation of a thin flame at low Mach number is listed in table V. The solution algorithm proceeds as follows: Eq. (4.36) is used to find the new coordinates of the flame front $\{\chi_f\}$ and the volume burnt at each point ΔV_b, while $(dV_b/dt)=\sum(\Delta V_b/\Delta t)$. The new value of the pressure is obtained by integrating Eq. (4.37). It should be noted that this equation can be strongly non-linear since both Q and S_u are dependent on the pressure P. Under certain circumstances, it can be replaced by the global conservation of mass and energy to avoid this non-linearity (Knio and Ghoniem [41]). The new value of the burnt gas temperature and density are computed from Eqs. (4.38) and (4.39), respectively. Eq. (4.40) is integrated to find the combustion-generated velocity field $\nabla\phi$ in terms of a distribution of volumetric sources along the flame front and a uniform distribution of volumetric sinks in the field.

The rotational component of the velocity field, produced by vorticity, is calculated from the integration of Eq. (4.41) using Eq. (2.21) that was derived in section 2. After adding the two components of the velocity, the vorticity field is transported according to Eq. (4.42) using Eq. (2.37) with $\zeta=0$. Extra vorticity is generated across the flame ω_f and along the boundaries of the domain according to Eq. (2.42). Finally, the flame burning speed and the heat release are updated using the appropriate laminar flame speed and heat release model.

TABLE V. GOVERNING EQUATIONS FOR THIN FLAME PROPAGATION AT THE LOW MACH NUMBER LIMIT

$$\frac{d\chi_f}{dt} = u(\chi_f) + S_u\, n(\chi_f) \tag{4.36}$$

$$\frac{dP}{dt} = \frac{\gamma Q \rho_u}{V} \frac{\partial V_b}{\partial t} \tag{4.37}$$

$$T_u(t) = s_i\, P^{\left(\frac{\gamma - 1}{\gamma}\right)} \tag{4.38}$$

$$\rho_u(t) = P(t)/T_u(t) \tag{4.39}$$

$$\nabla^2 \phi = - \frac{1}{\gamma P} \frac{dP}{dt} + \frac{QS_u}{T_u} \delta(x - \chi_f) \tag{4.40}$$

$$\nabla^2 \psi = -\omega \tag{4.41}$$

$$u = \nabla\phi + (\partial_2 \psi, -\partial_1 \psi)$$

$$\frac{D}{Dt}\left(\frac{\omega}{\rho}\right) = \omega_f\, \delta(x - \chi_f) \tag{4.42}$$

and

$$S_u = S_u(T_u, P)$$

$$Q = Q(T_u, P)$$

4.6. FLAME FRONT STABILITY

The outstanding question of existence, continuity and stability of a flame front when subjected to perturbations of wavelengths longer than its thickness is of obvious relevance to numerical work on turbulent flame propagation using reaction sheet models. In terms of the formulation presented above, the question can be posed in the following form: are solutions of Eq. (4.27), or equivalently Eq. (4.36), stable if perturbed by a sinusoidal wave at a small wavelength?

Recent analytical investigations on the subject have identified two primary mechanisms to cause flame front instability: hydrodynamic effects due to thermal expansion, and thermal-diffusive effects due to the dependence of the burning rate on the diffusion of heat and chemical species (Sivashinsky [67]). Other contributions of less importance can be attributed to changes of the chemical reaction rate due to perturbation of the temperature and composition, and to the effect of viscosity. Here we summarize qualitatively the most relevant conclusions of these studies, and their consequences on the numerical solution of Eqs. (4.36-42).

The study of the hydrodynamic instability of thin flames go back to the theory of Landau [45], which proved that flame fronts are absolutely unstable and that the rate of growth of perturbations is inversely proportional to its wavelength. (This situation is reminiscent of the Kelvin-Helmholtz instability of vortex sheets.) The mechanism of instability can be rationalized if one considers a plane stationary flame front with the reactants approaching at speed S_u, when its surface is perturbed by a sinusoidal wave. The perturbation causes the streamlines to diverge ahead of the convex portion and to converge ahead of the concave portion, both on the reactants side. Thus, the flow of reactants decelerates ahead of the peak of the sine wave while it accelerates in front of its valley. If the burning speed remains constant at S_u, the convex portion will bulge more into the reactants, while the concave portion will keep receding. Thus, small perturbations grow and the flame becomes more distorted. This is the basic result of Landau's linear theory.

Heat conduction improves flame front stability by changing the local value of the burning speed S_u. The convex segment of the perturbed flame heats up the reactants less than a plane flame, thus decreasing S_u below its value for a plane front. On the other hand, the concave segment heats the reactants more intensely, causing S_u to increase. Markstein [51] showed that if S_u is assumed to be proportional to the curvature of the flame front, with a positive constant of proportionality, a stabilizing mechanism is realized. However, experimental observations showed that the Lewis number L_e plays a major role in how the burning speed changes with curvature, and that S_u may be less for the concave portion than for the convex portion if $L_e<L_{ec}$. Thus, the stabilizing effect of heat conduction may be overcome by the destabilizing effect of mass diffusion. This phenomenon is known as the thermal-diffusive instability and is described next.

Consider the same plane flame with a sinusoidal perturbation. The bulge of the flame surface into the reactants increases the flux of chemical species onto any point on the front, hence increasing the burning speed S_u. The opposite happens on the concave part where the diffusive flux of species to the front is reduced, decreasing the local value of S_u. Thus the concave front moves into the reactants slower than a plane flame, while convex fronts move faster, i.e. mass diffusion acts in the opposite direction to heat conduction as far as stability is concerned. Depending on the value of $L_e><L_{ec}$, thermal-diffusive effects can stabilize or destabilize the flame even in the absence of hydrodynamic effects (Law [47]). L_{ec} is a critical value at which the two effects neutralize each other. Markstein's expression for the dependence of the burning speed on curvature should be modified to take the effect of Lewis number into consideration.

The results of the linear analysis apply only to small perturbations and cannot lead to steady solutions since perturbations grow indefinitely. The predicted exponential growth invalidates the theory after a short period of time. The non-linear theory is not yet well-developed, however, the following important observations can be made. The formation of structures of finite dimensions are expected as the growth of perturbations exceed the small linear amplitude (In the case of a vortex sheet, the growth of perturbations causes a roll-up, followed by the formation of a set of small scale eddies that interact in an organized pattern.) Cellular flames, which are observed in the laboratory, have been predicted to form as the thermo-diffusive instability grow into its non-linear range [15]. In this case, individual cells interact continuously in what appears to be a chaotic motion, while the flame remains as a coherent front.

The formation of stable curved flame fronts bulging into the reactants as a limit for the non-linear evolution of the hydrodynamic instability have been predicted by Emmons [26] and by Zeldovich [70], the latter on the basis of pure geometrical considerations. Within the non-linear range, one can use the Huygens' principle, which treats a continuous wave front in terms of a set of points acting as wave sources, to construct the flame front as it propagates in time. A sinusoidal wave perturbation with a large amplitude develops as follows: the concave part grows into a cusp and the convex part maintains its roundness bulging into the reactants (Sethian [65] showed these formations by numerical experiments using Huygens' rule.) Next, the angle between the two sides of the cusp will increase, i.e. the concave part will become flatter while the convex shape will persist. Thus, non-linear evolution will eventually lead to the smoothing out of the original perturbation and the development of a curved front bulging into the reactants (here we also mention that representing a vortex sheet by a set of vortex blobs leads to the formation of a well-defined set of eddies which merge into larger eddies.)

Zeldovich et al. [71] argue that the formation of a curved front as a result of the growth of the hydrodynamic instability improves the stability of the flame since perturbations will stretch as they move along the curved surface, hence their amplitudes will decrease. Thus, while the application of the

Huygens' principle leads to the formation of a curved flame bulging into the reactants, the stretching along the curved front acts to smooth out the perturbations, i.e. curved fronts are more stable than plane fronts. It is interesting to observe that this stabilizing mechanism is independent of the changes in the rate of burning with the front curvature due to thermal-diffusive effects, which leads to the conclusion that non-linear effects stabilize a front moving at a constant S_u.

The formation of cusps pose difficulties to the analytical or simple numerical solutions of Eq. (4.36) since the differentiability of the front is lost and the front tends to cave on itself. However, numerical solutions which can overcome these difficulties have been developed and are summarized in the following section.

4.7. NUMERICAL SCHEME

Numerical solutions of Eq. (4.36) have been obtained using an interface tracking algorithm (Noh and Woodward [59], and Chorin [20]), based on the volume-of-fluid scheme (Hirt and Nichols [38]) that minimizes the diffusion of the front and avoids the difficulties associated with the merging (collision) of several fronts. This algorithm is now summarized. The solution is obtained in terms of the volume fraction of products f in each square cell of an Eulerian grid, $0 \le f \le 1$ and the cell's size determines the resolution of the algorithm in terms of the smallest distance between two fronts. The material flux across the boundaries of a cell is determined by the advection field and the volume fractions in the neighboring cells, and the species equation is integrated to balance the change of f within a cell and the transport of products across its boundaries.

A pattern recognition algorithm is used to identify the most probable orientation of the local interface between reactants and products in each cell. In Cartesian coordinates, four possibilities are allowed: a vertical interface, two vertical interfaces forming a passage, a corner, and a horizontal interface. Besides allowing the reconstruction of the front from the concentration field, it also improves the accuracy of determining the transport flux between cells (Ghoniem [33]).

The evaluation of the rate of burning and the displacement of the flame surface due to the propagation of the flame normal to its surface at a speed S_u are evaluated by a direct extension of the advection algorithm. Exploiting the analogy between flame propagation and wave front motion, the advection field is replaced by the normal burning speed and the Huygens' rule is applied by maximizing the propagation in all possible directions. As shown by Chorin [20], eight directions are sufficient to obtain an accurate maximum. This algorithm does not require the evaluation of the normal to the surface, hence it can accommodate the existence of cusps and corners where the normal is not defined. Methods that employ marker particles, or discrete parameterization of fronts, have also been used to follow fronts similar in nature to flames. However, difficulties arise when concave surfaces fold on and form sharp angles, or when two initially-separated fronts merge together to form one continuous front. In both cases, the collision of particles involves some loss of information about the new configuration of the front (Sethian [66]). A case-in-point is the formation of islands of reactants within the products if two concave fronts collide side-by-side and their peaks merge while their valleys remain unburnt.

To check the validity of the solution algorithm described in section 4.5, the line interface scheme is incorporated to solve Eq. (4.36) and the numerical results are compared with an analytical solution for a flame propagating from the center of a circular chamber (Knio and Ghoniem [41]). The front propagates outwards as a circle, generating a radial velocity field obtained by integrating Eq. (4.40), while the pressure is recovered by integrating Eq. (4.37). Results for the flame front geometry, the pressure rise and rate of burning are plotted in Figures 9, 10, and 11, respectively. The accuracy of the algorithm is confirmed from the comparison between the predicted pressure rise and the analytical value, while the oscillations in the burning rate curve are due to the uncertainty in the interpretation of the line interface. The stability of the front is clearly visible in figure 9, where perturbations due to the numerical grid are not amplified along the flame surface.

5. TURBULENT FLAME PROPAGATION

Eqs. (4.36-4.42) were used to study turbulent flame propagation, implementing the vortex method in section 2, the flame model in section 4, and the numerical scheme in section 4.7, in different configurations. Here, we review the results for two important problems in the theory of turbulent flames: the effect of turbulence on the burning rate, and the stabilization of flames in a high speed velocity stream. Within the limitations imposed by the assumptions of this model, turbulence can affect combustion only by wrinkling the flame surface, hence generating extra burning area.

The wrinkling of a laminar flame due to stationary turbulence, modeled by a fixed array of vortices, was investigated in the study of Chorin [20]. A periodic array of vortices was used to construct the advection field, and combustion was assumed at constant density, i.e. $\rho_u = \rho_b$ and $\phi = 0$. The core radius of each vortex was taken as the characteristic microscale of turbulence, λ. The increase in the burning rate is due to the increase of the flame surface area by the advection field of the vortices. Chorin [20] found that the ratio of the turbulent burning speed S_T, defined as the total rate of burning per unit area of a plane flame passing through the center of the vortex, and the laminar burning speed is independent of S_u when the root mean square of the average turbulent fluctuations u' is much larger than S_u. By conducting numerical experiments at different λ, the results indicated that S_T/S_u is proportional to the scale λ.

Barr and Ashurst [7] employed a similar model of constant density combustion in which an array of vortices, with randomly-selected core size δ and strength Γ, is used to model the turbulent field. The implementation of the Huygens principle to compute the advancement of the flame normal to itself was replaced by the volume-of-fluid scheme [38]. The result of the computations show that S_T/S_u increases linearly with u' when $u' \sim O(S_u)$. The increase of S_T is due to the wrinkling of the flame front by the vortices, simulating the effect of a small scale turbulent field. The results of the thin flame model were then compared with the results of a direct computation of Eq. (4.21) for small ℓ_T and large T_a. In this case, turbulence affects burning by enhancing the mixing between reactants and products through extra transport fluxes, instead of extending the surface of the thin flame. The comparison showed that within the error bounds of the two models, the thin flame model can accurately account for the increase in the rate of burning due to turbulence.

Flames are stabilized in high velocity turbulent fields, in which the flow speed is higher than the laminar burning speed, by inducing flow separation hence establishing a hot zone of recirculation of products which supply energy to ignite the reactants. Successful stabilization of a flame is accomplished by adjusting the flow speed according to the fuel-air ratio, which governs S_u, to balance the convective transfer of heat from the recirculation zone by the reactants with the generation of heat by burning. If S_u is lower than a critical small value, reactants leave the ignition zone unburnt and the flame eventually extinct, i.e. blow off. On the other hand, if S_u is higher than an upper critical value, the flame propagates in the direction opposite to the flow direction into the incoming flow, i.e. flash back.

Figure 12 shows a numerical simulation that provides a qualitative interpretation of both processes. In this simulation, a fuel-air mixture is flowing downstream a rearward-facing step at R=500. The flow field is computed using the vortex method described in section 2 and the flame model in section 4 is implemented. A constant-density combustion model is used and the laminar burning speed S_u is varied to establish the limits of stabilization of the flame. Figure 12-a shows the blow-off condition, in which the flame surface is discontinuous and the products are driven away from the circulation zone at a rate faster than the rate of generation of products by the flame. The packet of products are separated from the recirculation zone when the passage between two packets is thinner than one cell size. Figure 12-b depicts a stable flame, where a front is established on the outer edges of the large scale eddies. Figure 12-c shows a flame propagating at a high laminar burning speed and approaching the inlet of the channel, indicating a tendency for flash-back.

The same problem was solved by Ghoniem et al. [28] and Hsiao et al. [39] for combustion with a density ratio of 5. The solution of Eq. (4.40) at constant pressure was simulated by a distribution of volumetric sources along the flame front. The flame front contour and the vorticity field are depicted in figure 13 for a flow at R=5000. The structure of the vorticity field, presented in terms of the computational vortex elements and their velocity, is formed of a set of large scale eddies that grow downstream of the step. The flame front, shown as the demarcation line between the reactants and the reacting mixture, is stabilized on the outer edges of these structures. The average velocity profiles for both the non-reacting and the reacting cases are shown in figure 14, computed at R=22,000. The results compare favorably with schlieren photographs of flame contours and measurements of the average velocity profiles. The volumetric expansion associated with the density jump across the flame produces an extra acceleration field that reduces the reattachment length.

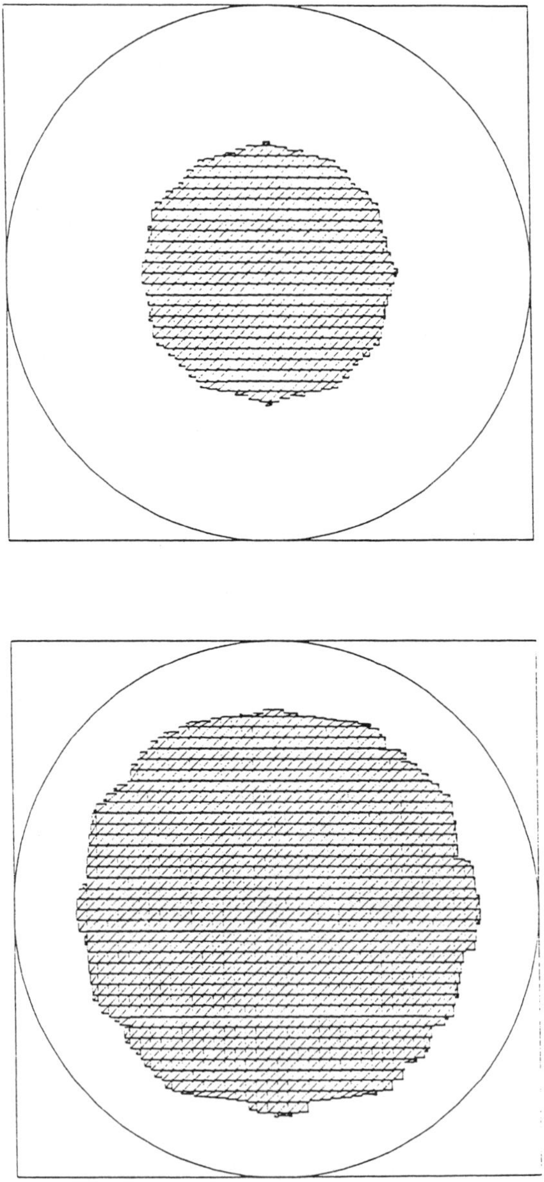

Figure 9. Representation of the flame front using the line interface propagation algorithm. The exact solution is a circle centered around the point of ignition at the center of the chamber.

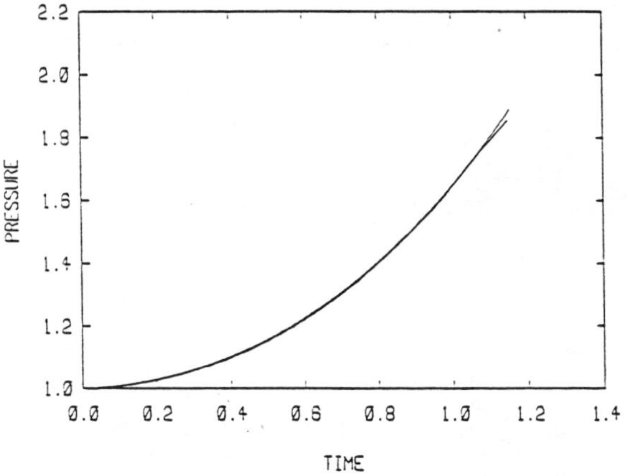

Figure 10. Numerical solution vs. the exact solution for the case in figure 9, presented in terms of the increase in pressure within the chamber.

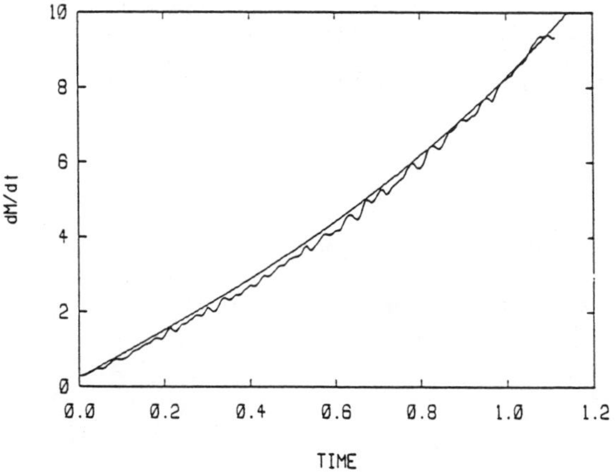

Figure 11. Comparison between the computed rate of burning and the exact values for the problem in figure 9.

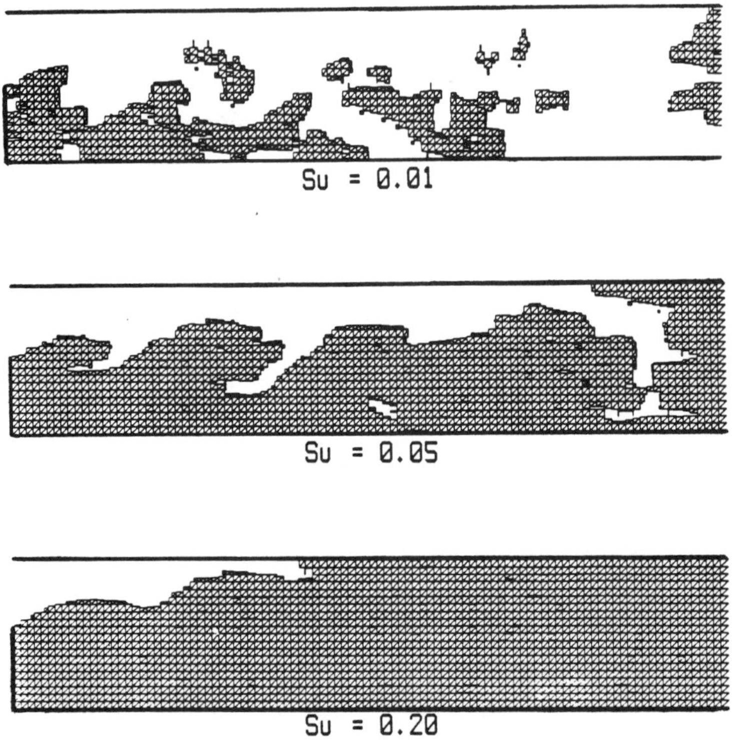

Figure 12. Flame front contours represented by the interface between reactants and products, in a flow of a fuel-air mixture at R=500, stabilized behind a rearward-facing step at values of S_u, showing the blow-off, a stable flame and the onset of flash-back.

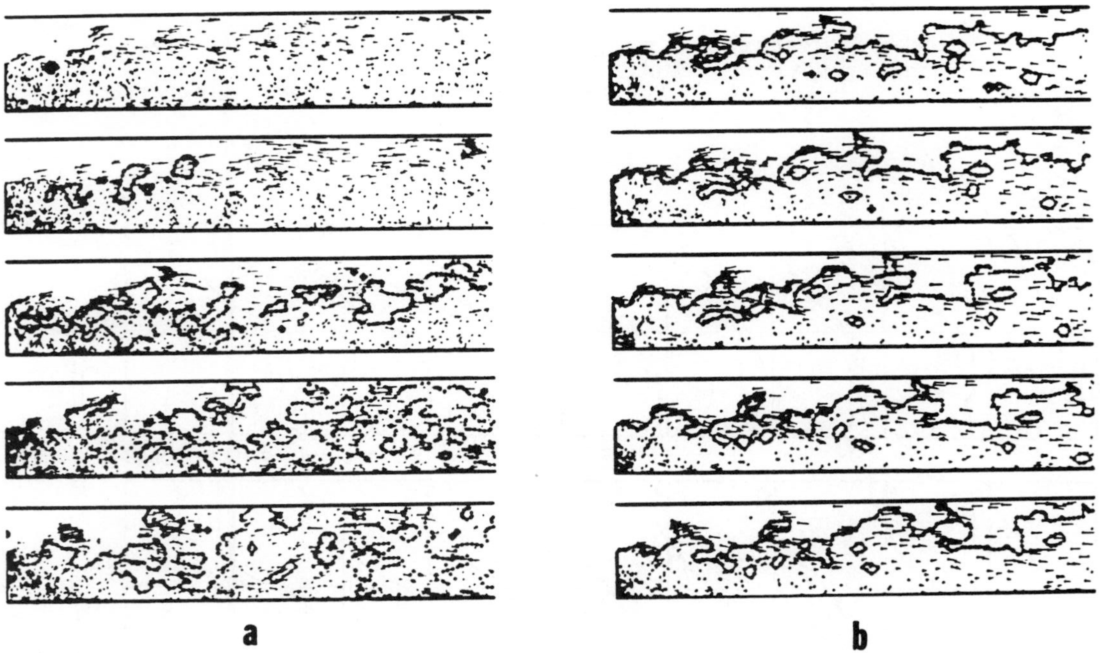

Figure 13. The flame front contours and the vorticity field in turbulent combustion behind a rearward-facing step at R=5000, S_{μ}=0.02 and density ratio of 4.25, showing (a) the ignition, and (b) the stationary state turbulent flame.

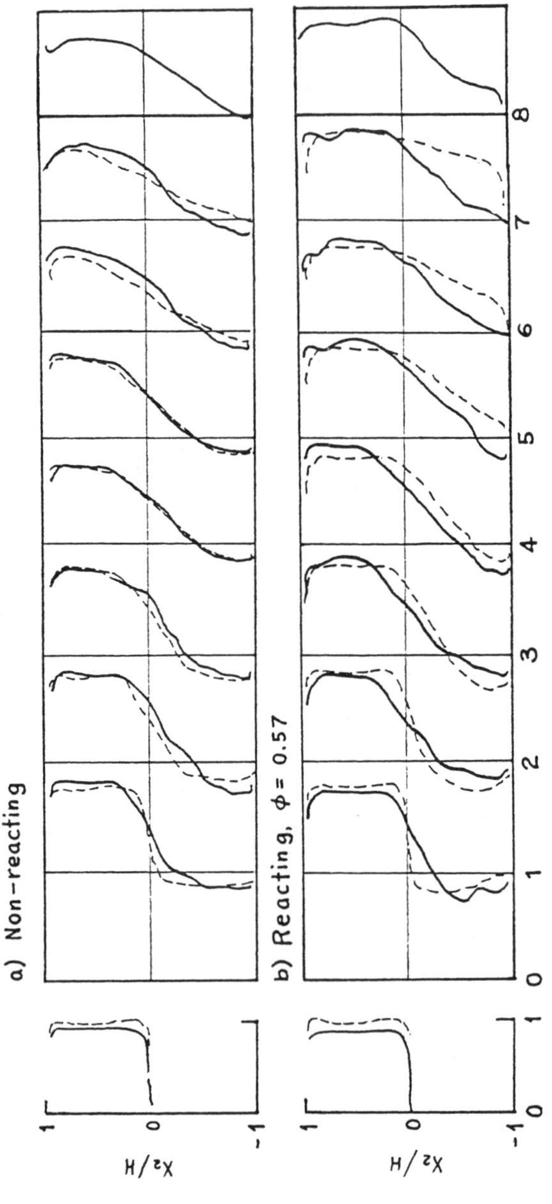

Figure 14. Average velocity profiles for a non-reacting and a reacting flow over a step at R=22000. Numerical results presented by solid lines and experimental data shown by broken lines. H is the step height.

BIBLIOGRAPHY

1. Anderson, C. and Greengard, C., "On vortex methods," Lawrence Berkeley Laboratory, University of California, Berkeley, LBL-16736, 1983.

2. Anderson, C., "Vortex methods for flows of variable density," submitted to J. Comp. Phys., 1985.

3. Aref, H., "Integrable, chaotic, and turbulent vortex motion in two dimensional flows," Ann. Rev. Fluid Mech., 15, 1983, pp. 345-389.

4. Aref, H. and Siggia, E., "Vortex dynamics of the two-dimensional turbulent shear layer," J. Fluid Mech., 100, 1980, pp. 705-737.

5. Ashurst, W., "Numerical simulation of turbulent mixing via vortex dynamics," Proc. 1st Symp. on Turbulent Shear Flow, (ed. by Durst et al.), Berlin, Springer-Verlag, 1979, pp. 402-413.

6. Ashurst, W., "Vortex simulation of a model turbulent combustor," Prog. Astro. Aero., 76, 1981, pp. 259-273.

7. Barr, P. and Ashurst, W., "An interface scheme for turbulent flame propagation," Sandia National Laboratories Report SAND82-8773, 1983.

8. Batchelor, G., An Introduction to Fluid Mechanics, Cambridge University Press, 1977.

9. Beale, J. and Majda, A., "Rates of convergence for viscous splitting of the Navier-Stokes equations," Math. Comp., 37, 156, 1981, pp. 243-260.

10. Beale, J. and Majda, A., "Vortex methods II: Higher order accuracy in two- and three-dimensions," Math. Comp., 39, 159, 1982, pp. 28-52.

11. Beale, J. and Majda, A., "Higher order accurate vortex methods with explicit velocity kernel," J. Comp. Phys., 58, 2, 1985, pp. 188-209.

12. Bray, K., "Turbulent flow with premixed reactants," Turbulent Reacting Flow, (ed. by P. Libby and F. Williams), Topics in Applied Physics, 4, Springer-Verlag, 1980, pp. 45-64.

13. Cantwell, B., "Organized motion in turbulent flow," Ann. Rev. Fluid Mech., 13, 1981, pp. 457-515.

14. Chandrasekhar, S., "Stochastic problems in physics and astronomy," Rev. Mod. Phys., 15, 1, 1943, pp. 1-89.

15. Chomiak, J., "Basic considerations in the turbulent flame propagation in premixed gases," Prog. Energy Combust. Sci., 5, 1979, pp. 207-221.

16. Chorin, A. and Bernard, P., "Discretization of a vortex sheet with an example of roll-up," J. Comp. Phys., 13, 1973, pp. 423-428.

17. Chorin, A., "Numerical study of slightly viscous flow," J. Fluid Mech., 57, 4, 1973, pp. 785-794.

18. Chorin, A., "Vortex sheet approximation of boundary layers," J. Comp. Phys., 27, 1978, pp. 423-442.

19. Chorin, A., "Vortex models and boundary layer instability," SIAM J. Sci. Stat. Comput., 1, 1980, pp. 1-21.

20. Chorin, A., "Flame advection and propagation algorithms," J. Comp. Phys., 35, 1980, pp. 1-11.

21. Christiansen, J., "Numerical simulation of hydrodynamics by the method of point vortices," J. Comp. Phys., 13, pp. 363-379.

22. Courant, R., Friedrichs, K. and Levy, H., "On the partial difference equations of mathematical physics, IBM Journal, March, 1967, pp. 215-234; originally published in Mathematische Annalen, 100, 1928, pp. 32-74,

23. Courant, R. and Hilbert, D., Methods of Mathematical Physics II, Intersciences New York, 1953, 830 p.

24. Damkohler, G., "The effect of turbulence on the flame velocity in gas mixtures," Z. Elekrochemie Angewandte Phys., 46, 1940, pp. 601-626, (NACA TM 1112, 1947).

25. Doob, J., "Probability approach to the heat equation," Trans. Amer. Math. Soc., 80, 1955, pp. 216-280.

26. Emmons, H., "Flow discontinuities associated with combustion," in High Speed Aerodynamics and Jet Propulsion, 1958, pp. 584-622.

27. Gagnon, Y. and Ghoniem, A., "Low frequency oscillations of recirculating flow at low Reynolds numbers" (in progress).

28. Ghoniem, A., Chorin, A. and Oppenheim, A., "Numerical modeling of turbulent flow in a combustion tunnel," Phil. Trans. R. Soc. Lond., A304, 1982, pp. 303-325.

29. Ghoniem, A. and Oppenheim, A., "Numerical solution of the problem of flame propagation by the use of the random element method," AIAA J., 22, 10, 1983, pp. 1429-1435.

30. Ghoniem, A., Marek, C. and Oppenheim, A., "Modeling interface motion of combustion (MIMOC), A computer code for two-dimensional unsteady turbulent combustion," NASA Technical Paper 2132, 1983.

31. Ghoniem, A. and Sethian, J., "Structure of turbulence in a recirculating flow; A computational study," AIAA-85-0146.

32. Ghoniem, A. and Sherman, F., "Grid-free simulation of diffusion using random walk methods," J. Comp. Phys., 1985 (in press).

33. Ghoniem, A., "Analysis of flame deformation in a turbulent field; effect of Reynolds number on burning rates," AIAA-85-0140.

34. Greengard, C., "Three-dimensional vortex methods," Ph.D. thesis, Department of Mathematics, University of California, Berkeley, 1984.

35. Hald, O., "Convergence of vortex methods for Euler's Equations, II," SIAM J. Num. Anal., 16, 1979, pp. 726-755.

36. Hald, O., "Convergence of a random method with creation of vorticity," Center for Pure & Applied Math., University of California, Berkeley, PAM-252, 1984.

37. Hald, O., "Convergence of vortex methods for Euler's equations III," Center for Pure & Applied Math., University of California, Berkeley, PAM-270, 1985.

38. Hirt, C. and Nichols, B., "Volume of fluid (VOF) method for the dynamics of free boundaries," J. Comp. Phys., 39, 1981, pp. 201-225.

39. Hsiao, C., Ghoniem, A., Chorin, A. and Oppenheim, A., "Numerical simulation of a turbulent flame stabilized behind a rearward-facing step," 20th Symposium (International) on Combustion, Ann Arbor, Michigan, 1984.

40. Inoue, O., "Vortex simulation of a turbulent mixing layer," AIAA J., 23, 1985, pp. 367-373.

41. Knio, O. and Ghoniem, A., "Numerical solution of flame propagation inside a confined circular chamber," in progress.

42. Kuwahara, K. and Takami, H., "Numerical studies of two-dimensional vortex motion by a system of point vortices," J. Phys. Soc. Japan, 34, 1973, pp. 247-252.

43. Kuwahara, K. and Takami, H., "Study of turbulent wake behind a bluff body by vortex method," Proceedings of IUTAM Symposium on Turbulence and Chaotic Phenomena in Fluids, ed. T. Tatsumi, North-Holland Publishing Corp., Amsterdam, 1983.

44. Krasney, R., "Desingularization of periodic vortex sheet roll up," submitted for publication.

45. Landau, L., "On the theory of slow combustion," J. Exp. Theor. Phys., 14, 1944, pp. 240.

46. Landau, L. and Lifshitz, E., Fluid Mechanics, Translated from Russian, Pergamon Press, New York, 1975, xii + 536 p. (first ed. 1959).

47. Law , C.K., "Heat and mass transfer in combustion: Fundamental concepts and analytical techniques," Prog. Energy Combust Sci., 10, 1984, pp. 295-318.

48. Lighthill, M., "Introduction: Boundary Layer Theory," Laminar Boundary Theory, ed. by L. Rosenhead, Oxford University Press, England, 1963, pp. 46-159.

49. Leonard, A., "Vortex methods for flow simulation," J. Comp. Phys., 37, 3, 1980, pp. 289-335.

50. Majda, A. and Sethian, J., "The derivation and numerical solution of the equations for zero Mach number combustion," Comb. Sci. and Tech., 42, 1985, pp. 185-205.

51. Markstein, G., Non-steady flame propagation, AGARDograph No. 75, Pergamom Press, 1964.

52. Marchioro, C. and Pulvirenti, M. "Hydrodynamics in two dimensions and vortex theory," Com. Math. Phys., 84, 1982, pp. 483-503.

53. McMurtry, P., Jose, W., Riley, J. and Metcalfe, R., "Direct numerical simulation of a reacting mixing layer with chemical heat release," AIAA-85-0143.

54. Milinazzo, F. and Saffman, P., "The calculation of large Reynolds number two-dimensional flow using discrete vortices with random walk," J. Comp. Phys., 23, 1979, pp. 380-392.

55. Monaghan, J., "Why particle methods work," SIAM J. Sci. Stat. Comp., 3, 1982, pp. 422-433.

56. Moore, D., "On the point-vortex method," SIAM J. Sci. Stat. Comp., 2, 1981, pp. 65-84

57. Nakamura, Y., Leonard, A. and Spalart, P., "Vortex simulation of an inviscid shear layer," AIAA/ASME 3rd Joint Thermophysics, Fluids, Plasma and Heat Transfer Conf., 1982, AIAA-82-0948, St. Louis, MO.

58. Ng, K. and Ghoniem, A., "Numerical simulation of a confined shear layer," 10th International Colloquium on Dynamics of Explosions and Reactive Systems, August 1985, Berkeley, CA.

59. Noh, W. and Woodward, P., "SLIC, simple line interface calculations," 5th International Conference on Numerical Methods in Fluid Dynamics (ed. A. Vooren and P. Zandbergen), Springer-Verlag, Berlin, 1976, pp. 330-339.

60. Peskin, C. and Wolfe, A., "The aortic sinus vortex," Fed. Proc., 37, 14, 1978, pp. 2784-2792.

61. Peyret, R. and Taylor, T., Computational Methods for Fluid Flow, Springer-Verlag, 1983.

62. Rehm, R., and Baum, H., "The equations of motion for thermally-driven, cognant flows," N.B.S.J. Res., 83, 3, 1978, pp. 297-308.

63. Saffman, P. and Baker, G., "Vortex Interactions," Ann. Rev. Fluid Mech., 11, 1979, pp. 95-122.

64. Sethian, J., "Turbulent combustion in open and closed vessels," J. Comp. Phys., 54, 3, 1984, pp. 425-456.

65. Sethian, J., "Curvature and the evolution of fronts," Comm. Math. Phys., 1985 (to appear).

66. Sivashinsky, G., "Hydrodynamic theory of flame propagation in an enclosed volume," Acta Astronautica, 6, 1979, pp. 631-645.

67. Sivashinsky, G., "Instabilities, pattern formation, and turbulence in flames," Ann. Rev. Fluid Mech., 15, 1983, pp. 179-199.

68. Spalart, P. and Leonard, A., "Computation of separated flows by a vortex tracing algorithm," AIAA-81-1246.

69. van der Vooren, A., "A numerical investigation of the rolling up of vortex sheets," Proc. R. Soc. Lon., A373, 1980, pp. 67-91.

70. Zeldovich, Ya., "Structure and stability of steady laminar flame at moderately large Reynolds numbers," Comb. Flame, 40, 1981, pp. 225-234.

71. Zeldovich, Ya., Istratov, A., Kichin, N., and Librovich, V., "Flame propagation in tubes: hydrodynamics and stability," Comb. Sci. and Tech., 24, 1980, pp. 1-13.

DEPARTMENT OF MECHANICAL ENGINEERING
MASSACHUSETTS INSTITUTE OF TECHNOLOGY
CAMBRIDGE, MASS. 02139

COMBUSTION

Lectures in Applied Mathematics
Volume 24, 1986

FLAME RESONANCE AND ACOUSTICS IN THE PRESENCE OF HEAT LOSS

A. C. McIntosh

ABSTRACT. A brief review is presented of some work on non-adiabatic laminar flame stability. The mathematical model used for the flame assumes a global Arrhenius type reaction scheme and is based upon the well-established method of large activation energy asymptotics. The global heat loss case is first considered and then the case of localised heat loss to a porous plug type burner.

The review is then extended to include the interaction of a low-level noise field with an anchored (and non-adiabatic) flame. The principles of matched asymptotic expansions are again applied using the fact that Mach number M is a small number. Long wavelength acoustic disturbances [$O(M^{-1})$ on a diffusion length scale] are linked to small perturbations in velocity within the combustion zone. The matching process yields frequency conditions governing the flame vibration which are strongly dependent on the acoustic boundary conditions. Acoustic emission alone is first of all considered and its consequent effect on stability. Finally results for the case of upstream acoustic feedback are presented. It is shown that considerable acoustic amplification is possible. In that the flame is non-adiabatic,the acoustic oscillations are strongly coupled not only with molecular and thermal diffusion balances (through Lewis number) but also to the fluctuating heat loss near the burner surface.

1. INTRODUCTION. The subject of laminar flame stability has received much attention by a considerable number of authors. Recently the book by Buckmaster and Ludford (1982) has brought together the main conclusions up to that date. A crucial tool

---

1980 Mathematics Subject Classification.

0075-8485/86 $1.00 + $.25 per page

in all the work has been the use of matched asymptotic expansions based on the largeness of activation energy. This tool was first successfully applied to modelling steady laminar flames, by Bush and Fendell (1970). The main result from the analysis of unsteady flames has been to predict the onset of pulsating instability for Lewis number less than unity (where Lewis number, Le is defined as mass diffusion divided by thermal diffusion) (Sivashinsky 1977).

For both regions of instability, bifurcations of the linear small amplitude solutions will occur, yielding pulsations or cellular patterns depending on the type of instability. Both phenomena have been observed in experiments and the matter is well reviewed by Sivashinsky (1983).

The subject of non-adiabatic flames, introduces a further parameter into the analysis. The heat loss can either be considered in a general manner (the 'global heat-loss' model) or to be lost to a localised region. The former approach is discussed in Section 3 and the latter approach in Section 4. Localised heat loss can, of course, be considered at various points near the flame. We have chosen to consider the case of heat loss to a porous-plug flame-holder. Such a situation can be readily set up by experiment and much work has been done using such flames (Botha and Spalding 1954, Kaskan 1957, Ferguson and Keck 1979).

The last section (5) considers the interaction of a low-level noise field with an anchored flame. This area is becoming of increasing interest since it is clear experimentally that the interactions can be severe (Oran and Gardner 1985). The latter work cited is a review of short time period acoustic coupling. The present work is restricted to long wavelength acoustic disturbances and thus the response is effectively quasi-steady. Acoustic emission is shown to affect flame stability for non-adiabatic flames and it is also shown that acoustic amplification can take place under certain conditions.

In that the paper is a brief review, only the main results of the theories considered are shown. For the detailed derivation of these results, the reader is referred to the original papers which are cited within each section. The next section lists the equations which are fundamental to any analysis of flame stability.

2. BASIC COMBUSTION EQUATIONS. Any theories of stability must effectively model the chemical, diffusive and thermal balances within the combustion zone. In the majority of works, for low speed flames, the effect of compressibility on a diffusion length scale is ignored, so that the equation of state is

$$\rho T = T_{01}, \tag{1}$$

where ρ (density), T (temperature) are non-dimensional quantities. All quantities, except temperature, in this mathematical treatment are non-dimensionalised with respect to upstream (initially) steady values. Temperature is non-dimensionalised with respect to its downstream (final) steady value. The equation of continuity for the two components of velocity (u and v) is:

$$\frac{\partial \rho}{\partial t} + \frac{\partial}{\partial x}(\rho u) + \frac{\partial}{\partial y}(\rho v) = 0. \tag{2}$$

The equation of energy balance assumes an overall Arrhenius type reaction rate involving θ_1, the non-dimensional activation energy (non-dimensionalised using the steady, burnt temperature):

$$\frac{\partial T}{\partial t} + u\frac{\partial T}{\partial x} + v\frac{\partial T}{\partial y} - \frac{1}{Le\rho}\frac{\partial}{\partial x}(\lambda\frac{\partial T}{\partial x}) - \frac{1}{Le\rho}\frac{\partial}{\partial y}(\lambda\frac{\partial T}{\partial y})$$
$$+ LeK(T-T_{01}) = \frac{Le(1+\sigma)Q_1}{\sigma}\Lambda_1 C_\ell C_r e^{\theta_1(1-1/T)}. \tag{3}$$

Occurring in this equation are Le (Lewis number defined as mass diffusion over thermal diffusion), λ (thermal conductivity), σ (ratio of molecular weights), Q_1 (heat of reactions) and C_ℓ, C_r which refer to lean and rich species mass fractions.

The term Λ_1 is the steady, pre-exponential eigenvalue and is typically (for far-from-stoichiometric conditions- see Clarke and McIntosh 1980) proportional to θ_1^2. The parameter K represents a global heat loss term, the effect of which is discussed in the next section. The equation of lean species conservation takes the form:

$$\frac{\partial C_\ell}{\partial t} + u \frac{\partial C_\ell}{\partial x} + v \frac{\partial C_\ell}{\partial y} - \frac{1}{\rho}\frac{\partial}{\partial x}(\lambda \frac{\partial C'_\ell}{\partial x}) - \frac{1}{\rho}\frac{\partial}{\partial y}(\lambda \frac{\partial C_\ell}{\partial y})$$

$$= -\mathrm{Le}\, \Lambda_1 C_\ell C_r e^{\theta_1(1-1/T)}, \qquad (4)$$

and the equations of momentum become,

$$\frac{\partial u}{\partial t} + u \frac{\partial u}{\partial x} + v \frac{\partial u}{\partial y} + \frac{1}{\gamma\rho}\frac{\partial p_f}{\partial x} = -g_r + \frac{S_c}{\rho}\,[\frac{\partial}{\partial x}(\lambda \frac{\partial u}{\partial x}) + \frac{\partial}{\partial y}(\lambda \frac{\partial u}{\partial y})$$

$$+ \frac{1}{3}\frac{\partial}{\partial x}(\lambda \frac{\partial u}{\partial x} + \lambda \frac{\partial v}{\partial y}) + \frac{\partial}{\partial y}(\lambda \frac{\partial v}{\partial x}) - \frac{\partial}{\partial x}(\lambda \frac{\partial v}{\partial y})], \qquad (5)$$

$$\frac{\partial v}{\partial t} + u \frac{\partial v}{\partial x} + v \frac{\partial v}{\partial y} + \frac{1}{\gamma\rho}\frac{\partial p_f}{\partial y} = \frac{S_c}{\rho}\,[\frac{\partial}{\partial x}(\lambda \frac{\partial v}{\partial x}) + \frac{\partial}{\partial y}(\lambda \frac{\partial v}{\partial y})$$

$$+ \frac{1}{3}\frac{\partial}{\partial y}(\lambda \frac{\partial u}{\partial x} + \lambda \frac{\partial v}{\partial y}) + \frac{\partial}{\partial x}(\lambda \frac{\partial u}{\partial y}) - \frac{\partial}{\partial y}(\lambda \frac{\partial u}{\partial x})]. \qquad (6)$$

In these last two equations γ is the ratio of specific heats and S_c is the flow Schmidt number ($= P_r/\mathrm{Le}$ where P_r is Prandtl number) and thus represents the balance of viscous diffusion with mass diffusion. Within the combustion zone, pressure variations are assumed to be of order M^2 and p_f represents the flow component. Thus,

$$p = 1 + M^2\, p_f, \qquad (7)$$

where M is flow Mach number. In the non-dimensionalising of all the equations, characteristic values of diffusion length and time have been used and, in particular, the gravitational acceleration term g_r in the longitudinal momentum balance is related to the dimensional acceleration g' by,

$$g_r \equiv g' \mathcal{D}'_{01}/u'^3_{01}, \tag{8}$$

where it is assumed the flow is upwards and $\mathcal{D}'_{01}$, u'_{01} represent the dimensional mass diffusion coefficient and initial upstream flow velocity respectively.

These equations are the starting point of all analyses, and can be simplified further by taking $\rho\lambda$ constant or indeed ρ constant (the latter to isolate diffusional thermal effects). Most analyses begin by thus decoupling the momentum equations (5) and (6). However, in reality there will be a most complex interaction between all six.

3. GLOBAL HEAT LOSS. One of the first papers to model global heat-losses from unsteady laminar flames was that by Joulin and Clavin (1979). In this work, the heat loss parameter K in equation (3), is given by

$$K = \frac{K'\lambda'_{01}}{c'^2_p \rho'^2_{01} u'^2_{01}}, \tag{9}$$

where c'_p is overall specific heat and K' is the (dimensional) volumetric heat loss coefficient. Writing

$$K = \frac{H}{2\theta}, \tag{10}$$

(where their activation energy θ is proportional to the θ_1 defined in the previous section) these authors then consider small $O(\theta^{-1})$ heat losses through H. For steady conditions, it is found

$$H = -\ell n\left(\frac{u'_{01}}{u'_{ad}}\right)^2, \tag{11}$$

where u'_{ad} is the adiabatic flame velocity. If one non-dimensionalises the heat loss with respect to fixed values at adiabatic conditions and define

$$\hat{H} = H\left(\frac{u'_{01}}{u'_{ad}}\right) = \frac{2\theta K'\lambda'_{01}}{c_p^{12}\rho_{01}^{12}u'^2_{ad}}\,, \tag{12}$$

then it is found (Fig. 1) that there is a max value of $\hat{H}$ beyond which no solution can be found. This quenching (q) value is $\hat{H}_q = 1/e$ and occurs at a value of velocity

$$\left(\frac{u'_{01}}{u'_{ad}}\right)_q = \frac{1}{\sqrt{e}}\,. \tag{13}$$

At this critical point ($\hat{H}_q = 1/e$), $H_q = 1$ and as indicated in the figure the diagram splits into "slow flames" with large heat losses (to the left) and "fast flames" with small heat losses (to the right). The results of the stability analysis show that all slow flames are unstable to planar disturbances but that the fast solutions can become stable beyond a certain point (labelled 'c' in Fig. 1). At this critical stability point, H_c is given by,

$$H_c = 2\sqrt{3-\ell_c} - (2-\ell_c), \tag{14}$$

where

$$\ell_c = \theta(\text{Le}-1). \tag{15}$$

For $H_c = 0$, the critical reduced Lewis number ℓ_c is $-4(1+\sqrt{3})$ (= -10.93...) which is inaccessible for most mixtures. Consequently adiabatic flames will be stable. However, the growing values of H_c, the corresponding value of ℓ_c reaches values up to -6. This means that the planar instability will be more readily accessible for flames with heat loss.

Figure 2 is taken from the work of Joulin and Sivashinsky (1983) and shows that the onset of cellular instability will occur for a reduced Lewis number ℓ_{cc} in the range,

$$0 < \ell_{cc} < 2, \tag{16}$$

with $\ell_{cc} = 2$ for $H = 0$ and $\ell_\infty = 0$ for $H = 1$. The general prediction from this result is that cellularity will be more readily observed near flammability limits (Markstein 1964). The work of Joulin and Sivashinsky (1983) extends the volumetric heat loss model to develop an equation for the evolution of the flame front by choosing suitable (new) time and space scales. This fourth order partial differential equation has led to predictions of the onset of various types of turbulence, such as stretched, oscillatory and spinning flames, and is dealt with in the review by Sivshinsky (1983).

Recently a very important paper by Pelcé and Clavin (1982) has successfully modelled the influence of hydrodynamics on the stability of freely propagating plane flames. This avoids the constant density assumption and clearly show that in reality both the diffusional and hydrodynamic (Landau) instability must be considered together. With a similar approach, Matalon and Matkowsky (1982) consider curved flame stability and Clavin and Williams (1982) using the full set of equations consider the onset of low-level intensity turbulence. These are only briefly referred to here as the subject of turbulence is really beyond the scope of this review.

4. LOCALISED HEAT LOSS.

4.1. STEADY RESULTS. We now address ourselves to the case of a flame near a porous plug type flame-holder (Fig. 3) where inlet conditions are such that temperature is held constant due to the large conductance of the holder and the fractional mass fluxes are fixed upstream. This is the Hirschfelder condition such that product species is not allowed to flow back into the holder. Referring to Clarke and McIntosh (1980) and McIntosh and Clarke (1984a), if a mass-weighted spatial coordinate is defined such that

$$x_1 \equiv \int_0^x \rho \, dx, \tag{17}$$

with the assumption

$$(\rho\lambda) = 1, \tag{18}$$

then the energy and species equations simplify to

$$\frac{dT}{dx_1} - \frac{1}{Le}\frac{d^2T}{dx_1^2} = \frac{Le(1+\sigma)Q_1}{\sigma}\Lambda_1 C_\ell C_r e^{\theta_1(1-1/T)}, \tag{19}$$

$$\frac{dC_e}{dx_1} - \frac{d^2C}{dx_1^2} = -\Lambda_1 C_\ell C_r e^{\theta_1(1-1/T)}, \tag{20}$$

with conditions at the holder surface:

$$T\Big|_{x_1=0} = T_{01}, \tag{21}$$

$$C_\ell\Big|_{x_1=0} - \left(\frac{dC_\ell}{dx_1}\right)\Big|_{x_1=0} = \text{constant}. \tag{22}$$

Using large activation energy asymptotics, where gradients in the pre-heat and equilibrium zones are matched with those within a small reaction zone, the main results obtained are:

$$\left(\frac{u'_{01}}{u'_{ad}}\right) = \left(\frac{T'_b}{T'_{ad}}\right)^2 \exp\left[-\frac{\theta_1}{2}\left(1 - \frac{T'_b}{T'_{ad}}\right)\right], \tag{23}$$

$$Lex_{|f|} = \ell n\left(\frac{1-T'_a/T'_{ad}}{1'T'_b/T'_{ad}}\right), \tag{24}$$

where $x_{|f|}$ is the non-dimensional (mass-weighted) stand off distance of the flame.

Figure 4 illustrates how $x_{f'}$ (no mass weighting) and $x_{|f|}$ (mass weighted) vary with T'_b/T'_{ad} for $\theta_1 T'_b/T'_{ad} = 10$ and $T'_{01}/T'_{ad} = 0.15$. For most realistic parameter values the minimum stand off distance occurs for $x_{|f|}$ in between 2 and 4 (McIntosh and Clarke, 1984a).

4.2. UNSTEADY RESULTS.

4.2.1. ONE DIMENSIONAL. Similar equations to (19) and (20) can be obtained with the addition of the time derivative of temperature and lean species. By making the assumption of small perturbations, e.g. for temperature,

$$T = T_s + \varepsilon T_u(x_1)e^{\omega t}, \tag{25}$$

where ω is complex frequency, the equations can be linearised. Using the same technique of matched asymptotic expansions, jump conditions are obtained in temperature and lean species. The result is a frequency condition for ω (see Margolis 1980, McIntosh and Clarke 1984b):

$$(s^2-\tfrac{1}{2}S+\tfrac{1}{2}R-RS) + \frac{2}{\theta_1 B_1}(\tfrac{1}{2}+R)(\tfrac{1}{2}-S)(s+S) = 0, \tag{26}$$

where

$$B_1 \equiv (1-T_{01})/(1-e^{-Lex_{|f|}}), \tag{27}$$

$$r \equiv \sqrt{\omega + \tfrac{1}{4}} \quad ; \quad s \equiv \sqrt{\omega/Le + \tfrac{1}{4}}\ , \tag{28a,b}$$

$$R \equiv \frac{r\cosh(rx_{|f|})}{\sinh(rx_{|f|})} \quad ; \quad S \equiv \frac{\cosh(Lesx_{|f|})}{\sinh(Lesx_{|f|})}. \tag{29a,b}$$

Result (26) is valid for near equi-diffusional flames and yields the stability diagram (Fig. 5). The overall effect of heat loss (as in the global heat loss case §3) is to make the planar instability more accessible. For most hydrocarbon fuels Lewis number greater (less) than unity corresponds roughly to fuel rich (lean) conditions. Thus the planar instability will generally be observed under fuel-lean conditions. As the activation energy gets large so the top neutral stability line is approached. Working independently Buckmaster (1983) pointed out that this curve will be more relevant for "limit" mixtures when flame speeds and flame temperatures are low.

Note from Fig. 4 below $x_{|f|} \approx 3{,}4$ the heat loss is such that the "slow" region from the previous section's analysis develops and Fig. 5 confirms such flames will be unstable for realistic values of activation energy.

4.2.2. TWO DIMENSIONAL - CONSTANT VELOCITY. The results of §4.2.1 can be simply extended if the assumption is made that density is constant. This somewhat artificial approach has the effect of highlighting diffusional thermal effects alone by imposing a constant velocity field on the temperature/lean species fluctuations. A limitied amount of physical understanding can be gained by such a move and using this technique the work of Buckmaster (1983) shows that the main effect of localised heat loss alone (in the absence of hydrodynamic effects) is to cause the cellular instability to be less accessible for flames with Lewis numbers near unity. Fig. 6 illustrates the form of these neutral stability curves in the present terminology. Equation (26) is replaced by one involving the wavenumber k,

$$\frac{k^2}{Le^2}(Le-1)+(s_1^2-\tfrac{1}{2}S_1+\tfrac{1}{2}R_1-R_1S_1)+\frac{2}{\theta_1 B_1}(\tfrac{1}{2}+R_1)(\tfrac{1}{2}-S_1)(s+S_1) = 0 \quad (30)$$

where,

$$r_1 \equiv \sqrt{\omega+k^2+\tfrac{1}{4}} \;; \qquad ; \; s_1 \equiv \sqrt{\omega/Le+k^2/Le^2+\tfrac{1}{4}}, \qquad (31a,b)$$

$$R_1 \equiv r_1 \frac{\cosh(r_1 x_{|f|})}{\sinh(r_1 x_{|f|})} \;; \; S_1 \equiv \frac{s_1\cosh(Lesx_{|f|})}{\sinh(Lesx_{|f|})} \,. \qquad (32a,b)$$

4.2.3. TWO DIMENSIONAL - NON CONSTANT DENSITY. The author has recently investigated the effect of gas expansion on the predictions of §4.2.2. Immediately one lifts the restriction of constant density, the problem becomes a lot more complicated. The reader is referred to McIntosh (1985a) for the full details of this work. The analysis is similar to that of Pelcé and Clavin (1982) in that hydrodynamic zones are considered before and after the flame. However, for the flame-holder case, the hydrodynamic zone upstream is within the holder itself in the form of Darcy type flow. Thus longitudinal (u_n) and transverse (v_n) components of velocity obey,

$$u_n = -\frac{\Omega_1}{\gamma}\frac{\partial p_{fn}}{\partial x_1}, \tag{33}$$

$$v_n = -\frac{\Omega_2}{\gamma}\frac{\partial p_{fn}}{\partial y}, \tag{34}$$

and a continuity condition,

$$\frac{\partial u_n}{\partial x_1} + \frac{\partial v_n}{\partial y} = 0, \tag{35}$$

where time variations of density are negligible due to the large conductance of the holder. First the transformation (17) is applied throughout, then the 2 - D equations (1) to (6) are linearised for small perturbations. Jump conditions are derived on the basis of large activation energy asymptotics which then leaves the somewhat intractable pre-heat equations to solve. These basically are second order differential equations with non-constant coefficients (depending on the steady solutions of temperature and lean species). The problem is made tractable by adopting the assumption that complex frequency and wavenumber are small. (The former assumption is valid near the cellular neutral stability boundary where $\omega \simeq 0$.) We define,

$$k = \overline{K}\, T_{01}\delta, \tag{36}$$

$$\omega = \overline{\omega}\delta, \tag{37}$$

and obtain series solutions in δ. After much tedious working a dispersion condition on the frequency ω is obtained of the form,

$$F_0+F_1\omega+F_2\left(\frac{k}{T_{01}}\right) + F_3\omega^2 + F_4\,\omega\left(\frac{k}{T_{01}}\right) + F_5\left(\frac{k}{T_{01}}\right)^2 +\ldots = 0, \qquad (38)$$

where $F_0, F_1, F_2, F_3, F_4, F_5$ all depend on the eight parameters:

$\theta_1 B_1^* \equiv \dfrac{E'_A}{R'T'_{b_1}}\left(1 - \dfrac{T'_{01}}{T'_{b1}}\right)$: Reduced activation energy

$Le \equiv \dfrac{\rho'_{01}\mathcal{D}'_{01}\, c'_p}{\lambda'_{01}}$: Lewis number

$T_{01} \equiv \dfrac{T'_{01}}{T'_{b1}}$: Inverse gas expansion

$x_{|f|}$: Mass weighted stand off distance

$g_r \equiv g' \dfrac{\mathcal{D}'_{01}}{u'^3_{01}}$: Gravity

$P_r \equiv \dfrac{\mu' c'_p}{\lambda'}$: Prandtl number (μ' = coefficient of dynamic viscosity)

Ω_1 : Longitudinal Darcy constant

Ω_2 : Transverse Darcy constant

The last parameter is effectively an imposed flame "stretch" ahead of the flame.

In reality the dependence of the neutral stability curves on these parameters is highly complicated. The theory can be tested for zero gas expansion against the exact results of the previous section, and it is found that agreement is good up to k values in the region of 0.2. This now gives confidence in considering the effect of gas expansion. As soon as $T_{01} \neq 1$ is introduced, all the other parameters listed above have an effect, mainly through the momentum equations. In particular the onset of cellularity is very much dependent on g_r, Ω_1 and Ω_2.

To demonstrate the overall effect on flame stability, we consider typical plots (Fig.7) of inlet speed against equivalence retio for a steady flat flame near a flame-holder where the adiabatic speed is taken as $u'_{01} = 20\text{cm sec}^{-1}$ and the dilution factor is taken to be 7. The stoichiometric adiabatic temperature is 1500°K, the gas expansion parameter T_{01} is 0.2 and the dimensional activation energy is 40,000 cal/mole. The inlet speed is plotted against equivalence ratio for $x_{|f|} = 2,3,4,5$ and ∞ (adiabatic conditions). The Lewis number is considered to vary with equivalence ratio from 0.58 (fuel lean) to 1.16 (fuel rich) and roughly represents a propane/air mixture. Lewis number is estimated between these extremes by using the formula suggested by Joulin and Mitani (1981) for order 2 reactions. It can be shown (Clarke 1984, McIntosh 1985a Appendix A) that the longitudinal Darcy constant Ω_1 is related approximately to pore radius (d') in the holder through,

$$\Omega_1 = \frac{8 d'^2 u'^2_{01}}{LeP_r K'^2_{01}} \tag{39}$$

where K'_{01} is the thermal diffusivity. Setting a value of $K'_{01} = 0.3\text{cm sec}^{-2}$ and d' = 0.01 cm with $P_r = 0.75$, $\Omega_2 = \Omega_1$, with a porosity of 0.75, the theory is used to establish the critical wavenumber for cellular instabilities to occur. The onset of such conditions is indicated in Fig. 7 by the shaded area just below adiabatic conditions.

This figure confirms the experimental observations of Botha and Spalding (1954) who found cellularity throughout the stoichiometric range as adiabatic conditions were approached. Botha and Spalding (p.94) came to the conclusion that preferential diffusion could not be the sole mechanism governing these flame irregularities and this work shows that it is primarily hydrodynamic influences within the holder which cause the cellularity to occur. An important finding was that the band of instability was very much the same as Ω_2/Ω_1 was reduced to values as low as 0.1. However, below this value the instability begins to disappear and finally is not present for $\Omega_2 = 0$. These findings agree very well with experiment - the effect of different types of porous plug burners will not be great but as soon as a hypodermic type holder is used (where transverse velocities in the holder are not allowed - Schimmer and Vortmeyer 1977) then the cellularity is not observed for sub-adiabatic conditions.

The foregoing work shows clearly the subtle part that hydrodynamic instabilites can play in flame behaviour, and in particular, the important effect that flame-holders can have on experimental results.

5. LONG-WAVE ACOUSTIC INTERACTION. This chapter extends the summary of non-adiabatic flame stability to include recent advances in the field of flame-acoustic interactions.

Flame-acoustic coupling is becoming a subject of increasing importance in combustion, particularly in understanding the stability and resonance of flames in enclosed regions where pressure waves interacting with flames can be amplified considerably (Oran and Gardner 1985). A very important parameter in any analysis is

$$\tau \equiv \frac{\text{diffusion time}}{\text{acoustic time}}, \tag{40}$$

as well as the gas expansion parameter,

$$B_1 \equiv 1 - \frac{\text{unburnt gas temperature}}{\text{burnt gas temperature}} = 1-T_{01}, \tag{41}$$

The work of Van Harten, Kapila and Matkowsky (1984) concentrates in the main on two cases, $\tau \ll 1$ and $\tau = 1$ with

$B_1 \to 0$ (i.e. little gas expansion). In the work of the author (McIntosh, 1984) $\tau = 1$ with no restriction on B_1 (i.e. $O(1)$ gas expansion is permitted). (In the work of Oran and others, dealing with very short wavelength disturbances, the relevant time ratio is chemical relaxation time/acoustic time).

What has led to the long-wave investigations has been due to applying momentum considerations to previous one dimensional stability theories. In these previous one-dimensional analyses of flame stability only energy and species equations have been the main governing principles employed. However, if one proceeds to reintroduce the decoupled momentum equation, there is a residual velocity perturbation which extends far upstream and downstream. On a diffusion length scale it is of constant magnitude, but at greater distances it has been shown (van Harten et. al. 1984 and McIntosh 1984) that it matches with a low level $O(M)$ acoustic field, where M is flow Mach number.

At large distances a new coordinate is defined,

$$\hat{x} \equiv Mx_1, \tag{42}$$

and the one-dimensional equations are considered with the full equation of state,

$$P = \frac{\rho T}{T_{01}}, \tag{43}$$

replacing equation (1), and the consequent addition of pressure spatial and time gradients in the energy equation (3). It is convenitn to still perform the transformation (17) in order to link any terms to the combustion zone results. The equations are linearised by considering small perturbations but in the acoustic zone the perturbations are themselves considered as a series in Mach number M. Thus,

$$p_u = M\bar{p}_u^{(a)} e^{\omega t} + \ldots \tag{44a}$$

$$T_u = M\bar{T}_u^{(a)} e^{\omega t} + \ldots \tag{44b}$$

$$\rho_u = M\rho_u^{(a)} e^{\omega t} + \ldots \tag{44c}$$

$$u_u = \bar{u}_u^{(a)} e^{\omega t} + \ldots \tag{44d}$$

The ordering of u_u ensures the development of the classical acoustic equations whilst enabling a balance with terms from the combustion zone. This is done using matched asymptotic expansions based on the smallness of Mach number M.

The author has included acoustic feedback (McIntosh, 1985b) by considering the flame to be anchored to a porous gauze upstream, beyond which there is (on an acoustic scale) a finite length tube open to atmospheric pressure (though such a boundary condition can be varied) (see Fig. 8). This, therefore, represents a very simplified model of a burner port, similar to that used by Roberts (1978) who has investigated noise amplification by flames. The physical set up also bears some resemblance to that of Madarame (1983). In the upstream acoustic zone (superscript "a1"), the solutons are of the form,

$$\bar{p}_u^{(a1)} = P_1^{(a1)} e^{\omega\hat{x}} + P_2^{(a1)} e^{-\omega\hat{x}}, \tag{45a}$$

$$\bar{\rho}_u^{(a1)} = \gamma^{-1}(P_1^{(a1)} e^{\omega\hat{x}} + P_2^{(a1)} e^{-\omega\hat{x}}), \tag{45b}$$

$$\bar{T}_u^{(a1)} = (1-\gamma^{-1})T_{01}(P_1^{(a1)} e^{\omega\hat{x}} + P_2^{(a1)} e^{-\omega\hat{x}}), \tag{45c}$$

$$\bar{u}_u^{(a1)} = -\gamma^{-1}(P_1^{(a1)} e^{\omega\hat{x}} - P_2^{(a1)} e^{-\omega\hat{x}}). \tag{45d}$$

Downstream, emitted waves alone are considered (zone "a2"):

$$\bar{p}_u^{(a2)} = P^{(a2)} e^{-\omega\hat{x}/\sqrt{T_{01}}}, \tag{46a}$$

$$\bar{\rho}_u^{(a2)} = T_{01}\gamma^{-1} P^{(a2)} e^{-\omega\hat{x}/\sqrt{T_{01}}} \tag{46b}$$

$$\bar{T}_u^{(a2)} = (1-\gamma^{-1}) P^{(a2)} e^{-\omega\hat{x}/\sqrt{T_{01}}} \tag{46c}$$

$$\bar{u}_u^{(a2)} = \frac{P^{(a2)}}{\gamma\sqrt{T_{01}}} e^{-\omega\hat{x}/\sqrt{T_{01}}} . \tag{46d}$$

Joining these with harmonic solutions in the combustion zone (where fluctuations in pressure are allowed of order (M) but only in time, not spatially with x_1), results in the conditions,

$$P_1^{(a1)} + P_2^{(a1)} = P^{(a2)} \tag{47}$$

$$(P_1^{(a1)} - P_2^{(a1)})V = \sqrt{T_{01}}\, P^{(a2)}, \tag{48}$$

where,

$$V = \frac{-\omega B}{LeG(\frac{1}{2}-s)} - T_{01} - \frac{A}{G}\left(\tfrac{1}{2}-S + \frac{se^{-Lex_{|f|}/2}}{shLesx_{|f|}}\right)$$

$$+ \frac{LeB_1}{\omega}\left[\left(\tfrac{1}{2}-S + \frac{se^{-Lex_{|f|}/2}}{shLesx_{|f|}}\right) - e^{-Lex_{|f|}}\left(\tfrac{1}{2}+S - \frac{se^{Lex_{|f|}/2}}{shLesx_{|f|}}\right)\right], \tag{49}$$

$$A = \frac{LeB_1}{\omega(\frac{1}{2}-S)}\left(\tfrac{1}{2}-S+\frac{se^{-Lex_{|f|}/2}}{shLesx_{|f|}}\right)(G-\frac{\omega}{Le}) + B_1\left(1 + \frac{re^{-x_{|f|}/2}}{shrx_{|f|}}\right), \tag{50}$$

$$B = \frac{2}{\theta_1}\left[\left(\tfrac{1}{2}-S + \frac{se^{-Lex_{|f|}/2}}{shLesx_{|f|}}\right) - (\tfrac{1}{2}-S)\left(1 + \frac{re^{-x_{|f|}/2}}{shrx_{|f|}}\right)\right], \tag{51}$$

$$G = D + \frac{2}{\theta_1 B_1}[(\tfrac{1}{2}+R)(s+S)+D](\tfrac{1}{2}-S), \tag{52}$$

$$D = s^2 - \frac{S}{2} + \frac{R}{2} - RS, \tag{53}$$

and B_1, r, s, R, S have the same definitions as in equations (27)-(29), and $x_{|f|}$ represents the initial (density-weighted) stand-off distance of the flame from the gauze. With a forced fluctuating pressure input at the upstream ($\hat{x} = -h_1$) end of the flow,

$$\bar{p}_u^{(a1)}(\hat{x} = -h_1) = P_i, \tag{54}$$

the relationship between the emitted pressure level $P^{(a2)}$ and input level P_i is given by

$$\left(\frac{P^{(a2)}}{P_i}\right) = \left(\frac{V}{V\text{ch}\omega h_1 - \sqrt{T_{01}}\ \text{sh}\omega h_1}\right). \tag{55}$$

5.1. STABILITY AND RESONANCE. Crucial to the understanding of equation (55) are the zeroes of the denominator, i.e. where,

$$V\text{ch}\omega h_1 - \sqrt{T_{01}}\ \text{sh}\omega h_1 = 0. \tag{56}$$

For $h_1 = \infty$ (i.e. acoustic emission alone), then the result of McIntosh (1984) is recovered viz:

$$V - \sqrt{T_{01}} = 0, \tag{57}$$

as the condition for resonance. This result leads to modification of the neutral stability curves in Le-$x_{|f|}$ space (see Fig. 9) from those obtained with no allowance made for acoustic emission. Acoustic emission has the effect of increasing the region of stability.

In the limit $x_{|f|} \to \infty$, it is found $V \to -T_{01}$ and the two results (47), (48) collapse down to those obtained by van Harten et. al. (1984). The effect of gas expansion is included through the factor $\sqrt{T_{01}}$. The structure of the flame has then no part to play on the acoustic disturbance and the flame becomes a contact discontinuity. The resonance condition (57) simplifies to

$$G(x_{|f|} = \infty) = 0, \tag{58}$$

where G is given by (52) and (equated to zero) is, in fact, the dispersion relation (26) with $x_{|f|} = \infty$, for flames without acoustic interference. It in fact simplifies to the classic result of Sivashinsky (1977):

$$(r-s) + \frac{4s}{\theta_1 B_1^*}(\tfrac{1}{2}+r) = 0, \tag{59}$$

where B_1^* is $1 - T_{01}$.

If one now considers the limit $B_1 \to 0$ ($x_{|f|}$ finite), it is found that $V \to -1$ and the results (47), (48) collapse down to the same simplified results for a contact discontinuity but with no effect from gas expansion. The resonance condition (57) then becomes,

$$G(x_{|f|} \text{ finite}) = 0, \tag{60}$$

(i.e. relation (26) of §4.2.1). These two limits clearly indicate that when considering flames with localised heat loss (i.e. $x_{|f|}$ finite) it is important to maintain the condition $B_1 \neq 0$ (i.e. $T_{01} \neq 1$), and that an equation (48) indicates, in reality the structure of the flame can have a considerable effect on the acoustic disturbance. Since T_{01} can in practice be as low as 0.2, flame/acoustic interference can be a large factor with non-adiabatic flames.

5.2. ACOUSTIC FEEDBACK. From equation (56) when h_1 is finite, there are a number of effects introduced which are explained in more detail in McIntosh (1985b). Firstly, tube modes are introduced (i.e. those present without the flame) and these are always stable. Secondly, heat-transfer modes are introduced which are only present for $x_{|f|}$ finite. Fig. 10 summarises the situation for $Le = 0.6$, indicating where regions of stability/instability are predicted for these types of modes.

There are, in fact, a number of curves. Those shown are for the first two modes. The lower curve is for frequencies in the range $\omega_i \simeq 0.2$ and values ±0.1 about this. The upper curve is for higher frequences (ω_i values up to about 0.7).

If we choose now a point where at least lower frequences are stable (we choose $Le = 0.6$, $h_1 = 7$, $x_{|f|} = 5$), then using equation (55) the sound pressure level can be calculated for a fluctuating upstream input P_i. We obtain

$$SPL = 20\log_{10}\left(\frac{p^{(a2)}}{P_i}\right) = 20\log_{10}\left|\frac{V}{V\mathrm{ch}\omega h_1 - \sqrt{T_{01}}\,\mathrm{sh}\omega h_1}\right|. \tag{61}$$

In Fig. 11, the SPL is plotted against frequency and shows for the low frequencies (which are stable) considerable amplification. As indicated above, the higher frequencies will, in this case, be unstable. The amplification dB level is about 18 near the lower (stable) peak (roughly corresponding to frequencies of the order of 100 Hz). The amplification will vary considerably with parameter values but overall, these results are in qualitative agreement with that of Roberts (1978). Further details are discussed in the author's work (McIntosh, 1985b).

6. CONCLUDING REMARKS. Considering the complexity of the subject of non-adiabatic laminar flames, it has only been possible to briefly review some of the results of current work in this area. In the material present, volumetric heat losses and localised heat loss have been considered and in the latter case the results of two dimensional instability analysis have been included allowing for gas expansion. It is now generally recognized that gas expansion and hydrodynamics have an important role to play in flame stability.

The reviw has been extended to include the effects of long-wave acoustic coupling for one-dimensional flames. In that this latter work is ongoing, the results are still preliminary in nature but serve to show how quite realistic boundary conditions can now be modelled and thus it is possible for theories to be applied to practical burners used in industry and laboratories.

REFERENCES

Botha, J.P. and Spalding, D.B. (1954) "The laminar flame speed of propane/air mixtures with heat extraction from the flame." Proc. Roy. Soc. A225, 71-96.

Buckmaster, J.D. and Ludford G.S.S. (1982) "Theory of Laminar Flames." Cambridge University Press.

Buckmaster, J.D. (1983) "Stability of the porous plug burner flame." SIAM J. Appl. Math. 43 (6), 1335-1349.

Bush, W.B. and Fendell, F.E. (1970) "Asymptotic analysis of laminar flame propagation for general Lewis numbers." Comb. Sci. Tech. 1, 421-428.

Clarke, J.F. (1984) "Regular reflections of a weak shock wave from a rigid porous wall." Q. Jnl. Mech. appl. Math. 37 (1), 87-111.

Clark, J.F. and McIntosh, A.C. (1980) "The influence of a flame-holder on a plane flame, including its static stability." Proc. Roy. Soc. A372, 367-392.

Clavin, P. and Williams, F.A. (1982) "Effects of molecular diffusion and thermal expansion on the structure and dynamics of premixed flames in turbulent flows of large scale and low intensity." Jnl. Fluid Mech. 116, 251-282.

Ferguson, C.R. and Keck, J.C. (1979) "Stand-off distances on a flat flame burner." Comb. and Flame 34, 85-98.

Joulin, G. (1982) "Flame oscillations induced by conductive losses to a flat burner." Comb. and Flame 46, 271-282.

Joulin, G. and Clavin, P. (1979) "Linear Stability Analysis of Non-adiabatic Flames: Diffusional-Thermal Model." Comb. and Flame 35, 139-153.

Joulin, G. and Mitani, T. (1981) "Linear stability analysis of two-reactant flames." Comb. and Flame 40, 235-246.

Joulin, G. and Sivashinsky, G.I. (1983) "On the dynamics of nearly-extinguished non-adiabatic cellular flames." Comb. Sci. and Tech. 31, 75-90.

Kaskan, W.E. (1957) "The dependence of flame temperature on mass burning velocity." 6th Symp. (Int.) on Comb., 134-143.

Madarame, H. (1983) "Thermally induced acoustic oscillations in a pipe." Bulletin of the JSME 26 (214), 603-610.

Margolis, S.B. (1980) "Bifurcation phenomena in burner-stablized premixed flames." Comb. Sci. and Tech. 22, 143-169.

Markstein, G.H. (1964) "Non-steady flame propagation." Pergamon Press, New York.

Matalon, M. and Matkowsky, B.J. (1982) "Flames as gasdynamic discontinuities." Jnl. Fluid Mech. 124, 239-259.

Matkowsky, B.J. and Olagunju, D.O. (1981) "Pulsations in a burner-stabilized premixed plane flame." SIAM J. Appl. Maths. 40 (3), 551-562.

McIntosh, A.C. and Clarke, J.F. (1984a) "A review of theories currently being used to model steady plane flames on flame-holders." Comb. Sci. and Tech. 37, 201-219.

McIntosh, A.C. and Clarke, J.F. (1984b) "Second order theory of unsteady burner-anchored flame with arbitrary Lewis number." Comb. Sci. and Tech. 38, 161-196.

McIntosh, A.C. (1984) "On the coupling of an anchored flame with an acoustic field." Proc. Inst. Acoustics 6 (1).

McIntosh, A.C. (1985a) "On the cellular instability of flames near porous-plug burners." To be published in Jnl. Fluid Mech.

McIntosh, A.C. (1985b) "The effect of upstream acoustic forcing and feedback on the stability and resonance behaviour of anchored flames." Proc. of 10th Int. Colloq. on Dynamics of Expl. and Reactive Syst., Berkeley, California.

Oran, E.S. and Gardner, J.H. (1985) "Chemical-Acoustic Interactions in Combustion Systems." Naval Res. Labs. Memo Rept. NRL.5554.

Pelce, P. and Clavin, P. (1982) "Influence of hydrodynamics and diffusion upon the stability limits of laminar pre-mixed flames." Jnl. Fluid Mech. 124, 219-237.

Roberts, J.P. (1978) "Amplification of an acoustic signal by a laminar premixed gaseous flame." Comb. and Flame 33, 79-83.

Schimmer, H. and Vortmeyer, D. (1977) "Acoustical oscillation in a combustion system with a flat flame." Comb. and Flame 28, 17-24.

Sivashinsky, G.I. (1977) "Diffusional-Thermal theory of Cellular flames." Comb. Sci. and Tech. 15, 137-146.

Sivashinsky, G.I. (1983) "Instabilities, pattern formation and turbulence in flames." Ann. Rev. Fluid Mech. 15. 179-199.

Van Harten, A., Kapila, A.K. and Matkowsky, B.J. (1984) "Acoustic coupling of flames." SIAM Jnl. Appl. Math. 44 (5), 982-995.

Luton College of Higher Education
Park Square, Luton, U.K.

Volumetric Heat Loss $\hat{H}$ non-dimensionalised with respect to adiabatic conditions.

$$\hat{H} = -\left(\frac{u_{o_1}'}{u_{ad}'}\right)^2 \ell_n \left(\frac{u_{o_1}'}{u_{ad}'}\right)^2$$

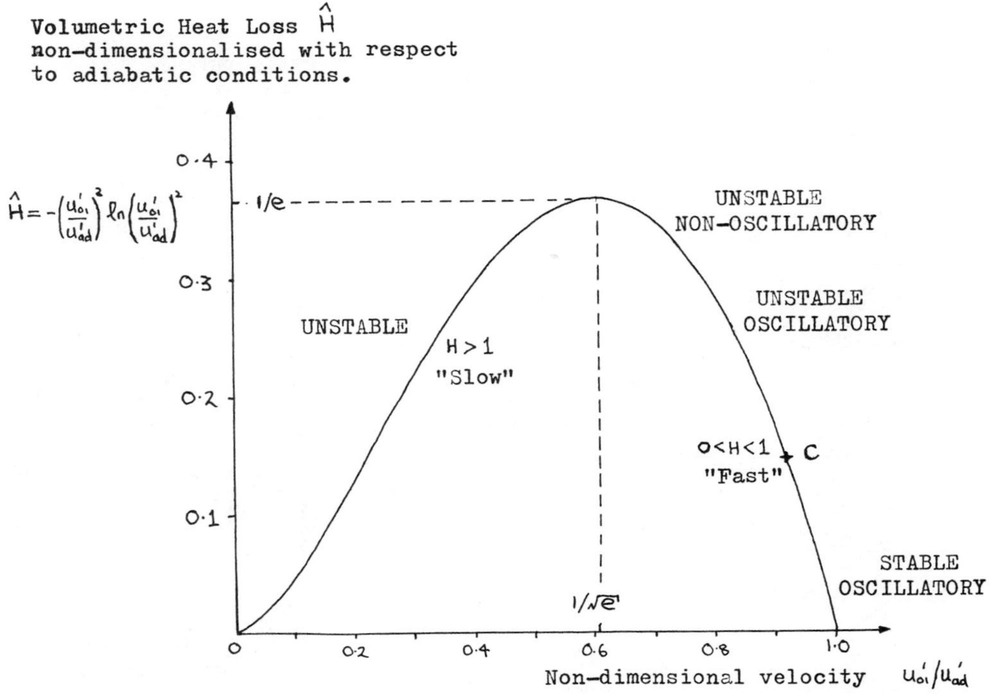

FIG.1   Volumetric Heat Loss parameter $\hat{H}$ versus non-dimensional velocity u_{o_1}'/u_{ad}' Typical stable/unstable regions marked for pulsating mode (after Joulin and Clavin, 1979).

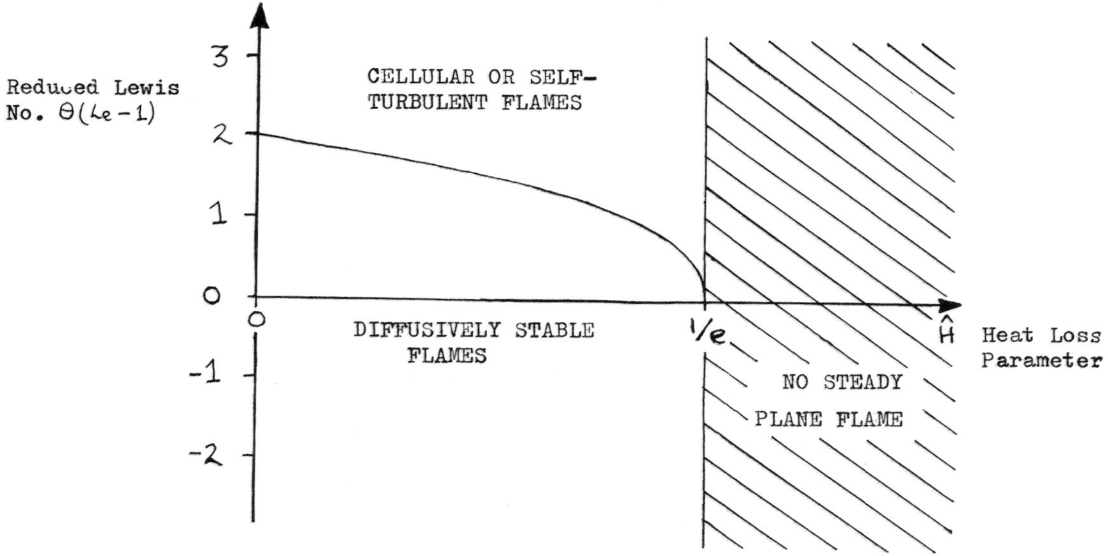

FIG. 2 Neutral two-dimensional stability boundary and stability/instability domains in the $(\Theta(Le-1), \hat{H})$ plane (after Joulin and Sivashinsky, 1983).

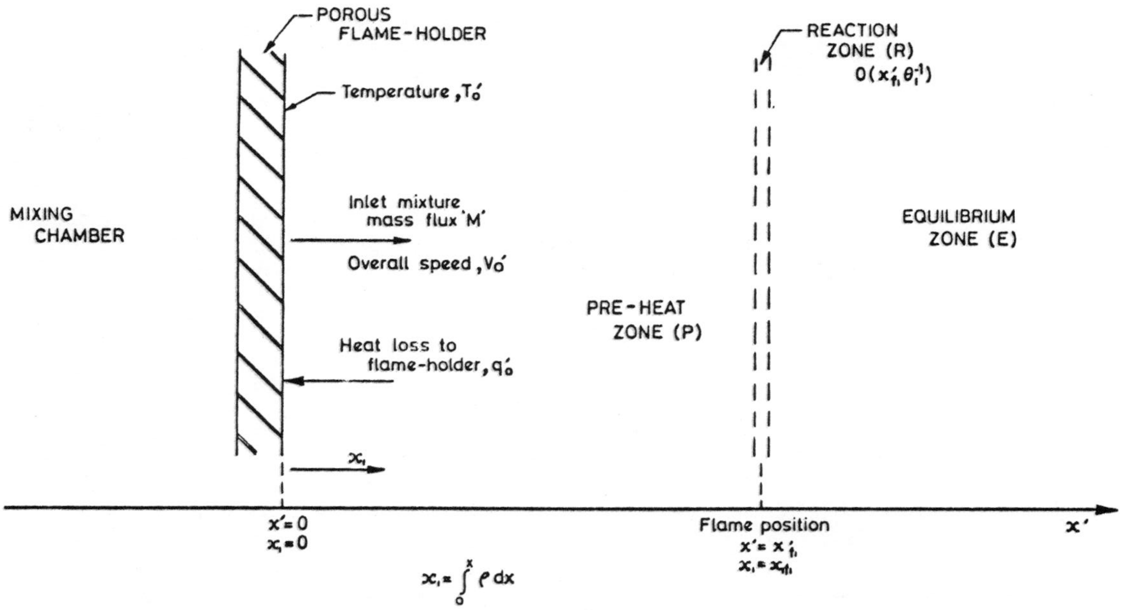

Fig. 3. Schematic of initially plane flame with porous plug flame-holder.

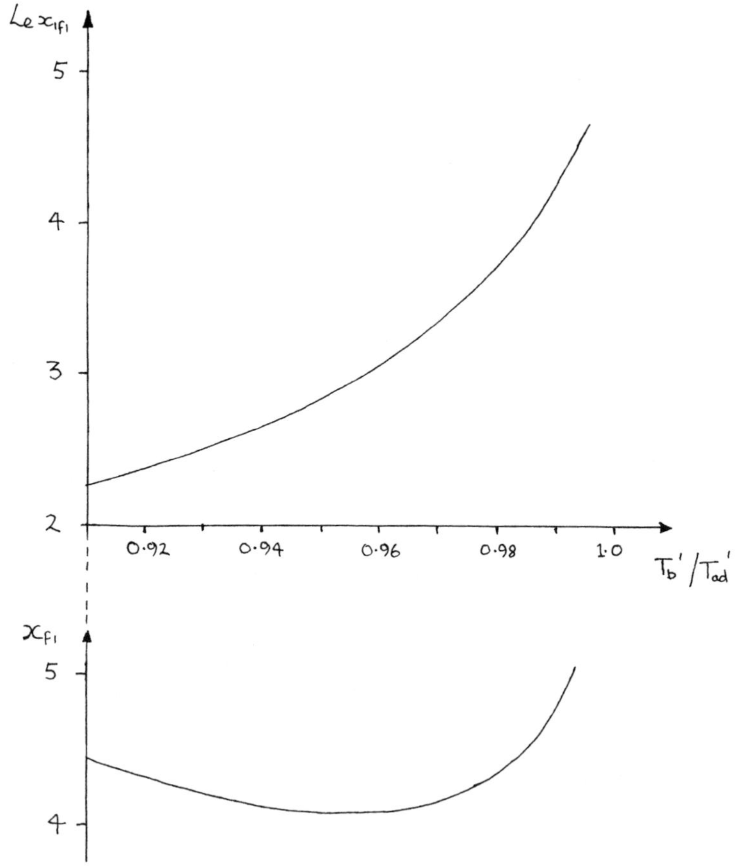

FIG. 4 Variation of mass-weighted stand-off distance ($Le\, x_{1f1}$) and actual stand-off distance (x_{f1}) with final temperature T_b'/T_{ad}' (both in units of $\lambda_{01}'/(\rho_{01}' u_{01}' c_p')$). For both plots $T_{01}'/T_{ad}' = 0.15$ and $\theta_1 T_b'/T_{ad}' = 10$.

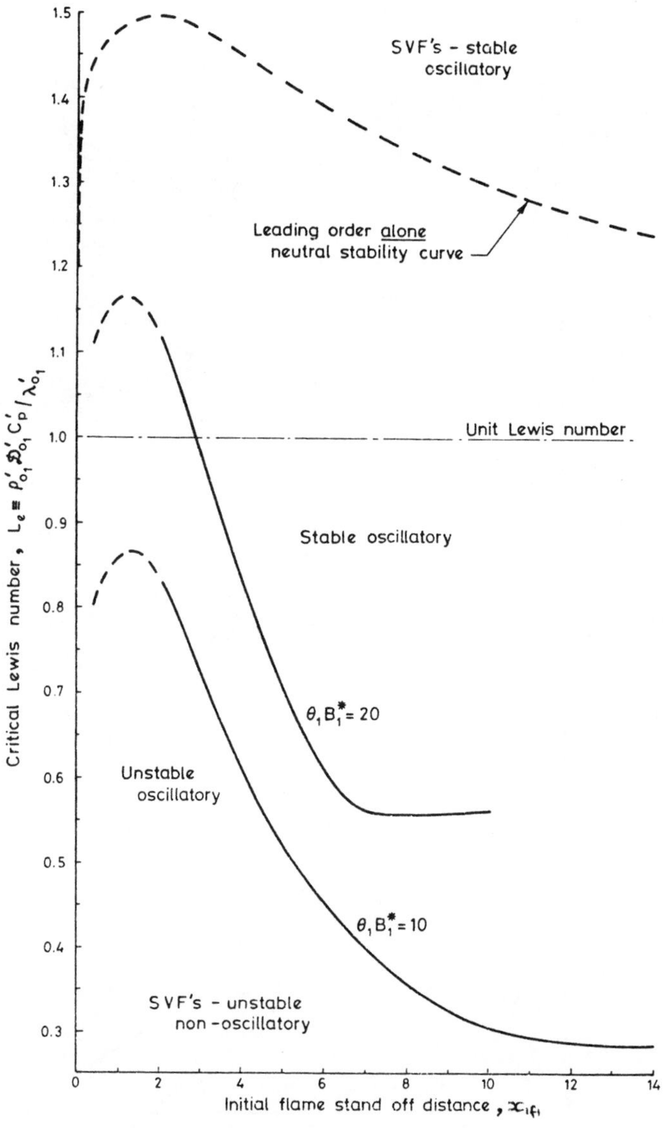

FIG. 5 Neutral stability curves for different values of activation energy $\theta_1 B_1^*$ ($B_1^* \equiv 1 - T'_{o1}/T'_{b1}$). (From McIntosh and Clarke 1984b).

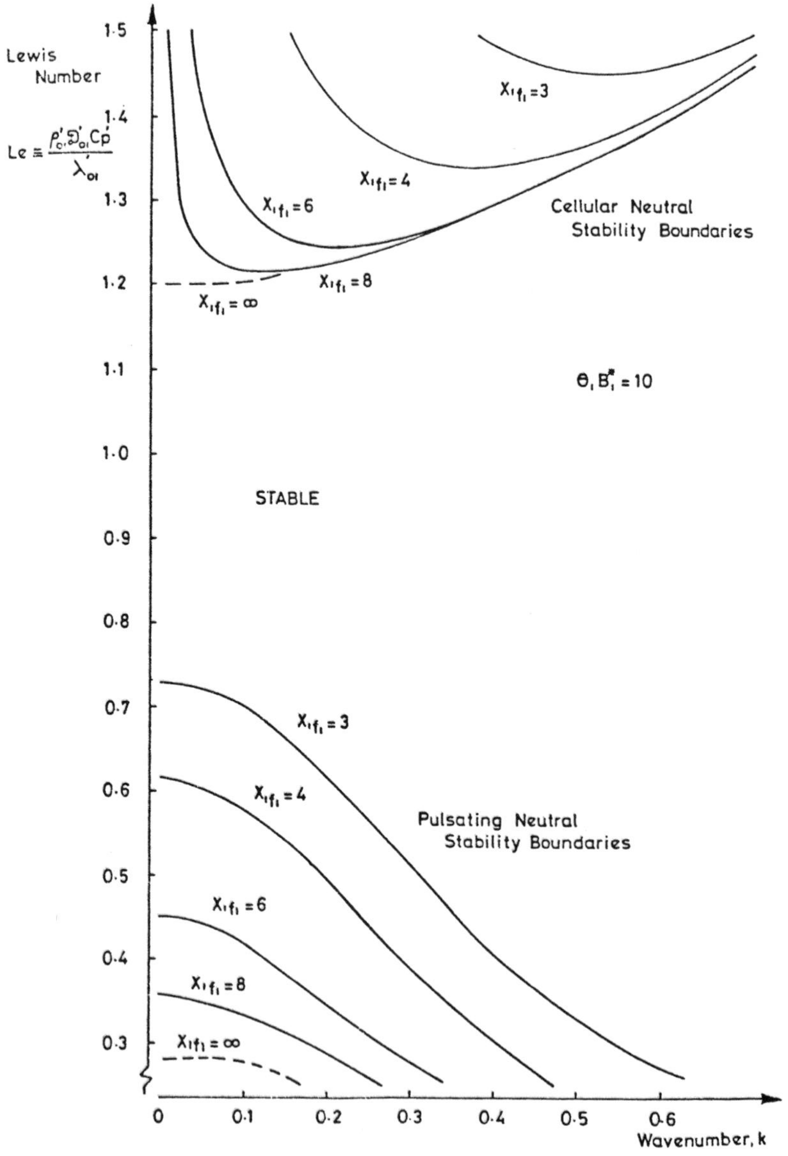

FIG. 6 Two dimensional neutral stability boundaries with zero gas expansion (ie ρ' =const.), activation energy $\theta_1 B_1^*=10$ ($B_1^*=1-T'_{o1}/T'_{b1}$). Stable response predicted within region enclosed by two curves for each x_{1f_1}. (From McIntosh I985a)

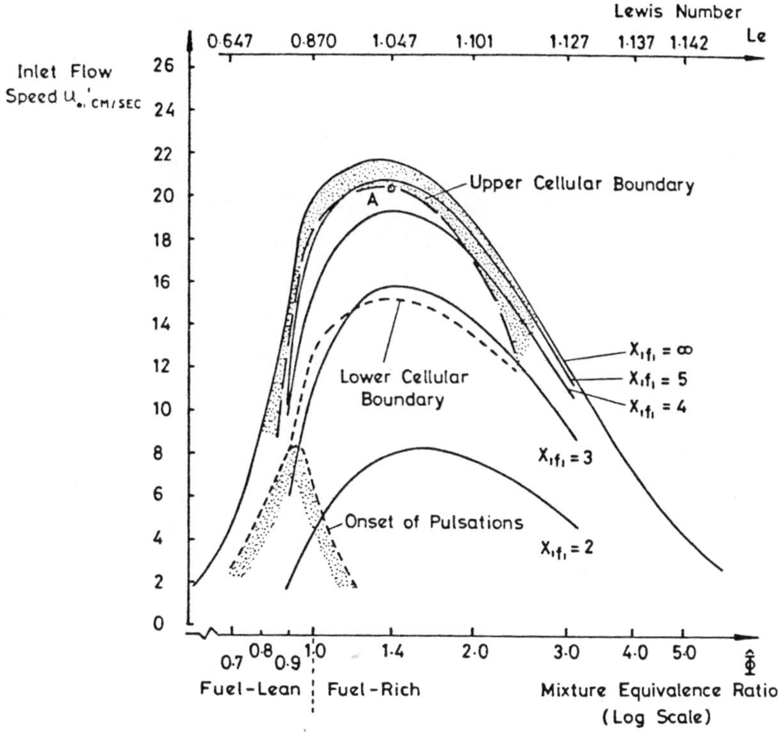

FIG. 7 Typical plot of inlet flow speed against equivalence ratio for various stand-off distances, x_{1f_1} . Curves roughly correspond to propane/air mixture burning on porous plug burner. Shaded regions correspond to cellular boundary (near adiabatic conditions) and onset of pulsations (at low speeds). Upper cellular boundary is shown to be accurate within limitations of approximate, small wavenumber theory, but lower cellular boundary is shown to be inaccurate by same limitations. Point A at $\hat{\Phi} = 1.4$, corresponds to where k_{crit} for cellularity = 0.2 . (From McIntosh 1985a).

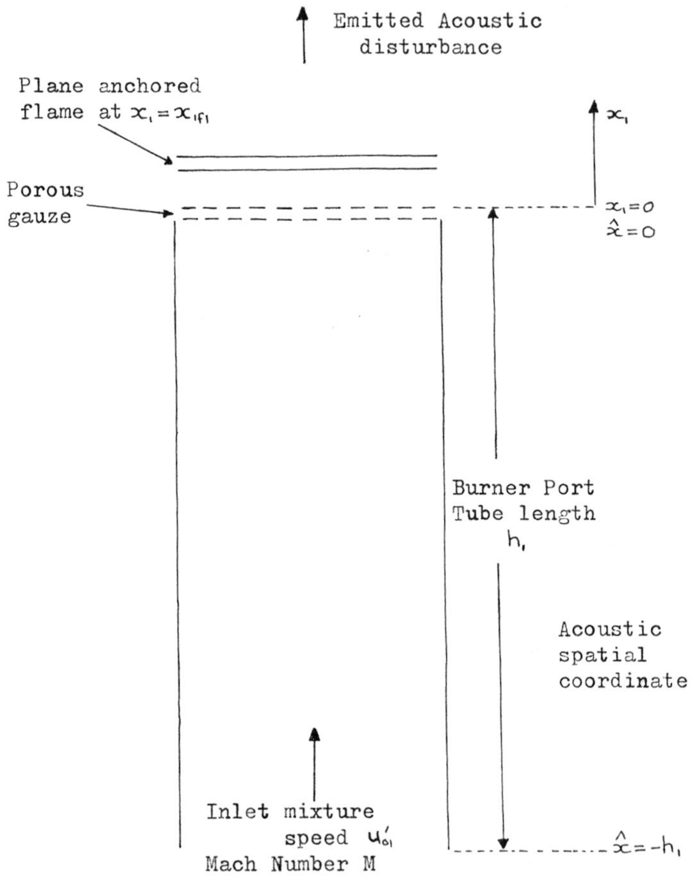

FIG. 8 Schematic of plane flame anchored on a porous gauze above an open burner port tube. (From McIntosh 1985b).

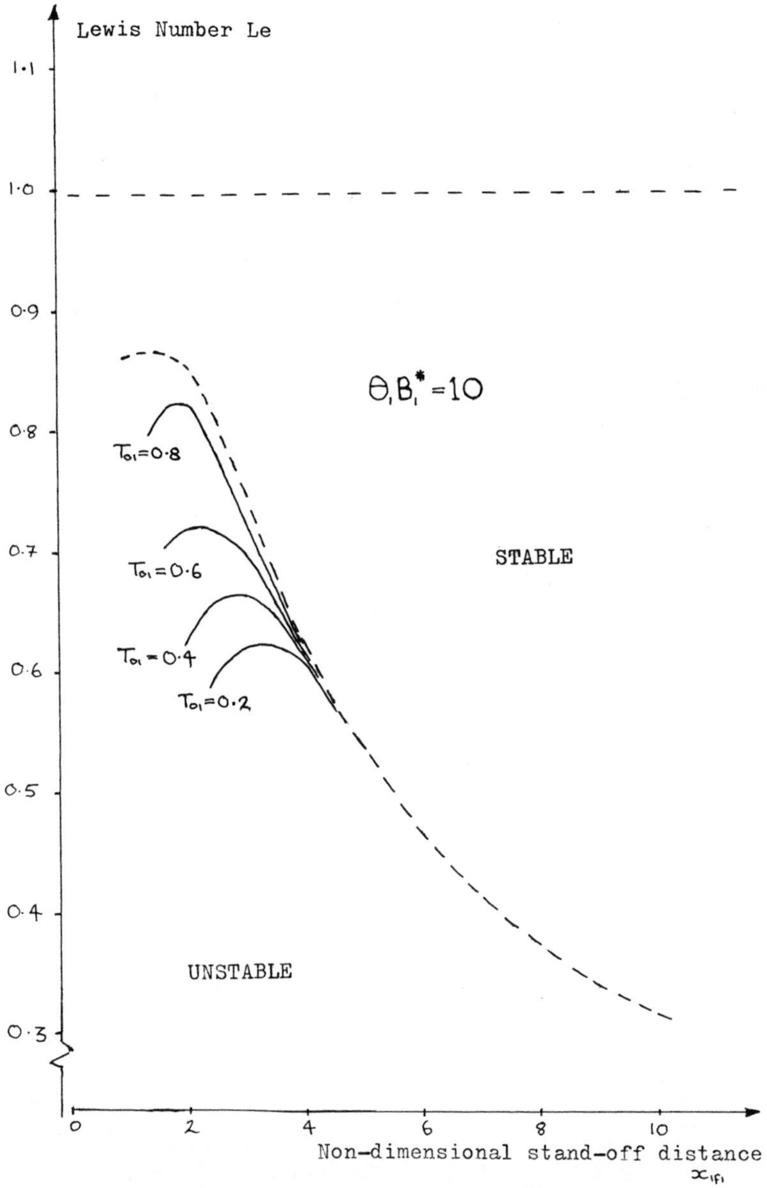

FIG. 9 Critical Lewis Number for neutral stability (with acoustic emission present) plotted against stand-off distance (mass weighted) of flame. The dotted curve is the non-acoustic curve for comparison. The family of curves are for varying temperature ratios (T_{01}'/T_{b1}') across the flame. (From McIntosh 1984)

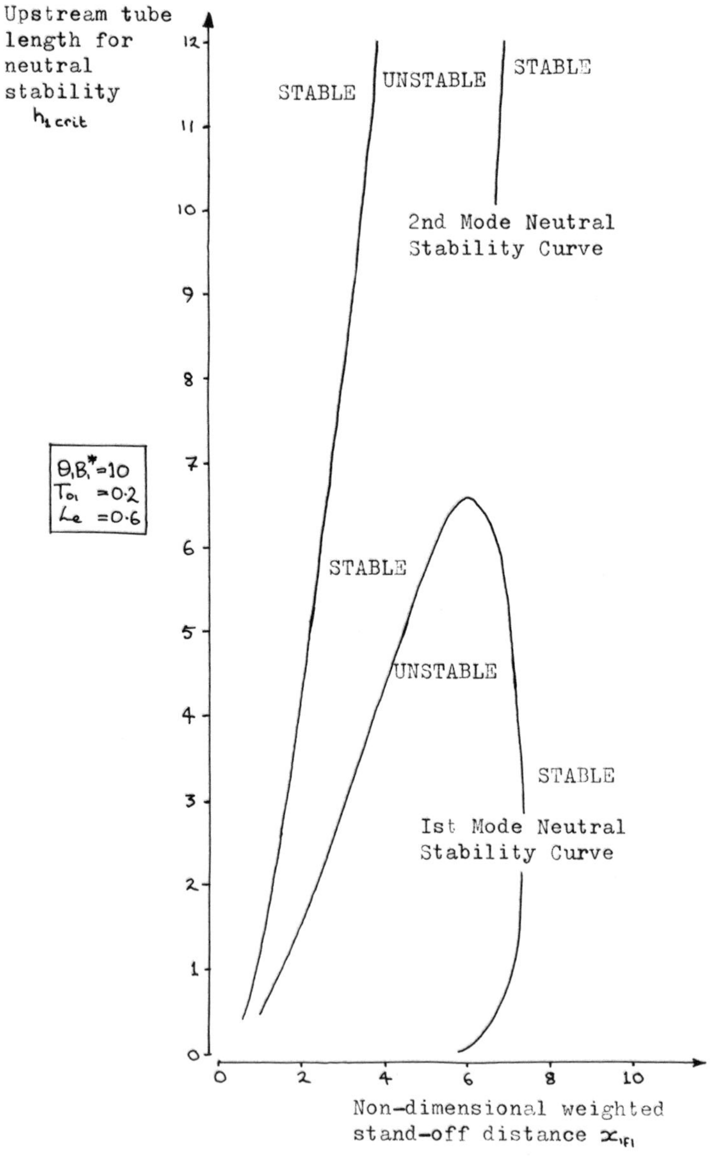

FIG. IO Critical Tube Length h_{1crit} for neutral stability plotted against stand-off distance x_{1f1} (mass weighted) of flame from porous gauze. Two modes are shown. Lewis Number, Le = 0.6 . (From McIntosh I985b).

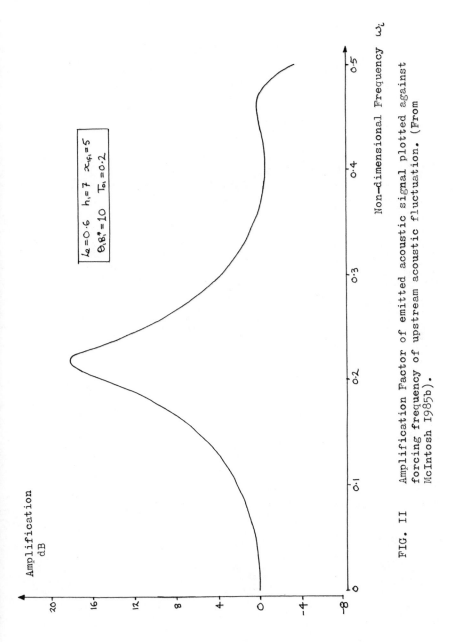

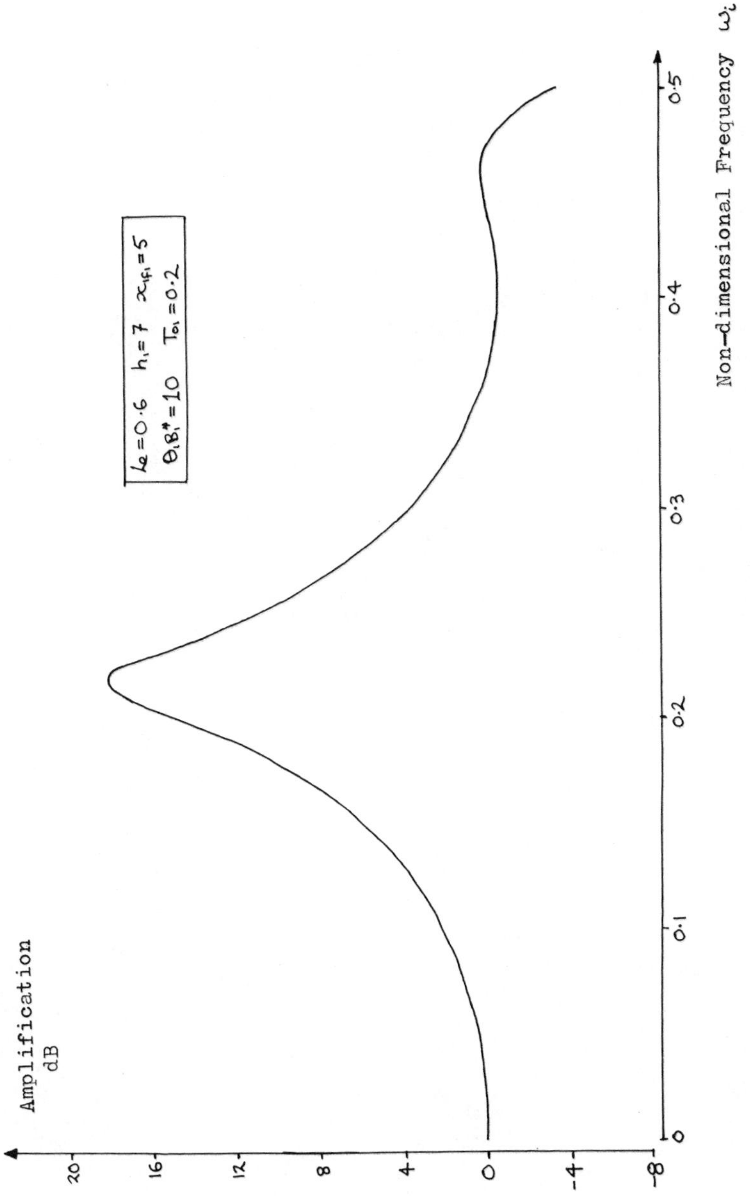

FIG. II Amplification Factor of emitted acoustic signal plotted against forcing frequency of upstream acoustic fluctuation. (From McIntosh 1985b).

Lectures in Applied Mathematics
Volume 24, 1986

THE ASYMPTOTIC ANALYSIS OF IGNITION EVENTS WITH DETAILED KINETICS

David W. Mikolaitis

ABSTRACT. A reaction mechanism patterned after elements of the hydrogen oxidation mechanism is analyzed with activation energy asymptotics. The reaction model incorporates a branched chain cycle, a recombination reaction, a chain breaking reaction, and a degenerate mechanism of much slower rate. This model is used to generate a qualitative description of ignition events near the second explosion limit. Since there are similarities between the model reaction scheme and the complete reaction mechanism for hydrogen oxidation, comparisons between the numerical analysis of the hydrogen system and the asymptotic analysis of the present system can be made under specialized conditions. It is seen that the behavior predicted by the asymptotic analysis is mirrored in the numerical analysis of the detailed system.

INTRODUCTION

Activation energy asymptotics (AEA) have been used for a variety of different problems concerning ignition processes. The analysis of Kassoy (1975) on the Frank-Kamenetskii problem is the pioneering effort at rational asymptotic representations of ignition solutions. These single active species problems are of some usefulness in understanding thermal explosions, but they cannot give any insight into the role of chemical kinetics in more realistic situations. Kapila (1978) as well as Sen and Law (1982) have analyzed a generalization of the Zeldovich-Liñán model (Zeldovich, 1948 and Liñán 1971) that has seen application in deflagration analyses (Hocks et al, 1981, Seshadri and

Peters, 1983, and Tam and Ludford, 1984, for example). Birkan and Kassoy (1983 and 1985) have used the Zeldovich-Liñán model for chain branching and termination and have added other features such as equilibrium dissociation effects.

These analyses are of interest not only for their predictions, but also for their mathematical methodology. It is unfortunate that many of the most interesting phenomena associated with ignition cannot be studied with such simple models. For example, the existence of the second explosion limit for many explosive systems is a manifestation of a competition between a chain branching and a chain breaking reaction. If the chain breaking reaction is faster, then the mixture is effectively inert. The explosion peninsula shown by many different reactant-oxidizer mixtures shows that the transition from explosive to non-explosive is very abrupt. Such behavior is well beyond the scope of the simple models that have been previously analyzed. Here we will analyze a fairly complicated model based on the chemistry of hydrogen oxidation. From this we will deduce the second explosion limit in its classical sense as a curve that separates regions of rapid reaction from regions of slow reaction. In this analysis of the evolution of the chemical state of the reacting mixture we will see that the second explosion limit is not a curve of zero thickness but rather a very thin region of transition. In this thin region a very small change in initial conditions can cause extraordinarily large changes in the time to thermal explosion.

REACTION MODEL

A reaction model that is sufficiently detailed for our purposes is given by

(1) $$A + B \rightarrow I_2 + I_4$$

(2) $$A + I_1 \rightarrow P + I_2$$

(3) $$B + I_2 \rightarrow I_1 + I_3$$

(4) $$A + I_3 \rightarrow I_1 + I_2$$

(5) $$I_1 + I_2 + M \rightarrow P + M$$

(6) $B + I_2 + M \rightarrow I_4 + M$

(7) $I_2 + I_4 \rightarrow I_1 + I_1$

analogous to the oxidation mechanism for hydrogen. If a strict analogy is sought then A and B can be taken as molecular hydrogen and oxygen, P is water, and the intermediate species I_1, I_2, I_3, I_4 are OH, H, O, HO_2. This mechanism cannot be complete since the species H_2O_2 (hydrogen peroxide) has not been included. It is hoped that reaction (7) can effectively model the degenerate (low temperature) mechanisms involving HO_2 and H_2O_2. Also, only one termination reaction has been considered. In very rich mixtures the reaction $H + H + M \rightarrow H_2 + M$ is an important part of the kinetics. This reaction is not a part of the model.

There are many aspects of the oxidation mechanism for hydrogen that are a part of the model, however. The main chain branching mechanism is given by reactions (2), (3), and (4). The most important initiation reaction is given by reaction (1). The essential competition responsible for the existence of the second explosion limit is the competition between reactions (3) and (6). All these features may allow at least a qualitatively correct description of the chemistry in the neighborhood of the second explosion limit, especially when the mixture is slightly on the "explosive" side of the curve where the slow reaction channels involving HO_2 and H_2O_2 are of lesser importance.

Since there are obvious difficiencies in using this model for a complete description of hydrogen oxidation, we will not attempt detailed numerical comparisons of our results with extensive numerical studies of hydrogen oxidation. We do feel, however, that there is a great enough similarity in the basic mechanisms to offer at least qualitative comparison and also the rate data for hydrogen-oxygen reactions can form a basis for assigning asymptotic orders to the sizes of the various rate constants in the modeling.

Keeping in mind that we are using hydrogen-oxygen data only as a guide, we see that the activation energies of reactions (1) to (4) are all large and of the same order of magnitude. The activation energy of reaction (3), in non-dimensional form, will be the large parameter in the problem and denoted by θ . The reference temperature throughout this analysis will be the nominal room temperature 293°K. θ is therefore $E_3/R(293°K)$. The initiation reaction (1) has an even higher activation energy, but it is only about 3.4 times the activation energy of reaction (3). In our model we will take the distinguished limit that the activation energy of reaction (1) is some O(1) number c times θ where c is greater than 1.

The activation energies of reactions (2) and (4) are also of the same order of magnitude as the activation energy of reaction (3). These activation energies in the hydrogen oxidation analogy are .307 and .530 times the activation energy of reaction (3). We will take these activation energies as $a\theta$ and $b\theta$ where $a<b<1$. The activation energy of the termination reaction (5) is identically zero in the rate constant data and the activation energies of reactions (6) and (7) are much smaller than those of (1) to (4). We will take the activation energy of (5) to be zero and those of (6) and (7) to be the O(1) numbers E_1 and E_2 when they are non-dimensionalized.

Appropriate asymptotic scales for the pre-exponential constants can be found using hydrogen oxidation reactions as a guide. If the constant for reaction (3) is absorbed into the time scaling then the appropriate non-dimensional preexponential constants are: $A_1 = .295$, $A_2 = .117$, $A_3 = 1$, $A_4 = .029$, $A_5 = .367$, $A_6 = T_B \exp(E_1/T_B)/p$ and $A_7 = 1.349$ where T_B is the value of the non-dimensionalized temperature where the competing reactions (3) and (6) have identical rates and p is the pressures given in atmospheres. From these numbers we can conclude that the so-called fast recombination limit is the most physically realistic asymptotic limit.

The value of θ using 293°K as the reference temperature is approximately 28.83, not an especially large number. Indeed, as the temperature of the mixture increases the "largeness" of θ is diminished. In the neighborhood of the second explosion limit for hydrogen-oxygen mixtures the value of θ/T is between 10 and 13. As θ/T is the parameter that must be large we have some confidence that meaningful results can be obtained at least through thermal induction. As the resulting thermal explosion occurs, we do not think that meaningful numerical results are possible but the asymptotic solutions may be of some qualitative interest. We will only pursue the problem up to thermal initiation.

THE GOVERNING EQUATIONS

As will be seen eventually, there will be only exponentially small deviations in concentrations and temperature until thermal induction begins. During thermal induction there will be reaction balancing among the intermediate species so that the entire mechanism behaves, in effect, like a simple $A \to B$ mechanism. We will therefore ignore confinement effects as they will be unimportant until appreciable quantities of product are made. Only then will the total number of moles change appreciably.

Under this restriction, assuming adiabatic conditions, no wall reactions, and homogeneous initial conditions, the governing equations are

$$\frac{dT}{dt} = q_1R_1 + q_2R_2 + q_3R_3 + q_4R_4 + q_5R_5 + q_6R_6 + q_7R_7 \tag{8}$$

$$\frac{d[A]}{dt} = -R_1 - R_2 - R_4 \tag{9}$$

$$\frac{d[B]}{dt} = -R_1 - R_3 - R_6 \tag{10}$$

$$\frac{d[P]}{dt} = R_2 + R_5 \tag{11}$$

$$\frac{d[I_1]}{dt} = -R_2 + R_3 + R_4 - R_5 + 2R_7 \tag{12}$$

$$\frac{d[I_2]}{dt} = R_1 + R_2 - R_3 + R_4 - R_5 - R_6 - R_7 \tag{13}$$

(14) $$\frac{d[I_3]}{dt} = R_3 - R_4$$

(15) $$\frac{d[I_4]}{dt} = R_6 - R_7$$

where the q's are the heat releases and the R's are given by

(16) $R_1 = A_1\ [A][B]\ \exp(-c\theta/T)$

(17) $R_2 = A_2\ [A][I_1]\ \exp(-a\theta/T)$

(18) $R_3 = A_3\ [B][I_2]\ \exp(-\theta/T)$

(19) $R_4 = A_4\ [I_3]\ T\ \exp(-b\theta/T)[A]$

(20) $R_5 = A_5\ p\ [I_1][I_2]/T^3$

(21) $R_6 = A_6\ p\ [B][I_2]\ \exp\ (-E_1/T - \theta/T_b)/T$

(22) $R_7 = A_7\ [I_2][I_4]\ \exp\ (-E_2/T)$.

The initial conditions are

(23) $T(0) = T_0$

(24) $[A](0) = A_0$

(25) $[B](0) = B_0$

(26) $[I_i](0) = 0, \quad i = 1, 2, 3, 4$

(27) $[P](0) = 0$.

We will be interested in solutions near the second explosion limit so that we will require that $T_B \sim T_0$. Our interest here is two-fold. First of all, in many applications it is convenient to have an ignition temperature. The detonation analysis of Kapila, Matkowsky, and van Harten (1983), for example, incorporates an ignition temperature as an $O(1)$ parameter in their study. It appears in such fundamentally important formulas as the wave speed formula. The simple $A \to B$ mechanism that they employed required the ad hoc addition of a cut-off temperature. In order to fully justify their study it becomes imperative that a physical mechanism be found for the cut-off temperature and a method for calculating that temperature must also be found. The original ideas of Belles (1959) on Detonation limits indirectly implies that the second explosion limit can be considered as an effective ignition temperature.

Another important reason for looking at the second explosion limit is that it has been extensively studied. We will use the results in Dougherty and Rabitz (1980) as representative and show that our results are qualitatively similar.

INITIATION STAGE

We can restrict attention to conditions near the second explosion limit by consdidering problems where $T_0 \sim T_B$. We will therefore write

(28) $T_B = T_o + a_o/\theta$, $a_o = 0(1)$

and also make the definitions

(29) $D_1 = A_1 [A]_0[B]_0$

(30) $D_2 = A_2[A]_0$

(31) $D_3 = A_3[B]_0$

(32) $D_4 = A_4T_0 [A]_0$

(33) $D_5 = A_5p_0/T_0^3$

(34) $D_6 = A_6p_0 [B]_0 \exp(-E_1/T_0 + a_0/T_0^2)/T_0$

(35) $D_7 = A_7 \exp(-E_2/T_0)$

in order to notationally simplify the reaction terms. All the D's are 0(1) numbers.

For small times we can write

(36) $\tau = \delta_t t$

(37) $[I_j] = \delta_j i_j$, $j = 1, 2, 3, 4$

(38) $[P] = \delta_p p$

where the δ's are gauge functions. We most now deduce the proper δ's.

At t = 0 the only reaction that can take place is the initiation reaction (1), so (1) must be important in the initial growth of species I_2 and I_4. $[I_3]$ can only grow through (3) and $[I_1]$ is first produced by reaction (3) as $[I_2]$ and $[I_3]$ grow. Species I_1 however is quickly consumed by reaction (2) so that

(39) $\delta_t \delta_2 \sim e^{-c\theta/T_0} \sim \delta_t \delta_4$

(40) $\delta_t \delta_1 \sim \delta_2 e^{-\theta/T_0} \sim \delta_1 e^{-a\theta/T_0}$.

The fastest P producing reaction in this regime is (2) so that

(41) $\delta_t \delta_p = \delta_1 e^{-a\theta/T_0}$

All of this implies that the proper scalings during the early moments of reaction are

(42) $\delta_t = \exp(-a\theta/T_0)$

(43) $\delta_1 = \delta_3 = \delta_p = \exp\left(-(c + 1 - 2a)\ \theta/T_0\right)$

(44) $\delta_2 = \delta_4 = \exp\left(-(c - a)\ \theta/T_0\right)$,

The leading order governing equations are

(45) $\frac{di_1}{d\tau} = D_3 i_2 - D_2 i_1$

(46) $\frac{di_2}{d\tau} = D_1$

(47) $\frac{di_3}{d\tau} = D_3 i_2$

(48) $\frac{di_4}{d\tau} = D_1$

(49) $\frac{dp}{d\tau} = D_2 i_1$

(50) $\frac{dT}{d\tau} = \frac{d[A]}{d\tau} = \frac{d[B]}{d\tau}$ = exponentially small terms (e.s.t.).

Although it is an easy task to calculate the small changes in T, [A], and [B], they will play no role in the solution. We will not report those calculations.

The solutions are straightforward and are

(51) $i_1 = \frac{D_1 D_3}{D_2^2}\left(D_2\tau - 1 + \exp(-D_2\tau)\right)$

(52) $i_2 = D_1\tau$

(53) $i_3 = D_1 D_3 \tau^2/2$

(54) $i_4 = D_1 \tau$

(55) $$p = \frac{D_1 D_3}{D_2^2}\left(1 - D_2\tau + \frac{1}{2} D_2^2 \tau^2 - \exp(-D_2\tau)\right)$$

As $\tau \to \infty$ we see that

$$i_1 \to D_1 D_3 \tau/D_2,\ i_2 \to D_1\tau,\ i_3 \to D_1 D_3 \tau^2/2$$

$$i_4 \to D_1\tau,\ p \to D_1 D_3 \tau^2/2\ .$$

As τ increases the rates of the various reactions will change. Reactions (2), (3), and (6) will increase linearly in time and reactions (4), (7), and (5) will increase quadratically. Reaction (1) approaches a constant value. For sufficiently large τ there must be a new ordering of the reaction rates so that a different set of leading order governing equations become valid. This will occur when either (2) becomes on the order of (1) or when (4) becomes on the order of (2). A very elementary analysis of the initiation solution shows that (4) becomes of the order of (2) well before (2) can become on the order of (1) since 1-a is O(1) greater than b-a. These reactions will be of similar rates when $t \sim \exp(b\ \theta/T_0)$.

INDUCTION TO BRANCHED CHAIN EXPLOSION

The previous solutions coupled with the estimate of when they are no longer valid indicate that the proper scalings in the next time regime are

(56) $\overline{\tau} = \exp(-b\theta/T_0)t$

(57) $[I_1] = \exp\left(-(c + 1 - a - b)\theta/T_0\right)\overline{i}_1$

(58) $[I_2] = \exp\left(-(c - b)\theta/T_0\right)\overline{i}_2$

(59) $[I_3] = \exp\left(-(c + 1 - 2b)\theta/T_0\right)\overline{i}_3$

(60) $[I_4] = \exp\left((c - b)\theta/T_0\right)\bar{i}_4$

(61) $[P] = \exp\left(-(c + 1 - 2b)\theta/T_0\right)\bar{P}$

Under these scalings the leading order governing equations are

(62) $0 = -D_2\bar{i}_1 + D_3\bar{i}_2 + D_4\bar{i}_3$

(63) $\dfrac{d\bar{i}_2}{d\bar{\tau}} = D_1$

(64) $\dfrac{d\bar{i}_3}{d\bar{\tau}} = D_3\bar{i}_2 - D_4\bar{i}_3$

(65) $\dfrac{d\bar{i}_4}{d\bar{\tau}} = D_1$

(66) $\dfrac{d\bar{P}}{d\bar{\tau}} = D_2\bar{i}_1$

(67) $\dfrac{dT}{d\bar{\tau}} = \dfrac{d[A]}{d\bar{\tau}} = \dfrac{d[B]}{d\bar{\tau}} = \text{e.s.t.}$

The solutions are

(68) $\bar{i}_1 = D_1D_3\left(-1 + 2D_4\bar{\tau} + \exp(-D_4\bar{\tau})\right)/D_2D_4$

(69) $\bar{i}_2 = D_1\bar{\tau}$

(70) $\bar{i}_3 = D_1D_3\left(-1 + D_4\,\bar{\tau} + \exp\,(-D_4\bar{\tau})\right)/D_4^2$

(71) $\bar{i}_4 = D_1\bar{\tau}$

(72) $\bar{P} = D_1D_3\,\left(1/D_4 - \bar{\tau} + D_4\,\bar{\tau}^{\,2} - \exp\,(-D_4\bar{\tau})/D_4\right)/D_4.$

after asymptotic matching to the previous stage. As $\bar{\tau}$ goes to infinity we see that now all the intermediate species grow linearly in time and only the product grows quadratically. Therefore for large enough time either the linear growth in the rates of reaction (2), (3), (4), and (6) will catch up to the rate of reaction (1) or the quadratic growth in reaction (7) will overcome the linear growth of reactions (2), (3), (4), and (6). Using these solutions and the scalings for this time regime we see that the first possiblility is the one that is achieved and this is accomplished when t is $O(\exp \theta/T_0)$.

THE BRANCHED CHAIN EXPLOSION

It is an easy task to see that the previous solutions demand that the proper scalings in the time regime appropriate for $t \sim O(\exp\theta/T_0)$ are

$$\tilde{\tau} = \exp(-\theta/T_0)t \tag{73}$$

$$[I_1] = \exp\left(-(c - a)\theta/T_0\right)\tilde{i}_1 \tag{74}$$

$$[I_2] = \exp\left(-(c - 1)\theta/T_0\right)\tilde{i}_2 \tag{75}$$

$$[I_3] = \exp\left(-(c - b)\theta/T_0\right)\tilde{i}_3 \tag{76}$$

$$[I_4] = \exp\left(-(c - 1)\theta/T_0\right)\tilde{i}_4 \tag{77}$$

$$[P] = \exp\left(-(c - 1)\theta/T_0\right)\tilde{P} \tag{78}$$

so that the leading order governing equations become

$$0 = -D_2\tilde{i}_1 + D_3\tilde{i}_2 + D_4\tilde{i}_3 \tag{79}$$

$$\frac{d\tilde{i}_2}{d\tilde{\tau}} = D_1 + D_2\tilde{i}_1 - D_3\tilde{i}_2 + D_4\tilde{i}_3 - D_6\tilde{i}_2 \tag{80}$$

$$0 = D_3\tilde{i}_2 - D_4\tilde{i}_3 \tag{81}$$

$$\frac{d\tilde{i}_4}{d\tilde{\tau}} = D_1 + D_6\tilde{i}_2 \tag{82}$$

$$\frac{d\tilde{P}}{d\tilde{\tau}} = D_2\tilde{i}_1 \tag{83}$$

$$\frac{dT}{d\tilde{\tau}} = \frac{d[A]}{d\tilde{\tau}} = \frac{d[B]}{d\tilde{\tau}} = \text{e.s.t.} \tag{84}$$

After elementary manipulations we find that equation (80) becomes

(85) $$\frac{d\tilde{i}_2}{d\tilde{\tau}} = D_1 + (2D_3 - D_6)\tilde{i}_2 \, .$$

When $2D_3 - D_6 > 0$ then I_2 grows exponentially but if $2D_3 - D_6 < 0$ then it approaches a constant value. This critical condition $2D_3 = D_6$ is attained when the rate of reaction (6) is precisely twice the rate of reaction (3). This is the classical calculation of the second explosion limit in hydrogen-oxygen mixtures.

When the initial conditions are very close to the classical second explosion limit but still on the "explosive" side of the curve, we would find that the value of $\mathcal{D}$ defined as $2D_3 - D_6$ would be a small positive number. Under such conditions the growth in intermediate species concentrations is not too large but for conditions not very far removed at all from the second explosion limit the growth rates become very large. This very rapid change in growth rate is much quicker than that associated with Arrhenius kinetics only. Here we will see that an $O(1/\theta)$ change in initial conditions will be able to cause an exponentially large change in reaction rate. In Arrhenius kinetics with large activation energy such changes in initial conditions would only account for $O(1)$ changes in rate for a simple $A \to B$ type mechanism.

Such considerations in conjunction with the fast wave analyses of Clarke (1983 a, b) where the chemical structure of the detonation wave is seen to be a "convected explosion" gives additional support to Belles' model of detonation limits as calculated from the second explosion limit. Additional work by Kassoy and Clarke (1985) lends further support to the idea that the chemistry behind the shock can be profitably considered as a "convected explosion".

The solutions for $\mathcal{D} \neq 0$ are

(86) $$\tilde{i}_1 = 2D_1D_3\,(\exp(\mathcal{D}\tilde{\tau}) - 1)/D_2\mathcal{D}$$

(87) $$\tilde{i}_2 = D_1\,(\exp\,(\mathcal{D}\tilde{\tau}) - 1)/\mathcal{D}$$

(88) $$\tilde{i}_3 = D_1D_3\,(\exp\,(\mathcal{D}\tilde{\tau}) - 1)/D_4\mathcal{D}$$

(89) $$\tilde{i}_4 = D_1D_6\,(\exp\,(\mathcal{D}\tilde{\tau}) - \tilde{\tau}\mathcal{D}^{-1} - 1) + D_1\,\tilde{\tau}$$

(90) $\tilde{P} = 2D_1D_3\left(\exp(\mathcal{D}\tilde{\tau}) - \tilde{\tau}\mathcal{D}^{-1} - 1\right)$.

I) $2D_3 - D_6 < 0$, The Slow Branch

When the initial conditions are slightly on the "non-explosive" side of the classical second explosion limit curve we see that the concentrations of intermediate species I_1, I_2, and I_3 all approach constant values on the $t \sim 0\,(\exp \theta/T_0)$ time scale. I_4 continues to grow linearly in time so that eventually reactions (1) and (7) will have similar rates. This will occur when $t \sim 0\,(\exp\,(c-1)\theta/T_0)$. The previous solutions demand that the proper scalings in this time regime are

(91) $\hat{\tau} = \exp\left(-(c-1)\theta/T_0\right)t$

(92) $[I_1] = \exp\left(-(c-a)\theta/T_0\right)\hat{i}_1$

(93) $[I_2] = \exp\left(-(c-1)\theta/T_0\right)\hat{i}_2$

(94) $[I_3] = \exp\left(-(c-b)\theta/T_0\right)\hat{i}_3$

(95) $[I_4] = \exp\left(-\theta/T_0\right)\hat{i}_4$

(96) $[P] = \exp\left(-\theta/T_0\right)\hat{P}$

so that the leading order governing equations become

(97) $0 = -D_2\hat{i}_1 + D_3\hat{i}_2 + D_4\hat{i}_3 + 2D_7\hat{i}_2\hat{i}_4$

(98) $0 = D_1 + D_2\hat{i}_1 - D_3\hat{i}_2 + D_4\hat{i}_3 - D_6\hat{i}_2 - D_7\hat{i}_2\hat{i}_4$

(99) $0 = D_3\hat{i}_2 - D_4\hat{i}_3$

(100) $\dfrac{d\hat{i}_4}{d\hat{\tau}} = D_1 + D_6\hat{i}_2 - D_7\hat{i}_2\hat{i}_4$

(101) $\dfrac{d\hat{P}}{d\hat{\tau}} = D_2\hat{i}_1$

(102) $\dfrac{dT}{d\hat{\tau}} = \dfrac{d[A]}{d\hat{\tau}} = \dfrac{d[B]}{d\hat{\tau}} = \text{e.s.t.}$

The solution for $\hat{i}_4$ after a symptotic matching is given implicitly by

(103) $\hat{i}_4 + (\mathcal{D} + D_6)\ \ell n|1 + 2D_7\hat{i}_4/(\mathcal{D} - D_6)|/2 = 2D_1\hat{\tau}$

and the other concentrations in terms of $\hat{i}_4$ are

(104) $\hat{i}_2 = -D_1/(\mathcal{D} + D_7\hat{i}_4)$

(105) $\hat{i}_3 = D_3\hat{i}_2/D_4$

(106) $\hat{i}_1 = D_3\hat{i}_2/D_2 + D_4\hat{i}_3/D_2 + 2D_7\hat{i}_2\hat{i}_4/D_2$.

As i_4 goes to $-\mathcal{D}/D_7$ we see that the other intermediate species concentrations become unbounded. This occurs when

(107) $\hat{\tau}_e \simeq \dfrac{-\mathcal{D}}{2D_1D_7} + \dfrac{\mathcal{D} + D_6}{4\ D_1D_7}\ \ell n|1 - \dfrac{2\mathcal{D}}{\mathcal{D} - D_6}|$.

Of course, the growth in the intermediate species concentrations cannot continue unabated. This type of growth rate is eventually terminated when the rate of reaction (7) becomes of the order of the reactions (2) through (6). This implies that the proper scalings in the next time regime are

(108) $t = \exp\left((c - 1)\theta/T_0\right)\hat{\tau}_e + \exp(\theta/T_0)\ \tau^*$

(109) $[I_1] = \exp(-\theta/T_0)i_1^*$

(110) $[I_2] = \exp(-a\theta/T_0)i_2^*$

(111) $[I_3] = \exp\left(-(1 + a - b)\theta/T_0\right)i_3^*$

(112) $[I_4] = \exp(-\theta/T_0)i_4^*$

with the resulting leading order equations

(113) $0 = -D_2i_1^* + D_3i_2^* + D_4i_3^* - D_5i_1^*i_2^* + 2D_7i_2^*i_4^*$

(114) $\dfrac{di_2^*}{d\tau^*} = D_2i_1^* - D_3i_2^* + D_4i_3^* - D_5i_1^*i_2^* - D_6i_2^* - D_7i_2^*i_4^*$

(115) $0 = D_3i_2^* - D_4i_3^*$

(116) $0 = D_6i_2^* - D_7i_4^*i_2^*$

These equations do admit exact solution (in implicit form) but the most important aspect of these solutions is that the long time behavior is given by

(117) $i_{1_*}^{\ *} \to D_3/D_5$
(118) $i_{2_*}^{\ *} \to D_2D_3/(D_5(D_3 + 2D_6))$
(119) $i_{3_*}^{\ *} \to D_2D_4/(D_5(D_3 + 2D_6))$
(120) $i_4^{\ *} \to D_6/D_7$

II) $2D_3 - D_6 > 0$, The Fast Branch

If $\mathfrak{D} > 0$ then the solutions display exponential growth as mentioned earlier. This growth cannot continue forever and indeed reaction (7) eventually becomes of the order of reaction (1). In this regime, attained when

(121) $\tilde{\tau} \sim (c - 2)\ \theta/ T_0$,

the appropriate scalings are

(122) $t = (c - 2)\theta \exp(\theta/T_0)/\ T_0 + \hat{\tau} \exp(\theta/T_0)$

(123) $[I_1] = \exp(-(2 - a)\ \theta/T_0)\ \hat{i}_1$

(124) $[I_2] = \exp(-\theta/T_0)\hat{i}_2$

(125) $[I_3] = \exp(-(2 - b)\theta/T_0)\hat{i}_3$

(126) $[I_4] = \exp(-\theta/T_0)\hat{i}_4$

(127) $[P] = \exp(-\theta/T_0)\hat{P}$

with leading order equations

(128) $0 = -D_2\hat{i}_1 + D_3\hat{i}_2 + D_4\hat{i}_3 + 2D_7\hat{i}_2\hat{i}_4$

(129) $\dfrac{d\hat{i}_2}{d\hat{\tau}} = D_2\hat{i}_1 - D_3\hat{i}_2 + D_4\hat{i}_3 - D_6\hat{i}_2 - D_7\hat{i}_2\hat{i}_4$

(130) $0 = D_3\hat{i}_2 - D_4\hat{i}_3$

(131) $\dfrac{d\hat{i}_4}{d\hat{\tau}} = D_6\hat{i}_2 - D_7\hat{i}_2\hat{i}_4$

(132) $\dfrac{d\hat{P}}{d\hat{\tau}} = D_2\hat{i}_1$

The solutions for large time ($\hat{\tau} \to \infty$) are

(133) $\hat{i}_1 \sim 2(D_3 + D_6)\, \hat{i}_2/D_2$

(134) $\hat{i}_3 \sim D_3 \hat{i}_2/D_4$

(135) $\hat{p} \sim (D_3 + D_6)\, \hat{i}_2/D_3$

(136) $\hat{i}_4 \sim D_6/D_7$

(137) $\hat{i}_2 \;\alpha\; \exp\,(2D_3\, \hat{\tau})$.

At an even later time given by

(138) $t \simeq \left(\frac{1-a}{2D_3} + \frac{c-2}{\mathcal{D}}\right)\, \theta \exp\,(\theta/T_0)$

reactions (2) and (5) become of comparable magnitudes. The resulting transition problem is identical to the last stage of the slow branch analysis. This final transition occurs at this time if $\mathcal{D} < 0$ as opposed to

(139) $t \simeq \left(\frac{\mathcal{D} + D_6}{4D_1D_7}\, \ell n \left|\frac{1 - 2\mathcal{D}}{\mathcal{D} - D_6}\right| - \frac{\mathcal{D}}{2D_1D_7}\right) \exp\,((c - 1)\, \theta/T_0)$.

for $\mathcal{D} > 0$. The ratio of these times is approximately

(140) $\frac{t_{slow}}{t_{fast}} \simeq \frac{1}{\theta} \exp\,((c - 2)\, \theta/T_0)$.

It is now clear that the second explosion limit gives a curve that separates regions of fast from slow reaction and that the transition between the two regimes is rapid but not discontinuous. We can therefore conclude that the second explosion limit defines an effective ignition temperature for at least detonations and that the cut-off temperature used in the analysis of Kapila, Matkowsky, and van Harten (1983) has a physical meaning. If the temperature and pressure upstream of the shock front in the detonation wave is specified then there is a minimum Mach number possible for the detonation wave and at the minimum Mach number the conditions just behind the shock correspond to a point on the second explosion limit curve. For stronger shocks the conditions just behind the shock are situated in the "explosive" region and for weaker shocks the conditions just behind the shock are situated in the "non-explosive" region. Since an analysis similar to Clarke's (1983 a,b) convected explosion analysis will apply with this more complicated chemical mechanism it is obvious that the behavior shown here will correspond to an effective ignition temperaure that is a function of the upstream pressure.

The second explosion limit does terminate in reaction vessel studies when it is intersected by the so-called third explosion limit. The third explosion limit is not an abrupt transition like the second explosion limit but rather a gradual transition from effectively "inert" initital conditions to effectively "reactive" initial conditions. This gradual transition is caused by the increase in rate of the recombination reactions involving HO_2. At high enough pressures there are appreciable reaction rates among the "degenerate" mechanisms found on the left side of the second explosion limit curve. It has been argued that the correspondence between detonation limits for hydrogen-oxygen and hydrogen-air mixtures and the extrapolation of the second explosion limit curve are completely fortuitious because no accounting for the third explosion limit is made. This analysis shows that there is still an abrupt change in rate across that extrapolation of the second explosion limit curve but that the amount of change decreases for increasing temperature. On the left side of the curve in pressure-temperature space the reaction mechanism is dominated by the low temperature processes. These processes may give rise to appreciable reaction in an adiabatic reaction vessel but it is very possible that they are too slow (and hence too sensitive to heat loss) to be the main heat liberating mechanism driving a detonation front. More work must be done before a definative answer can be given but the results given here are encouraging.

CONCLUDING REMARKS

This analysis shows a number of features not found in previous asymptotic analyses of branched chain explosions. In the studies of Kapila (1978), and Birkan and Kassoy (1985) particular attention was paid to the "transition" limit between slow and fast recombination. Sen and Law (1982) looked at the slow recombination limit exclusively. Here we have presented evidence that fast recombination limit is the limit of physical interest and have presented solutions for that limit. In addition we have represented the chain branching cycle through its three constituent reactions as opposed to modeling the cycle with a single reaction as in the previous studies. By using all three reactions in the cycle we have shown that reaction balancing does indeed occur among the three reactions so that their net effect can be represented as a single reaction during the branched chain explosion. Before the explosion, however, their rates are not balanced. Only through an analysis such as that given here can the initial growth of intermediates be observed.

If the analysis started here is carried through to completion, then we would find that the intermediate species concentrations are always exponentially small. The largest intermediate species concentration would be given by $[I_2]$ and it would be on the order of exp $(-a\theta/T)$ during the thermal explosion. In the real physical system the concentrations of the intermediate species appear to become on the order of $1/\theta$ lending support to the notion that transitional recombination rate limits are the most appropriate. This idea is incorrect as far as hydrogen oxidation is concerned. First of all, using the rate data for the elementary reaction steps in the hydrogen oxidation mechanism clearly shows that the fast recombination limit is the one that is physically correct. Also, the transitional recombination limit requries that the final temperature of the mixture be within $O(1/\theta)$ of a specified temperature where the propagating and recombination reactions will be of similar rate and the concentrations of the radicals will be of a desired size picked a priori. $O(1)$ changes in the initial temperature will cause the reacting mixture to become either a fast or slow recombination limit-type mixture. This type of behavior is not seen in real physical systems.

How is it then that the fast recombination limit is appropriate and yet there is appreciable radical concentrations during thermal explosion? The answer is that the large activation energy limit is not sufficiently accurate at the elevated temperatures found near the end of the thermal explosion process. Asymptotically it is required that exp $(-a\theta/T)$ be not only very much less than 1 throughout the portion of the ignition event that is being studied, but it must also be small relative to T/θ as terms of this size are retained relative to terms of the order of exp $(-a\theta/T)$. Using hydrogen oxidation data and a typical hot temperature of 2000°K, we find that exp $(-a\theta/T)$ is about .27 whereas T/θ is about .24. At such elevated temperatures the asymptotic ordering of the sizes of the terms in the governing equations breaks down. The θ's found in practice are just not large enough to generate numerically accurate solutions all the way through the entire ignition event. The early stages can be accurately determined through asymptotic techniques for two reasons. The first reason is that the initial temperaure of the mixture is relatively low. This makes the quantity T/θ sufficiently small. The second and more important reason is that the early events in the ignition process are dominated by the initiation reaction. The activation energy of the initiation reaction is the real large parameter for the first few stages in the ignition sequence and it is typically much larger than the activation energies found in the branched chain cycle. The combination of low temperatures and large activation energies then guarantee accurate solutions with asymptotic techniques. Solutions for events past thermal induction are probably not numerically accurate but may be of interest for their qualitative behavior.

The analysis presented here is analogous to the numerical results of Dougherty and Rabitz (1980) as given in their Figure 7(b) for conditions very close to the second explosion limit but slightly on the "explosive" side of the curve. The orders of the concentrations as well as the coupling of the growth rates among the various constituents is exactly as predicted here. Furthermore, it is very clearly seen that the thermal explosion process begins when the concentrations of the intermediate species are at least four orders of magnitude below the concentrations of the major species. This is additional evidence that the fast recombination limit is the most physically realistic asympototic limit.

When conditions are well removed for the second explosion limit and in the explosive regime, then the analysis of Mikolaitis (1985) will apply corresponding to the numerical solutions given in Figure 7(c) of Dougherty and Rabitz (1980). In that study, the reaction channels involving H_2O_2 and HO_2 were neglected. The resulting solutions were seen to be qualitatively similar to the explosion events for conditions removed from the second explosion limit condition.

BIBLIOGRAPHY

1. Belles, F. E., (1959), Seventh (International) Symposium on Combustion, Butterworths Scientific Publications, London, p. 745.
2. Birkan, M. A. and Kassoy, D. R., (1983), Comb. Sci. Tech., 33, 135.
3. Birkan, M. A. and Kassoy, D. R., (1985), The Unified Theory for Chain Branching Thermal Explosions with Dissociation - Recombination and Confinement Effects, to appear in Combustion Science and Technology.
4. Clarke, J. F., (1983a), Comb. Flame, 50, 125.
5. Clarke, J. F., (1983b), JFM, 136, 139.
6. Dougherty, E. and Rabitz, H., (1980), J. Chem. Phys., 72, 6571.
7. Hocks, W., Peters, N., and Adomeit, G., (1981), Comb. Flame, 41, 157.
8. Kapila, A. K., (1978), J. Eng. Math., 12, 221.
9. Kapila, A. K., Matkowsky, B. J., and van Harten, A., (1983), Siam J. Appl. Math., 43, 491.
10. Kassoy, D. R., (1975) Comb. Sci. Tech., 10, 27.
11. Kassoy, D. R. and Clarke, J. F., (1985), JFM, 150, 253.
12. Liñán, A., (1971), A Theoretical Analysis of Premixed Flame Propagation with an Isothermal Chain Reaction, unpublished technical report.
13. Mikolaitis, D., (1985), in preparation.
14. Sen, A. and Law, C. K., (1982), Comb. Sci. Tech., 28, 75.

15. Seshadri, K. and Peters, N., (1983), Comb. Sci. Tech., 33, 35.

16. Tam, R. and Ludford, G. S. S., (1984), Comb. Sci. Tech., 40, 303.

17. Zeldovich, Y. B, (1948), Zhar. Fizi. Khi., (USSR), 22, 27, (English Translation in NACA TM 1282, 1951).

DEPARTMENT OF ENGINEERING SCIENCES
UNIVERSITY OF FLORIDA
GAINESVILLE, FL 32611

Lectures in Applied Mathematics
Volume 24, 1986

THERMAL EXPANSION EFFECTS ON PERTURBED PREMIXED FLAMES: A REVIEW

T. L. Jackson and A. K. Kapila

ABSTRACT. Thermal expansion is the means by which a flame is coupled to the flow through which it propagates. Although it has long been recognized that this coupling may vitally influence the response of a flame to perturbations, self-consistent analytical treatments of perturbed flames which take this coupling into account have only recently been developed. Even there, coupling is only admitted as a perturbation. Analysis is either confined to long-wave disturbances, which amounts to regarding hydrodynamics as the primary effect, or one adopts the constant-density approximation, where diffusion plays the dominant role. This paper reviews recent numerical work on the perturbed-flame problem in a framework where diffusion and thermal expansion are considered equally important. In the near-equidiffusional formulation the full range of wavelengths, including long and short waves, is explored. Stability results are presented for nonadiabatic flames propagating downwards under gravity. Flame response to high-frequency acoustic oscillations is also included.

1. INTRODUCTION. When steady and planar, the premixed flame is essentially a reactive-diffusive entity, and the role of hydrodynamics is minimal. The situation changes, however, if the flame becomes corrugated because of a disturbance. The flame profile now shifts differentially in the direction of propagation, creating transverse gradients of temperature and concentration which induce diffusive fluxes along the flame surface. At the same time, thermal expansion of the gas

This work was supported by the Army Research Office.

through the corrugated flame produces a refracted streamline pattern, thereby generating transverse convective fluxes. These two mechanisms alter the local balance of energy and reactant fraction, thus modifying the speed and structure of the flame. Deflection of streamlines at the flame also influences the essentially incompressible flow upstream. Thus, diffusive-hydrodynamic coupling not present in the steady, planar mode is restored.

Although the importance of this coupling has long been recognized, self-consistent analytical treatments of perturbed flames which take it into account have only recently been attempted. Even there, either hydrodynamics or diffusion is considered as the dominant effect, and coupling is admitted only as a perturbation. For disturbances whose wavelength is much larger than the flame thickness a hydrodynamic description is employed, wherein the flame is a nearly plane and nearly steady discontinuity in an ideal gas flow, and the effect of diffusion is lost. For near-unity Lewis numbers, diffusion can be reinstated as a perturbation in the framework of Slowly Varying Flames (SVFs). The first attempt of this kind, due to Sivashinsky, employed the small heat release assumption (1977), but that restriction was later removed by Pelce and Clavin (1982), Matalon and Matkowsky (1982) and Frankel and Sivashinsky (1983), working independently. These authors demonstrated that for a given heat release, all wave numbers larger than a critical value are stable, the critical value being a monotonically decreasing function of the Lewis number. The collective effect of heat loss and gravity in this context was examined by Clavin and Nicoli (1985).

The above analysis, it must be reiterated, is chiefly hydrodynamic, and does not apply to wavelengths as short as the flame thickness, because then the flame ceases to be slowly varying. One turns, therefore, to the theory of Near-Equidiffusional Flames (NEFs) for which flame thickness and diffusion time are the relevant scales. Unlike the SVFs, the NEFs are unsteady and fully three-dimensional, so that even the linearized perturbation equations are analytically intractable because of the presence of variable coefficients. Further progress in the literature is based on the so-called Constant Density Approximation (CDA) which neglects thermal expansion and treats diffusion as the primary effect. The coupling is lost again, but can be reinstated in a weak form by making the small heat release assumption (Sivashinsky, 1983).

The present work reviews the recent numerical work of Jackson and Kapila (1984, 1985) who have re-examined the stability problem by according equal importance to hydrodynamics and diffusion. This is done by adopting the NEF formulation without invoking the CDA. The full range of wave lengths is considered and stability boundaries for nonadiabatic flames are obtained, including the effects of gravity and heat loss. Results are displayed in the form of neutral stability curves in the Lewis number-wave number plane. The final section deals with flame response to planar, high-frequency oscillations.

2. GOVERNING EQUATIONS. A nonadiabatic premixed flame propagating down a vertical tube can be modelled by the following equations:

$$\frac{D\rho}{Dt} + \rho\nabla\cdot\underline{v} = 0, \quad \rho\frac{D\underline{v}}{Dt} + \nabla p = \kappa\left[\frac{1}{3}\nabla(\nabla\cdot\underline{v}) + \nabla^2\underline{v}\right] - \rho G\underline{i},$$

$$\rho C_p\frac{DT}{Dt} = \lambda\nabla^2 T + Qw - B\left(T - T0\right), \quad \rho\frac{Dy}{Dt} = \mu\nabla^2 Y - w,$$

$$p_o = \rho R^o T/W, \quad w = A\rho Y e^{-E/R^o T}.$$

Here ρ is the density, $\underline{v}$ the velocity and T the temperature of the gas mixture. The reactant mass fraction is Y and p denotes pressure deviation from the ambient value p_0. The ambient temperature T_0 is also that of the fresh mixture. The other quantities appearing above are the specific heat c_p, viscosity κ, thermal conductivity λ, species diffusion coefficient μ, gravitational acceleration G, chemical heat release Q, molecular weight W, heat-loss coefficient B, pre-exponential factor A, Universal gas constant R^o and the activation energy E. The unit vector pointing vertically upwards is denoted by $\underline{i}$.

These equations are nondimensionalized by selecting the fresh-mixture values p_0, ρ_0, T_0 and Y_0 as reference quantities. The nonadiabatic steady flame speed v_n is chosen as the reference speed, rather than the adiabatic value v_0; their ratio is denoted by ψ, i.e.,

$$\psi = v_n/v_0 \ .$$

The characteristic length and time scales are taken, respectively, to be

$$l_n = l_0/\psi \ , \quad t_n = t_0/\psi^2 \ ,$$

where l_0 and t_0 are the adiabatic scales

$$l_0 = \lambda/(\rho_0 c_p v_0) \ , \quad t_0 = \lambda/(\rho_0 c_p v_0^2).$$

It is convenient to define the adiabatic and nonadiabatic flame Mach numbers, according to the relations

$$M_0 = v_0/c_0 \ , \ M_n = v_n/c_0 = \psi M_0 \ ,$$

where c_0 is the frozen sound speed in the fresh mixture. Finally, pressure deviations from the ambient are referred to $\gamma M_n{}^2 p_0$, γ being the specific-heats ratio. The governing equations then take the dimensionless form

$$\frac{D\rho}{Dt} + \rho\nabla\cdot\underline{v} = 0, \ \rho\frac{D\underline{v}}{Dt} + \nabla p = Pr\left[\frac{1}{3}\nabla(\nabla\cdot\underline{v})+\nabla^2\underline{v}\right] - \frac{\rho}{\psi^3\phi}\underline{i} \ , \ \rho T = 1 \quad (1)$$

$$\rho\frac{DT}{Dt} = \nabla^2 T + \alpha\Omega - \frac{\varepsilon\beta}{\alpha\psi^2}(T-1), \quad \rho\frac{DY}{Dt} = L^{-1}\nabla^2 Y - \Omega \ ,$$

$$\Omega = \left(\frac{D}{\psi^2 M_0{}^2}\right)\rho Y e^{-\theta/T} \ . \quad (2)$$

The yet undefined nondimensional parameters appearing above are:

$Pr = \kappa c_p/\lambda$, Prandtl number ,
$\phi = v_0^2/(l_0 G)$, Froude number ,
$\alpha = QY_0/(c_p T_0)$, Heat release parameter ,
$\beta = \alpha B\lambda/(\rho_0^2 c_p^2 v_0^2 \varepsilon)$, Heat loss parameter ,
$\theta = E/(R^0 T_0)$, Activation energy ,
$\varepsilon = (1+\alpha)^2/\theta$, Reciprocal activation energy ,
$L = \lambda/(\mu c_p)$, Lewis number ,
$D = \lambda A/(\rho_0 c_p c_0^2)$, Damkohler number .

A straightforward analysis of the governing equations for steady, planar propagation (Buckmaster and Ludford, 1982) in the limit

$$\theta \to \infty \text{ , i.e. } \varepsilon \to 0 \tag{3}$$

shows that the adiabatic flame Mach number is given by

$$\frac{D}{M_0^2} = \frac{\alpha^2(1+\alpha)}{2L\varepsilon^2}\left[1+O(\varepsilon)\right] \exp\left[\theta/(1+\alpha)\right] \tag{4}$$

while the nonadiabatic-to-adiabatic flame-speed ratio depends upon the heat-loss parameter according the the relation

$$2\beta + \psi^2 \ln \psi^2 = 0 \text{ .} \tag{5}$$

This relation is graphed in Fig. 1.

Additional simplification of the full, unsteady equations can be obtained if one is interested only in special perturbations of an established flame. These correspond to the NEF framework which, in addition to large activation energy, is characterized by near-unity Lewis number and nearly-uniform enthalpy, i.e.,

$$L^{-1} = 1 - \varepsilon l/\alpha \text{ , } T + \alpha Y \equiv H = 1 + \alpha + \varepsilon h \text{ .} \tag{6}$$

Of course, this framework is valid only over O(1) lengths behind the flame; the $O(\theta^{-1})$ heat loss acting over $O(\theta)$ lengths will cause O(1) enthalpy variations.

In view of (6), an asymptotic analysis in the limit $\varepsilon \to 0$ leaves eqs. (1) unaltered while eqs. (2) reduce, to leading order in ε (see Buckmaster and Ludford, 1982), to

$$\rho \frac{DT}{Dt} - \nabla^2 T = 0 \text{ in the unburnt region ,}$$

$$T = 1 + \alpha \text{ in the burnt region ,} \tag{7}$$

$$\rho \frac{Dh}{Dt} = \nabla^2 h + (1/\alpha)\, \nabla^2 T - \frac{\beta}{\alpha\psi^2}(T-1) \, . \tag{8}$$

Equations (1), (7) and (8) now constitute the basic set. It must be supplemented by appropriate initial and boundary conditions as well as jump conditions across the reaction zone, which is now an interface separating the burnt and unburnt regions.

For a two-dimensional flow, let the reaction-zone location be given by $x = x_f \equiv F(y,t)$. It is convenient to shift to a reference frame in which the reaction zone is stationary. To that end, let a new coordinate ξ be defined by the transformation

$$x = F(y,t) + \xi \, , \tag{9}$$

and let the unburnt and burnt regions be confined, respectively, to $\xi < 0$ and $\xi > 0$. Let S denote the longitudinal velocity relative to the moving frame, defined by

$$S = u - F_t - v\, F_y \tag{10}$$

where

$$\underline{v} = \langle u,v \rangle \tag{11}$$

and suffixes denote partial derivatives. Then, Eqs. (1), (7) and (8) transform into

$$\rho_t + (\rho S)_\xi + (\rho v)_y = 0 \, , \tag{12}$$

$$\rho(u_t+Su_\xi+vu_y) + p_\xi = \Pr\{\nabla^2 u + \tfrac{1}{3}(S_{\xi\xi}+v_{\xi y})\} - \frac{\rho}{\psi^3\phi}\ , \tag{13}$$

$$\rho(v_t+Sv_\xi+vv_y) + p_y - F_y p_\xi =$$
$$\Pr\{\nabla^2 v + \tfrac{1}{3}(\frac{\partial}{\partial y} - F_y\frac{\partial}{\partial \xi})(S_\xi+v_y)\}\ , \tag{14}$$

$$\rho(T_t+ST_\xi+vT_y) = \nabla^2 T\ ,\ \xi < 0\ ;\quad T = 1+\alpha\ ,\ \xi > 0\ , \tag{15}$$

$$\rho(h_t+Sh_\xi+vh_y) = \nabla^2 h + (1/\alpha)\nabla^2 T - \frac{\beta}{\alpha\psi^2}(T-1)\ , \tag{16}$$

$$\rho T = 1\ , \tag{17}$$

where now,

$$\nabla^2 \equiv (1+F_y{}^2)\frac{\partial^2}{\partial\xi^2} + \frac{\partial^2}{\partial y^2} - 2F_y\frac{\partial^2}{\partial\xi\partial y} - F_{yy}\frac{\partial}{\partial\xi}\ . \tag{18}$$

The appropriate boundary conditions for the stability problem are

$$\rho = u = T = 1,\ p_\xi + \frac{\rho}{\psi^3\phi} = 0,\ \text{and}\quad v = h = 0\ ,\ \text{at}\ \xi = -\infty\ ;$$
$$\text{no exponential growth at}\ \xi = \infty\ . \tag{19}$$

The jump conditions across the reaction zone (see Buckmaster and Ludford, 1982, Chap. 8) at $\xi = 0$ are

$$\delta T = \delta h = \delta u = \delta v = \delta\rho = \delta S = 0\ , \tag{20}$$

$$\delta p = \tfrac{4}{3}\Pr\rho S\,\delta T_\xi\ , \tag{21}$$

$$\delta u_\xi = \frac{\rho S}{1+F_y^2}\delta T_\xi\ ,\ \delta v_\xi = -\frac{\rho S}{1+F_y^2}F_y\delta T_\xi\ ,\ \delta S_\xi = \rho S\delta T_\xi\ , \tag{22}$$

$$\delta T_\xi = -(\alpha/\psi)(1+F_y{}^2)^{-1/2}\exp(h/2)\ , \tag{23}$$

$$\delta h_\xi = -(1/\alpha)\delta T_\xi\ , \tag{24}$$

where

$$\delta\phi = \phi\Big|_{\xi=0+} - \phi\Big|_{\xi=0-} .$$

Equations (10)-(24) describe completely the NEF problem.

3. STEADY FLAME, AND LINEARIZATION.

The steady, plane solution of the above equations, denoted by the superscript s, is given by

$\xi < 0$:

$$T^s = 1 + \alpha e^{\xi} , \quad h^s = -1\xi e^{\xi} + \frac{\beta}{\psi^2} (\xi-2)e^{\xi} ,$$

$$p^s = -\alpha e^{\xi} + \frac{4}{3} \Pr \alpha e^{\xi} + \frac{1}{\psi^3 \phi} \ln \left\{ \frac{1+\alpha e^{\xi}}{(1+\alpha)e^{\xi}} \right\} ,$$

$\xi > 0$:

$$T^s = 1 + \alpha, \quad h^s = -\frac{\beta}{\psi^2} (\xi+2), \quad p^s = -\alpha - \frac{\xi}{\psi^3 \phi(1+\alpha)} ,$$

and

$$\rho^s = 1/T^s, \quad u^s = S^s = T^s , \quad F^s = -t . \tag{25}$$

It is convenient to introduce the longitudinal mass flux m, defined by

$$m = \rho S ,$$

whose steady-state value is given by

$$m^s = 1 .$$

Small perturbations of the steady flame are governed by linearized equations, which can be obtained by setting $\psi = \psi^s + \sigma\tilde{\psi}(\xi)e^{iky+\omega t}$ for all variables and taking the limit $\sigma \to 0$. Details are given in Jackson and Kapila (1984, 1985); suffice it to say that a linear, 8th-order eigenvalue problem with variable coefficients emerges in the region $\xi \gtrless 0$, which has a solution only if the growth rate ω and the transverse wave number k are appropriately related.

4. STABILITY RESULTS. For nonzero α the eigenvalue problem has to be treated numerically, and this was done by appending a rootfinder based on Muller's procedure to the boundary-value solver COLSYS. Computations were performed only for unit Prandtl number although more general values of Pr can be introduced without additional complications.

The results are displayed on the kl-diagram in the form of neutral curves, on which either $\omega = 0$ or Re $\omega = 0$, and which separate the stable from the unstable regions. In the interest of expositional clarity and ease of comparison with existing results, several limiting cases are treated in order. For each case, the CDA limit $\alpha \to 0$ is considered first because (i) it can be extracted analytically, (ii) it serves as an initial guess for the numerical procedure, and (iii) it clarifies the role of thermal expansion, which corresponds to $\alpha \neq 0$. Case I below is discussed in detail in Jackson and Kapila (1984), the subsequent cases are treated more completely in Jackson and Kapila (1985).

I. Adiabatic flame without gravity ($\psi = 1$, $G = 0$)

In the limit $\alpha \to 0$ one obtains the dispersion relation

$$16\omega^3 + \left[48k^2+8+2l-l^2/4\right]\omega^2 + (1+12k^2)(1+4k^2+l/2)\omega + k^2(1+4k^2+l/2)^2 = 0$$

which generates the well-known Sivashinsky diagram (Sivashinsky, 1977; Buckmaster and Ludford, 1982) sketched in Fig. 2. There, ω vanishes on the left boundary

$$2 + l + 8k^2 = 0$$

and on the k = 0 axis, while Re ω = 0 and Im $\omega \neq 0$ on the right boundary

$$1 = 4(1+8k^2)(1+12k^2)^{-1}\left[1+\{3(1+8k^2)\}^{1/2}\right] .$$

The left boundary corresponds to cellular instability and the right boundary to pulsatile instability. The results for $\alpha \neq 0$, computed numerically, are displayed in Fig. 3 for four different values of α. The right boundary does not appear to differ very much from the corresponding CDA curve. The left boundary also retains the shape of its CDA counterpart, except near small values of k, where it veers to the right and intersects the right boundary. Thus, increase in thermal expansion causes a corresponding reduction in the region of stability, and allows cellular instability to occur at smaller excursions of the Lewis number from the value of unity. This effect is most pronounced at smaller values of k; in fact, for each α however small there is a minimum k below which the flame is unstable for all Lewis numbers, suggesting the predominance of hydrodynamics over diffusion for long waves. This last result is in complete accord with the SVF analyses mentioned in the Introduction. Thus, the dot-dash curve of Fig. 4 corresponds to Matalon and Matkowsky's dispersion relation (1982)

$$\omega_o(\alpha) + k\,\omega_1(\alpha,1) = 0$$

where ω_0 and ω_1 are given by their Eq. (6.22) after q and l in their paper are replaced by the respective values α and $1/\alpha$, and α is assigned the value one. Fig. 4 shows good agreement of the SVF curve with our numerical results for small k. It must be pointed out, however, that the SVF analysis has nothing to say about the right boundary.

II. Nonadiabatic flame without gravity ($\psi < 1$, $G = 0$)

The corresponding CDA limit, investigated by Joulin and Clavin (1979), yields the dispersion relation

$$2(\nu-1)\left[\nu^2 + (\nu+1)\ln\psi\right] = 1\left[1-2\nu+2\omega\right], \quad \nu = (1+4k^2+4\omega)^{1/2}, \quad \mathrm{Re}\ \nu > 1,$$

which must be examined in conjunction with the plot of flame-speed ratio ψ against heat-loss parameter β sketched in Fig. 1. Only the upper branch of this plot will be considered henceforth because flames corresponding to the lower branch are found to be unstable to planar perturbations at all Lewis numbers.

Stability curves corresponding to the above dispersion relation are shown in Fig. 5 for ψ lying between the adiabatic value unity and the flammability-limit value $e^{-1/2}$. One finds that as ψ decreases (i.e. β increases), both the cellular branch and the pulsatile branch move inward, thereby shrinking the region of stability. Thus heat loss promotes both cellular and pulsatile combustion. The latter is especially significant because, as is often mentioned in the literature, pulsating adiabatic flames correspond to unrealistically large values of Lewis number. The same displacement of the right boundary is achieved by heat loss to the burner in anchored flames (Buckmaster, 1983) although there, when heat loss is increased beyond a critical vlaue, the right boundary reverses its movement and begins to proceed rightwards. In burner-supported flames the left boundary also moves to the <u>left</u> with increasing heat loss, especially at low wave numbers, thereby acting as a stabilizer against cellular instability.

Figs. 6 and 8 show the combined effects of thermal expansion and heat loss. For fixed thermal expansion ($\alpha=1$) and increasing heat loss (Fig. 6), both boundaries move inwards for moderate k, faster than they did for $\alpha = 0$. Thus heat loss encourages instability. For the right branch the same holds true for k small as well, as shown in Fig. 7, which displays the behavior of the critical Lewis number for planar pulsations as a function of ψ for various α. For k small the opposite is true for the cellular boundary, for which heat loss acts as a stabilizing factor. This numerical feature is confirmed analytically by the limit $\omega = 0$, $1 = O(1)$ and $\alpha \to 0$, which, when specialized to k small, leads to the asymptotic result

$$\ell \sim \left(1+2\ \ell n\ \psi\right)\left(-2 + \frac{\alpha}{k}\right)$$

The effect of nonadiabaticity is now clear; as ψ decreases below the adiabatic value of unity, the first factor on the right is negative, and hence stabilizing, for k sufficiently small. Fig. 8 is plotted at fixed heat loss (ψ=0.8) but at different values of thermal expansion. The effect of increasing α is similar to that for the adiabatic case (Fig. 5); the pulsating boundary moves very little while cellular instability is significantly enhanced.

III. Adiabatic flame propagating downward under gravity ($\psi = 1$, $G = 0$)

For adiabatic flames the effect of buoyancy is inversely proportional to the Froude number $\phi = v_o^2/(l_oG)$, and is therefore more pronounced on slower flames. Typical values of ϕ range from 0.5 for the slowest flames to 10^6 for the fastest. Influence of buoyancy is, of course, nil in the CDA limit.

For nonzero α, the results are shown in Fig. 9 for three different values of ϕ. The right branch is essentially unaffected by buoyancy and the same holds true for the left branch at moderate k. For k small buoyancy opposes the hydrodynamic instability and can suppress it for ϕ small enough. This effect is almost completely captured by the SVF analyses, as shown by the dotted curves in Fig. 9 which correspond to Clavin and Pelce's results (1982). Also, an asymptotic analysis of the left branch in the limit $\alpha \to 0$ reveals that for small k,

$$1 \sim -2 + \frac{\alpha}{k} - \frac{\alpha}{\phi}\left(\frac{1}{k^2} + \frac{1}{k}\right) .$$

The negative sign of the last term for finite ϕ indicates its stabilizing effect.

In Fig. 10 results are displayed at fixed $\phi(=20)$ and increasing α. As α is raised, the Lewis number stability band for which all waves are stable continues to shrink, primarily due to the movement of the cellular branch. At a critical α the band disappears, and for larger α there always exists an intermediate range of unstable wave numbers at all values of ℓ.

The combined effect of buoyancy and nonadiabaticity is shown in Fig. 11 which is drawn for $\psi = 0.8$, $\phi = 20$ and at three different values of α. Comparison with Fig. 10 shows that for fixed α and ϕ, heat loss promotes pulsating instability at all k and cellular instability at moderate k, while the latter is suppressed at small k (cf. discussion of Fig. 6, corresponding to zero buoyancy). Comparison with Fig. 8 shows that buoyancy only influences the lower end of the left boundary.

IV. Nonadiabatic flame propagating downward under gravity ($\psi < 1, G \neq 0$)

This dominance of buoyancy for long waves is confirmed by examining, for k small, the asymptotic limit $\alpha \to 0$, $1 = O(1)$:

$$\ell \sim -2 + \frac{\alpha}{k} + 2\left(-2 + \frac{\alpha}{k}\right)\ell n\ \psi - \frac{\alpha}{\psi^3\phi}\left(\frac{1+2\ \ell n\ \psi}{k^2}\right) + \frac{1}{k}\Big) .$$

For $e^{-1/2} < \psi < 1$ and finite ϕ, the last term is dominant and stabilizing as $k \to 0$.

4. FLAME-ACOUSTIC COUPLING. In a recent paper, Van Harten, Kapila and Matkowsky (1984) studied the interaction of a slender flame with an applied acoustic field. Their analysis is based on the limit $\theta \to \infty$ and assumes that the strength of the acoustic disturbance is $O(\varepsilon)$, that the acoustic time and

and diffusion time are of the same order, and that the characteristic length of wrinkles in the flame is comparable to the acoustic wave length. Then the flame behaves essentially as a density discontinuity convected by the fluid, across which pressure and normal velocity are continuous to leading order. The jump in the tangential velocity across the flame and the flame speed are both determined by analyzing the flame structure which is now unsteady and one-dimensional. For the special case of a plane disturbance incident normally on a plane flame the situation is simpler since the tangential velocity vanishes. The structure equations then reduce, in our notation, to

$$\frac{\partial \rho}{\partial t} + \frac{\partial m}{\partial \xi} = 0 \ , \quad \rho T = 1 \ ,$$

$$\rho \frac{\partial T}{\partial t} + m \frac{\partial T}{\partial \xi} = \frac{\partial^2 T}{\partial \xi^2} \quad \text{for} \quad \xi < 0 \ , \ T = 1+\alpha \quad \text{for} \quad \xi > 0 \ ,$$

$$\rho \frac{\partial h}{\partial t} + m \frac{\partial h}{\partial \xi} - \frac{\gamma-1}{\gamma} \frac{dp}{dt} = \frac{\partial^2 h}{\partial \xi^2} + \frac{\ell}{\alpha} \frac{\partial^2 T}{\partial \xi^2} \ . \tag{64}$$

The only change is that p now denotes the $O(\varepsilon)$ perturbation in the ambient pressure and the mass flux m, defined by

$$m = \rho S$$

is employed in favor of the relative speed S. The equations are basically those governing the NEFs, except that an extra pressure term now appears in the enthalpy equation. Jump conditions for T, h, T_ξ and h_ξ are the same as those in the set (20-24), once F_y is set to zero there for the case of the planar flame. In addition, as $\xi \to \pm\infty$, the solution must match with the acoustic zone outside.

The above equations constitute a difficult, nonlinear problem which can, however, be linearized for the special, but important, case of the applied acoustic amplitude being smaller than $O(\varepsilon)$. Then, one seeks solutions of the form

$$f(\xi,t) = f^s(\xi) + \sigma\bar{f}(\xi)e^{i\omega t}$$

where the superscript s denotes the steady state, given by the relevant expressions in (25), and ω is the applied frequency. Substitution into (64) followed by the limit $\sigma \to 0$ leads to the linearized problem which, as before, can be solved analytically for $\xi > 0$. Only the problem for $\xi < 0$ needs to be considered, and it consists of the equations

$$\bar{m}' = \frac{i\omega\bar{T}}{(1+\alpha e^{\xi})^2}, \quad \bar{T}'' - \bar{T}' - \frac{i\omega\bar{T}}{1+\alpha e^{\xi}} = \alpha e\xi \bar{m},$$

$$\bar{h}'' - \bar{h}' - \frac{i\omega\bar{h}}{1+\alpha e^{\xi}} = -\frac{\ell}{\alpha}\bar{T}' - \frac{i\omega\ell\bar{T}}{\alpha(1+\alpha e^{\xi})} - \ell(2+\xi)e^{\xi}\bar{m} - \frac{\gamma-1}{\gamma} i\omega$$

with boundary conditions

$$\bar{T} = 0, \quad \bar{h} = \frac{\gamma-1}{\gamma} \quad \text{at} \quad \xi = -\infty,$$

$$\bar{T} = 0, \quad \bar{T}' = \frac{\alpha\bar{h}}{\alpha}, \quad \bar{h}' + \frac{\ell}{\alpha}\bar{T}' = \lambda\left(\bar{h} - (1+\alpha)\left(\frac{\gamma-1}{\gamma}\right)\right) \quad \text{at } \xi = \infty,$$

where

$$\lambda = \frac{1}{2}\left[1 - \left\{1 + \frac{4i\omega}{1+\alpha}\right\}^{1/2}\right]$$

and the applied acoustic pressure at the flame has been taken to be

$$\bar{p} = e^{i\omega t}.$$

In Van Harten et al (1984) the above problem was solved analytically in the limit $\alpha \to 0$. Then, in the above equations, T must be replaced by αT, with the result that $\bar{m}$ becomes independent of ξ and to leading order, the problem becomes one with constant coefficients. Details are given in Van Harten et al (1984) and one finds, in particular, the following expression for $\bar{m}$:

$$\bar{m} = \frac{\gamma-1}{\gamma} \frac{(1+4i\omega)\{1+(1+4i\omega)^{1/2}\}}{1+4i\omega - \frac{\ell}{4}(1+4i\omega)^{1/2} + \frac{\ell}{4}}.$$

A graph of the modulus of m against ω is shown in Fig.12. It exhibits resonance at the critical value l_c of 1, where

$$\ell_c = 4\left(1 + \sqrt{3}\right)$$

corresponding to the lowest point of the right boundary in the CDA diagram of Fig. 2.

For $\alpha = 0$ the problem must be solved numerically, and this was done for various values of α and for $\ell = 10$. The mass flux perturbation m is now a function of ξ. Fig. 12 also shows the graph of the modulus of $m(-\infty)$ against ω for various x. One finds that the effect of thermal expansion is to increase the amplitude at a given frequency and to lower the resonant frequency.

BIBLIOGRAPHY

1. Buckmaster, J. D. (1983) "Stability of the porous plug burner flame". SIAM J. Appl. Math., 43, 1335.

2. Buckmaster, J. D. and Ludford, G. S. S. (1982) Theory of Laminar Flames, Cambridge University Press.

3. Clavin, P. and Nicoli, C. (1985) "Effects of heat losses on the limits of stability of premixed flames propagating downwards. Combustion and Flame, to appear.

4. Frankel, M. L. and Sivashinsky, G. I. (1983) "The effect of viscosity on the hydrodynamic stability of a plane flame front". Comb. Sci. Tech., 19, 207.

5. Jackson, T. L. and Kapila, A. K. (1984) "Effect of thermal expansion on the stability of plane, freely propagating flames". Comb. Sci. Tech., 41, 191.

6. Jackson, T. L. and Kapila, A. K. (1985) "Effect of thermal expansion on the stability of plane, freely propagating flames. Part II: Incorporation of gravity and heat loss". Submitted for publication.

7. Joulin, G. and Clavin, C. (1979) "Linear stability analysis of nonadiabatic flames: diffusional thermal model". Combustion and Flame, 35, 139.

8. Matalon, M. and Matkowsky, B. J. (1982) "Flames as gasdynamic discontinuities". J. Fluid Mech., 124, 239.

9. Pelce, P. and Clavin, P. (1982) "Influence of hydrodynamics and diffusion upon the stability limits of laminar premixed flames".

10. Sivashinsky, G. I. (1977) "Diffusional-thermal theory of cellular flames". Comb. Sci. Tech., 15, 137.

11. Van Harten, A., Kapila, A. K. and Matkowsky, B. J. (1984) "Acoustic coupling of flames". SIAM J. Appl. Math., 44, 982.

Department of Mathematical Sciences
Rensselear Polytechnic Institute
Troy, New York 12180-3590.

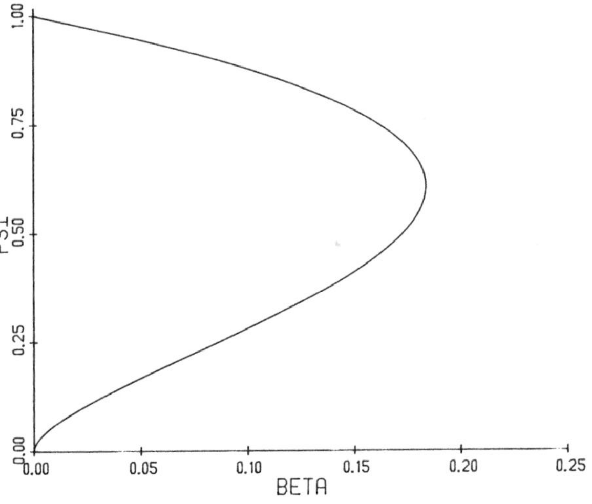

Fig. 1. Nonadiabatic-to-adiabatic flame-speed ratio ψ vs. heat-loss parameter β.

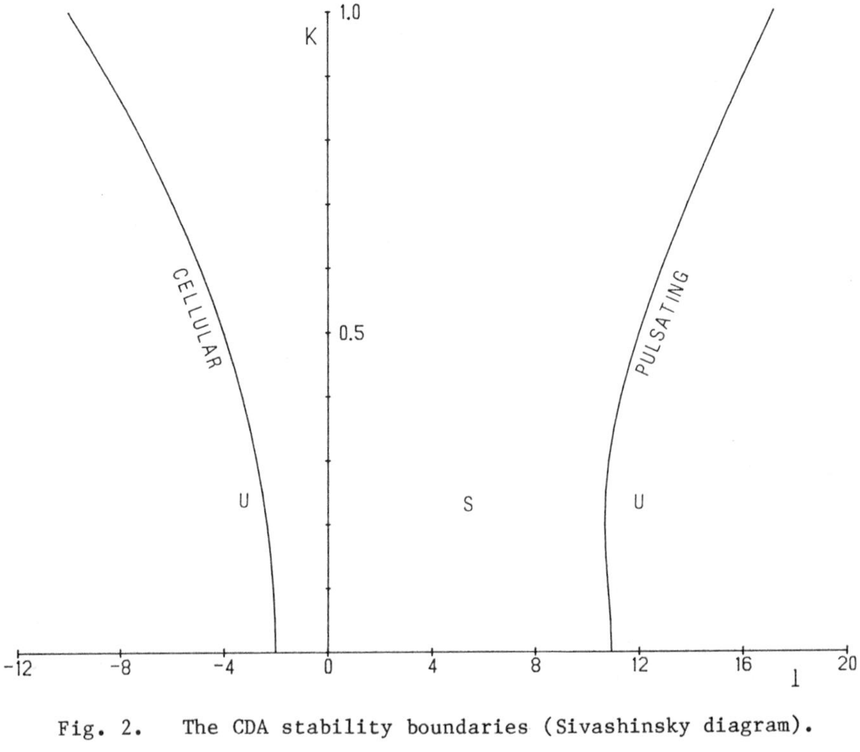

Fig. 2. The CDA stability boundaries (Sivashinsky diagram).

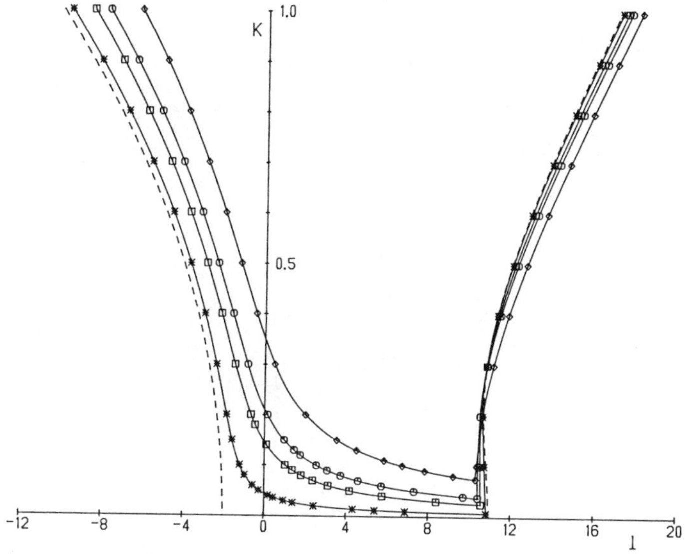

Fig. 3. Stability boundaries for no heat loss (ψ=1), no gravity (ϕ=∞) and different values of thermal expansion: α = 0.1(*), α = 0.5(□), α = 1(O), α = 5(◇). The dashed curves correspond to CDA.

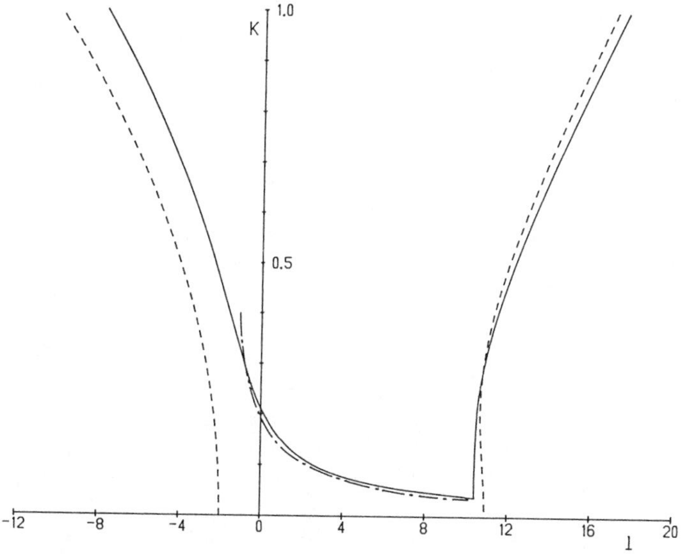

Fig. 4. Stability boundaries for no heat loss (ψ=1), no gravity (ϕ=∞) and α = 1. The dashed curve corresponds to CDA, and the dot-dash curve to the SVF analysis.

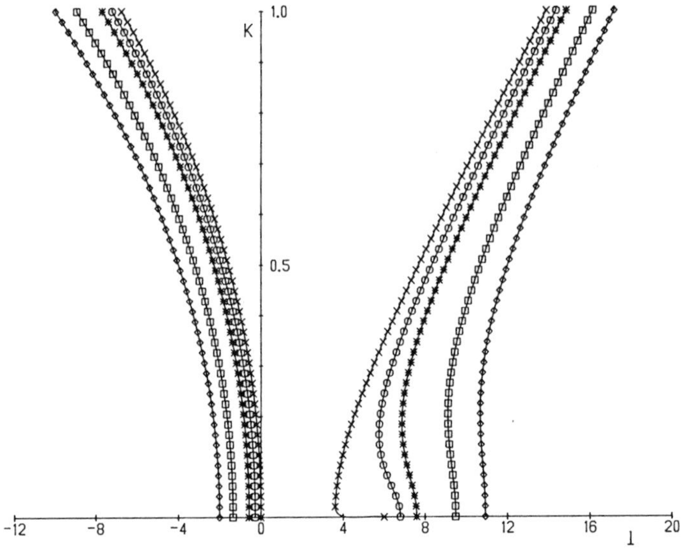

Fig. 5. CDA stability boundaries with heat loss: $\psi = 1(\diamond)$, $\psi = 0.85(\square)$, $\psi = 0.7(*)$, $\psi = 0.65(0)$, $\psi = e^{-1/2}(X)$. These results are due to Joalin and Clavin (1979).

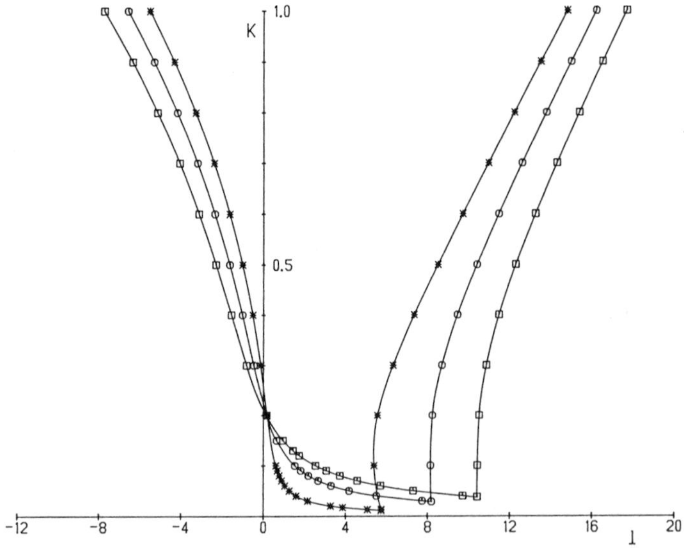

Fig. 6. Stability boundaries for no gravity ($\phi=\infty$), fixed thermal expansion ($\alpha=1$) and different values of heat loss: $\psi = 1(\square)$, $\psi = 0.8(0)$, $\psi = 0.65(*)$.

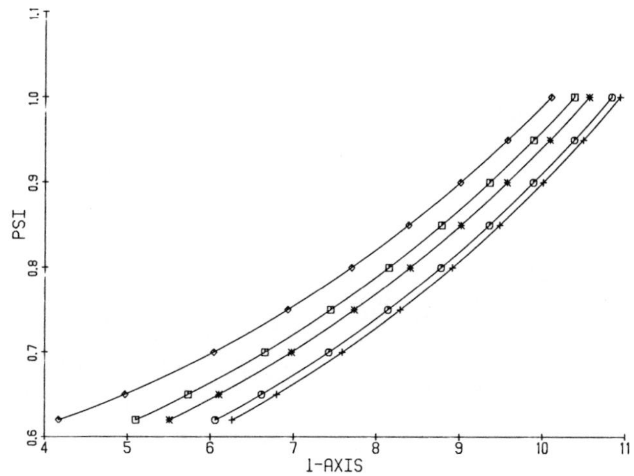

Fig. 7. Critical Lewis number ℓ_c for planar pulsations as a function of ψ for various α: CDA(+), α = 0.1(O), α = 0.5(*), α = 1(□), α = 5(◇).

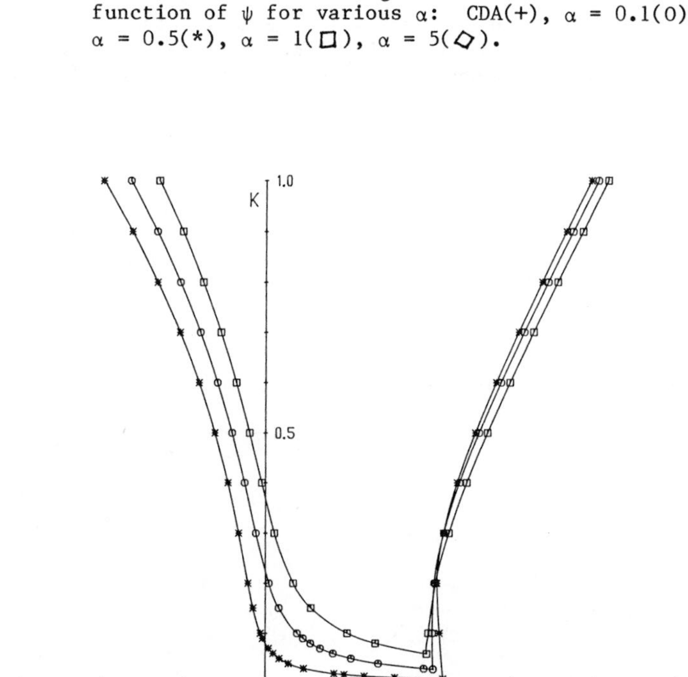

Fig. 8. Stability boundaries for no gravity ($\phi=\infty$), fixed heat loss (ψ=0.8) and different values of thermal expansion: α = 0.2(*), α = 1(O), α = 5(□).

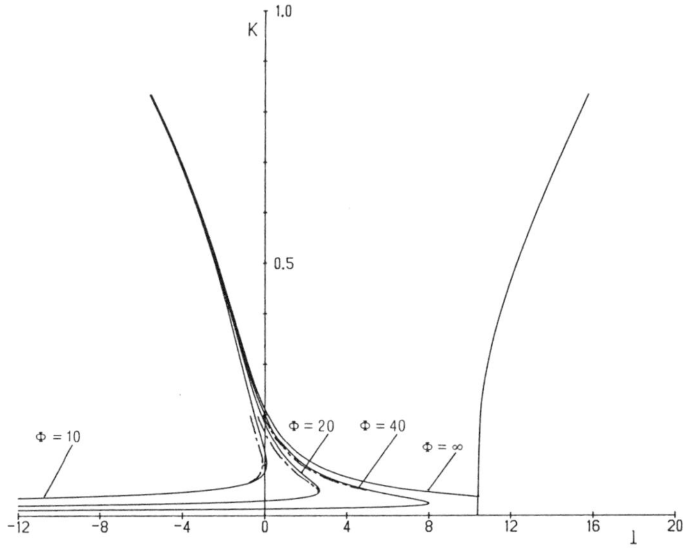

Fig. 9. Stability boundaries for no heat loss (ψ=1), given thermal expansion (α=1) and different Froude numbers ϕ. The dashed curves correspond to the SVF results of Pelce and Clavin.

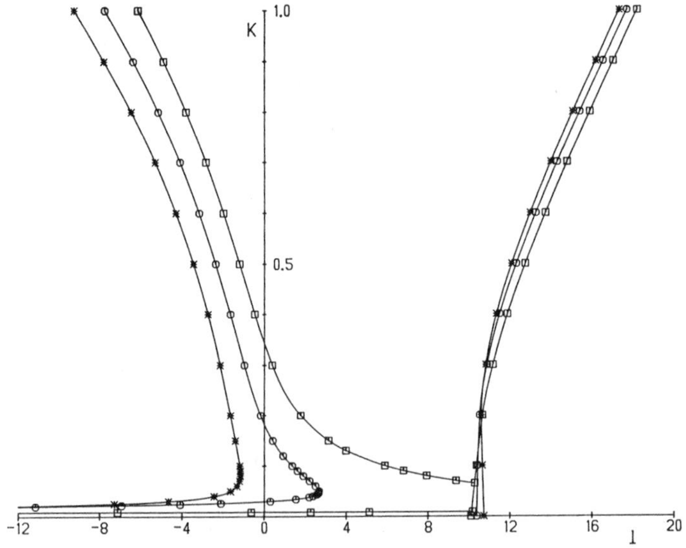

Fig. 10. Stability boundaries for no heat loss (ψ=1), given buoyancy (ϕ=20) and different values of thermal expansion: α = 0.2(*), α = 1(O), α = 5(□).

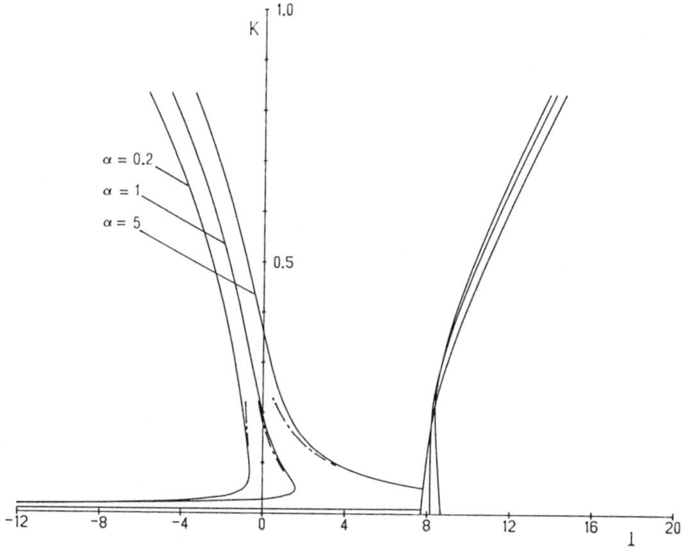

Fig. 11. Stability boundaries for given heat loss (ψ=0.8), given buoyancy (ϕ=20) and different values of thermal expansion. The dashed curves correspond to the SVF results of Clavin and Nicoli.

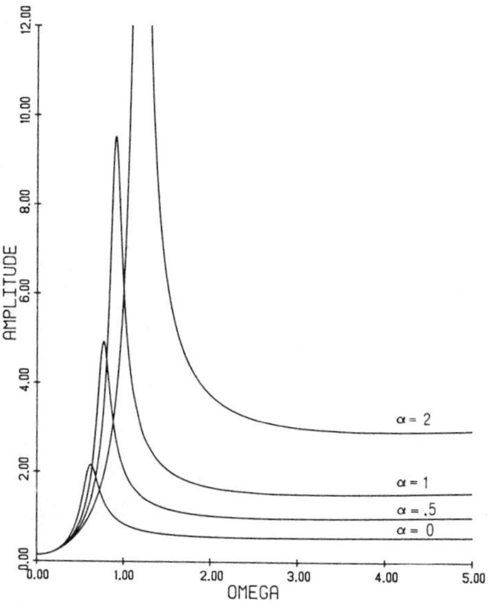

Fig. 12. Modulus of $\overline{m}(-\infty)$ vs. ω at ℓ = 10 for various α.

Lectures in Applied Mathematics
Volume 24, 1986

MODELING END-GAS KNOCK IN A RAPID-COMPRESSION MACHINE: REVIEW AND EXTENSION

W. B. Bush, F. E. Fendell, and S. F. Fink[1]

ABSTRACT. A rapid-compression machine is a laboratory apparatus to study aspects of the compression stroke, the combustion event, and the expansion stroke of an Otto-cycle engine. As a simple model of such a machine, unsteady one-dimensional nonisobaric laminar flame propagation through a combustible mixture, enclosed in a variable volume, is examined in the asymptotic limits of the Arrhenius-activation temperature large relative to the conventional adiabatic-flame temperature and the reference diffusion length small relative to the reference enclosure length. In these limits, a thin propagating flame separates the nondiffusive expanses of the burned and the unburned gases. The pressure throughout the enclosure is spatially homogeneous for smooth flame propagation. However, expansion of the hot burned gas results in the compressional heating of the remaining unburned gas, and, in fact, the spatially homogeneous gas may undergo autoconversion prior to the arrival of the propagating flame. If such an explosion is too rapid for acoustic adjustment, large spatial differences in the pressure arise, and the resulting nonlinear waves produce audible knock. Here, attention is concentrated on what fraction, if any, of the total charge may undergo autoconversion for a given operating condition, and what enhanced heat transfer from the end gas would preclude autoconversion--although too great a heat transfer from the end gas could result in flame quenching (unburned residual fuel).

1980 Mathematics Subject Classification. 80A32.

[1]Portions of this study were supported by the Army Research Office under contract DAAG29-83-C-0010 and by the Department of Energy under contract DE-AC04-78ET-13329.

0075-8485/86 $1.00 + $.25 per page

1. INTRODUCTION. During nonisobaric flame propagation through an initially homogeneous premixture of combustible gases in an (in general, variable-volume) enclosure, end-gas knock may occur. That is, the residual unburned gas may be preheated, prior to flame arrival, by compression, owing to the expansion of the already burned gas and possibly also owing to the movement of the walls of the enclosure. If this preheating results in a high-enough unburned-gas temperature, virtually the entire residual gas may undergo autoignition and rapid conversion to product in a spatially homogeneous explosion. This phenomenon is known as end-gas knock. Because the temporal scale of the abrupt local autoconversion may be shorter than the acoustic-wave speed, the pressure field (that, during normal flame propagation, is well approximated to be a function of time only) becomes spatially nonuniform, and the resulting nonlinear waves interact with the walls to produce loud noise. In view of the current capacity to treat the exhaust of Otto-cycle engines, the potential damage to engine components from end-gas knock is a major obstacle to operation at high thermal efficiency, i.e., at high compression ratios [1, 2].

In this paper, further insight into end-gas knock is sought from a mechanical-engineering viewpoint, as distinct from a chemical-kinetic one. That is, a Shvab-Zeldovich-type formulation of the one-dimensional unsteady laminar interplay of species and heat diffusion, exothermic chemical reaction, compressional heating, and heat loss is solved for an impervious noncatalytic variable-volume enclosure. Such a model may suffice for a (so-called) rapid-compression machine for cases in which ignition is arranged such that a planar flame propagates parallel to one movable (piston-like) end wall. Often, such a machine serves as a conveniently instrumented laboratory apparatus for detailed examination of compression and reaction (and perhaps subsequent expansion as well). In contrast, in an Otto-cycle engine, the range of parametric variation may, in practice, be constrained to a relatively narrow range by cost and available equipment [3, 4].

Accordingly, the following questions are addressed:

(1) Under what conditions of operation is the onset of knock to be expected in a rapid-compression machine?

(2) If knock does occur, what fraction of the initial premixture undergoes autoconversion to product?

(3) What heat transfer from the end gas would counteract the compressional heating to produce smooth flame propagation across the entire combustible mixture?

Specifically, to answer these questions, the aforementioned formulation is examined in the asymptotic limits of

(1) the Arrhenius-activation temperature large relative to the initial adiabatic-flame temperature;

(2) the reference diffusion length (i.e., the initial mass-transfer coefficient/adiabatic-flame speed ratio) small relative to the reference enclosure length.

In these limits, analyses are presented for

(1) the unburned and burned external regions of the flow, where the diffusion and reaction-rate contributions to the species and energy equations are negligible;

(2) the propagating flame zone--consisting of a relatively thin convective-diffusive subregion and a still thinner diffusive-reactive one--that separates the external regions.

2. FORMULATION. A low-Mach-number laminar unsteady one-dimensional flow, in which fuel F and oxygen O pass exothermically to product P (in the presence of nitrogen N) via the direct one-step bimolecular second-order Arrhenius-type global reaction

$$\nu_F F + \nu_O O + \nu_N N \rightarrow \nu_P P + \nu_N N, \qquad (2.1)$$

where ν_i is the stoichiometric coefficient of species i, is considered. For this flow, the dimensional conservation equations for mass, species, and energy, and the state equation are taken, in the domain $0 < x^* < L^*(t^*)$, $t^* > 0$, to be [5, 6]:

$$\frac{\partial \rho^*}{\partial t^*} + \frac{\partial(\rho^* u^*)}{\partial x^*} = 0: \quad \frac{\partial \psi^*}{\partial x^*} = \rho^*, \quad \frac{\partial \psi^*}{\partial t^*} = -\rho^* u^*; \tag{2.2}$$

$$\rho^*\left(\frac{\partial Y_i}{\partial t^*} + u^* \frac{\partial Y_i}{\partial x^*}\right) - \frac{\partial}{\partial x^*}\left((\rho^* D^*)\frac{\partial Y_i}{\partial x^*}\right) = -w^*,$$

$$\text{for } i = F, O, \text{ with } Y_i = \frac{m^* \tilde{Y}_i}{m_i^* \nu_i}; \tag{2.3}$$

$$\rho^*\left(\frac{\partial T^*}{\partial t^*} + u^* \frac{\partial T^*}{\partial x^*}\right) - \frac{\partial}{\partial x^*}\left((\lambda^*/c_p^*)\frac{\partial T^*}{\partial x^*}\right) = \left(\frac{Q^*}{c_p^*}\right) w^* + \frac{1}{c_p^*}\frac{dp^*}{dt^*}; \tag{2.4}$$

$$p^* = \rho^* R^* T^*. \tag{2.5}$$

Here, t^* and x^* are the time and Cartesian spatial coordinates, respectively; u^* is the gas speed and ψ^* is the stream function; ρ^* and T^* are the density and temperature; R^* is the gas constant for the mixture, taken to be composed of species of comparable molecular weights; p^* is the pressure, taken to be a function of time only from consideration of the equation of conservation of momentum; $\tilde{Y}_i$ and m_i^* are the mass fraction and molecular weight, respectively, of species i; $m^* = \nu_F m_F^* + \nu_O m_O^* + \nu_N m_N^*$; D^* and λ^* are the mass-transfer and thermal-conductivity coefficients, respectively, and c_p^* is the (constant universal) heat capacity at constant pressure; Q^* is the heat of combustion per mass of premixture; and w^* is the reaction rate. The Lewis-Semenov number is taken as a constant of order unity, i.e.,

$$Le = \frac{(\lambda^*/c_p^*)}{(\rho^* D^*)} = \text{const} \sim O(1). \tag{2.6}$$

It is also taken that

$$\rho^{*2} D^* \sim p^{*n} = \text{fnc}(t^*). \tag{2.7}$$

Further, consistent with the nature of the formulation, the reaction rate is taken to be

$$w^* = B'^* T^{*\alpha} \rho^{*\nu_F + \nu_O} Y_F^{\nu_F} Y_O^{\nu_O} \exp\left\{-\frac{T_a^*}{(T^* - T_{u0}^*)}\right\}, \tag{2.8}$$

where B'^* is the (constant) frequency factor; T_a^* is the (constant) Arrhenius activation temperature; T_{u0}^* is the uniform premixture temperature, introduced to modify the Arrhenius factor, so as to preclude the "cold-boundary difficulty"; and α characterizes the pre-exponential thermal dependence of the reaction rate. Here, $\alpha = \{(\nu_F + \nu_0) - 1\}$ is adopted.

It is convenient to adopt the von Mises transformation $(x^*,t^*) \to (\psi^*,t^*)$, where the stream function ψ^* is defined in (2.2) Under this transformation, the species and energy conservation equations, (2.3) and (2.4), become

$$\frac{\partial Y_i}{\partial t^*} - p^{*n}\frac{\partial^2 Y_i}{\partial \psi^{*2}}$$

$$= -B'^*(p^*/R^*)^\alpha Y_F^{\nu_F} Y_O^{\nu_O} \exp\left\{-\frac{T_a^*}{(T^*-T_{u0}^*)}\right\}, \text{ for } i = F, O, \tag{2.9}$$

$$\frac{\partial T^*}{\partial t^*} - Le\, p^{*n}\frac{\partial^2 T^*}{\partial \psi^{*2}}$$

$$= (Q^*/c_p^*)B'^*(p^*/R^*)^\alpha Y_F^{\nu_F} Y_O^{\nu_O} \exp\left\{-\frac{T_a^*}{(T^*-T_{u0}^*)}\right\} + \frac{T^*}{(c_p^*/R^*)}\frac{1}{p^*}\frac{dp^*}{dt^*}. \tag{2.10}$$

The mass conservation and mapping equations are

$$\frac{\partial u^*}{\partial \psi^*} = \frac{\partial}{\partial t^*}\left(\frac{1}{\rho^*}\right), \quad \frac{\partial x^*}{\partial \psi^*} = \frac{1}{\rho^*}. \tag{2.11a,b}$$

For nondimensionalization,

$$t = \frac{t^*}{(L_0^*/u_{u0}^*)}, \quad \psi = \frac{\psi^*}{(\rho_{u0}^* L_0^*)}; \tag{2.12}$$

$$Y(\psi,t) = \frac{Y_F(\psi^*,t^*)}{Y_{Fu}} = \frac{(1-\phi)}{\phi}\left\{\frac{Y_O(\psi^*,t^*)}{Y_{Ob}} - 1\right\}, \tag{2.13a}$$

$$T(\psi,t) = \frac{T^*(\psi^*,t^*) - T_{u0}^*}{(T_{b0}^* - T_{u0}^*)}; \tag{2.13b}$$

$$p(t) = \frac{p^*(t^*)}{p_0^*}; \tag{2.14}$$

$$\rho(\psi,t) = \frac{\rho^*(\psi^*,t^*)}{\rho_{u0}^*}, \quad u(\psi,t) = \frac{u^*(\psi^*,t^*)}{u_{u0}^*}, \quad x(\psi,t) = \frac{x^*(\psi^*,t^*)}{L_0^*}. \tag{2.15a-c}$$

Here, the subscript *u* denotes unburned-premixture-state conditions, the subscript *b* denotes burned-state conditions, and the subscript *0* denotes initial conditions. Specifically, L_0^* is the initial separation distance between the walls, u_{u0}^* is the initial adiabatic-flame speed, and T_{b0}^* is the initial adiabatic-flame temperature. Also, $\phi = (Y_{Fu}/Y_{Ou})$ is the equivalence ratio, with $\phi < 1$ for the fuel-lean-stoichiometry cases of interest; thus, $Y_{Fb} = 0$. Hence, with $\Phi = \{\phi/(1-\phi)\}$ of order unity, $Y_{Ob} = (Y_{Ou}-Y_{Fu}) = Y_{Fu}/\Phi$, and $T_b^* = \{T_u^* + (Q^*/c_p^*)Y_{Fu}\}$ (including $T_{b0}^* = \{T_{u0}^* + (Q^*/c_p^*)Y_{Fu}\}$). Note also that $p_0^* = \rho_{u0}^* R^* T_{u0}^*$.

Thus, the nondimensional species and energy equations are, for $\alpha, n = 1$ and $\nu_F, \nu_O = 1$,

$$\frac{\partial Y}{\partial t} - \varepsilon p \frac{\partial^2 Y}{\partial \psi^2} = -\frac{1}{\varepsilon}\Lambda p Y(1+\Phi Y)\exp\left\{-\beta\frac{(1-T)}{T}\right\}, \tag{2.16}$$

$$\frac{\partial T}{\partial t} - \varepsilon Le\, p \frac{\partial^2 T}{\partial \psi^2} = \frac{1}{\varepsilon}\Lambda p Y(1+\Phi Y)\exp\left\{-\beta\frac{(1-T)}{T}\right\} + \frac{(\gamma-1)}{\gamma}\frac{(1+KT)}{Kp}\frac{dp}{dt}, \tag{2.17}$$

where

$$\varepsilon = \frac{(D_{u0}^*/u_{u0}^*)}{L_0^*} \ll 1, \tag{2.18}$$

$$\beta = \frac{T_a^*}{(T_{b0}^* - T_{u0}^*)} \gg 1,\quad \Lambda = \left(\frac{B'^* D_{u0}^*}{u_{u0}^{*2}}\right)\left(\frac{p_0^*}{R^*}\right)\left(\frac{Y_{Fu}}{\Phi}\right)\exp(-\beta) \gg 1, \tag{2.19a,b}$$

$$K = \frac{(T_{b0}^* - T_{u0}^*)}{T_{u0}^*} = \frac{(Q^*/c_p^*)Y_{Fu}}{T_{u0}^*} > 1,\quad \gamma = \frac{c_p^*}{c_v^*} = \frac{c_p^*}{(c_p^* - R^*)} > 1. \tag{2.20a,b}$$

The nondimensional state, continuity, and mapping equations are

$$\rho = \frac{p}{(1+KT)},\quad \frac{\partial u}{\partial \psi} = \frac{\partial}{\partial t}\left(\frac{1}{\rho}\right) = \frac{\partial}{\partial t}\left(\frac{(1+KT)}{p}\right),\quad \frac{\partial x}{\partial \psi} = \frac{1}{\rho} = \frac{(1+KT)}{p}. \tag{2.21a-c}$$

From steady (isobaric) fuel-lean laminar flame propagation, for $\beta \gg 1$, the only case of practical interest in an automotive context, it has been determined that, for $\nu_F = 1$,

$$\Lambda = \frac{\beta^2}{2\,Le^2}\left[1 + O(\bar{\beta}^{1})\right], \tag{2.22}$$

where the higher-order terms are known but are not utilized here.

In this formulation, a finite spatial domain $0 < x < L(t)$ is examined, where the (cylinder-head-like) wall, at $x = 0$, is fixed, and the (piston-like) wall, $x = L(t)$, moves via a prescribed rule (see Sec. 3), with $L(0) = 1$. Under the von Mises transformation, with the fixed wall located at $\psi = 0$, the moving wall is located at $\psi = \Psi(t)$, with $\Psi(0) = \Psi_0 = \text{const.} > 0$, i.e.,

$$\psi = \int_0^x \rho\, dx_1 = p\int_0^x \frac{dx_1}{(1+KT)}\,;\quad \Psi = \int_0^L \rho\, dx_1 = p\int_0^L \frac{dx_1}{(1+KT)}\,. \tag{2.23a,b}$$

In general, the initial conditions for this flow geometry are

$$Y \to 1,\ T \to 0\ (p \to 1)\ \text{as}\ t \to 0\ \text{for}\ \psi > (\psi_{f0})_+ \to 0; \tag{2.24a}$$

$$Y \to 0,\ T \to 1\ (p \to 1)\ \text{as}\ t \to 0\ \text{as}\ \psi \to (\psi_{f0})_-\,. \tag{2.24b}$$

Note that, subject to (2.24), it follows from (2.23) that $\Psi_0 = 1$. In general, if both walls are impermeable noncatalytic adiabatic ones,[2] the boundary conditions are

$$\frac{\partial Y}{\partial \psi}, \frac{\partial T}{\partial \psi} \to 0,\quad x, u \to 0\ \text{as}\ \psi \to 0\ (t > 0); \tag{2.25a}$$

$$\frac{\partial Y}{\partial \psi}, \frac{\partial T}{\partial \psi} \to 0,\quad x \to L(t),\ u \to L'(t)\ \text{as}\ \psi \to 1\ (t > 0). \tag{2.25b}$$

---

[2] On physical grounds, since the pressure is spatially invariant, the results depend on the wall *separation* as a function of time. As long as the temporal history of the wall separation is the same, the results are invariant to which end wall moves or both end walls move. Also, from (2.19) and (2.22), altering some of the parameters not only alters the dimensionless groups, but also alters the derived quantity u^*_{u0}, so that the temporal and stream-function scales used for nondimensionalization are altered. Finally, in a straightforward generalization of (2.24), sparking need not occur necessarily at time $t = 0$; e.g., compressional heating, owing to wall motion, may occur for a finite time interval before sparking initiates flame propagation.

The jump conditions at the flame, $\psi = \psi_f(t)$, are

$$[\![Y]\!]_f = \{Y_b(\psi_f(t),t) - Y_u(\psi_f(t),t)\} = \{Y_{bf}(t) - Y_{uf}(t)\} = -1; \qquad (2.26a)$$

$$[\![T]\!]_f = \{T_b(\psi_f(t),t) - T_u(\psi_f(t),t)\} = \{T_{bf}(t) - T_{uf}(t)\} = 1. \qquad (2.26b)$$

3. EXTERNAL HYDRODYNAMICS OF FLAME PROPAGATION. Exterior to the thin-flame structure, there is an unburned region, defined by $\psi_f(t) < \psi < 1$, and a burned region, defined by $\psi_f(t) > \psi > 0$. In both of these unburned and burned regions, the diffusion and reaction-rate contributions are taken to be negligible, so that the species and energy equations, (2.16) and (2.17), reduce to

$$\frac{\partial Y}{\partial t} = 0, \quad \frac{\partial T}{\partial t} - \frac{(\gamma-1)}{\gamma}\frac{(1+KT)}{Kp}\frac{dp}{dt} = 0. \qquad (3.1,2)$$

Here, the unburned region, $\psi_f(t) < \psi < 1$, is considered to consist of a bulk-gas region, $\psi_f(t) < \psi < \psi_i$, and an end-gas region, $\psi_i < \psi < 1$, where ψ_i (= given const) denotes the interface, or contact surface, between the bulk-gas and end-gas regions. This division of the unburned region holds for $0 < t < t_i$. When $t = t_i$, the last of the unburned bulk gas is now burned (bulk) gas, i.e., $\psi_f(t_i) = \psi_i$. For $t > t_i$, the end gas is the sole component of the unburned region. For $0 < t < t_i$, the bulk gas is the sole component of the burned region, $\psi_f(t) > \psi > 0$. However, for $t > t_i$, the burned region consists of a burned end-gas region, $\psi_f(t) > \psi > \psi_i$, and a burned bulk-gas region, $\psi_i > \psi > 0$. To distinguish among these regions, the following notation is introduced:

unburned bulk gas $(\psi_f(t) < \psi < \psi_i)$

$$Y(\psi,t) = Y_u(\psi,t), \quad T(\psi,t) = T_u(\psi,t), \quad \gamma = \kappa; \qquad (3.3a)$$

unburned end gas $(\psi_i < \psi < 1)$

$$Y(\psi,t) = X_u(\psi,t), \quad T(\psi,t) = \Theta_u(\psi,t), \quad \gamma = \mu; \qquad (3.3b)$$

burned bulk gas $(\psi_i > \psi_f(t) > \psi > 0)$

$$Y(\psi,t) = Y_b(\psi,t), \quad T(\psi,t) = T_b(\psi,t), \quad \gamma = \sigma; \tag{3.3c}$$

burned end gas $(\psi_f(t) > \psi > \psi_i)$

$$Y(\psi,t) = X_b(\psi,t), \quad T(\psi,t) = \Theta_b(\psi,t), \quad \gamma = \chi; \tag{3.3d}$$

For the unburned bulk gas, (3.1) and (3.2) become

$$\frac{\partial Y_u}{\partial t} = 0, \quad \frac{\partial T_u}{\partial t} - \frac{(\kappa-1)}{\kappa}\frac{(1+KT_u)}{Kp}\frac{dp}{dt} = 0. \tag{3.4a,b}$$

The solutions of (3.4), subject to the initial conditions for this region,

$$p(0) = 1, \; Y_u(\psi,0) = 1, \; T_u(\psi,0) = 0, \text{ and } \psi_f(0) < \psi < \psi_i, \tag{3.5}$$

with $\psi_f(0) = \psi_{f0} \to 0$, ψ_i = const (to be specified), are

$$Y_u(\psi,t) = 1, \quad T_u(\psi,t) = T_u(t) = \frac{1}{K}\left[\{p(t)\}^{(\kappa-1)/\kappa} - 1\right]. \tag{3.6a,b}$$

In a similar manner, for the unburned end gas, the solutions of

$$\frac{\partial X_u}{\partial t} = 0, \quad \frac{\partial \Theta_u}{\partial t} - \frac{(\mu-1)}{\mu}\frac{(1+K\Theta_u)}{Kp}\frac{dp}{dt} = 0, \tag{3.7a,b}$$

subject to the initial conditions

$$p(0) = 1, \; X_u(\psi,0) = 1, \; \Theta_u(\psi,0) = 0 \text{ and } \psi_i < \psi < 1, \tag{3.8}$$

are

$$X_u(\psi,t) = 1, \quad \Theta_u(\psi,t) = \Theta_u(t) = \frac{1}{K}\left[\{p(t)\}^{(\mu-1)/\mu} - 1\right]. \tag{3.9a,b}$$

For the burned-bulk-gas region, (3.1) and (3.2) become

$$\frac{\partial Y_b}{\partial t} = 0, \quad \frac{\partial T_b}{\partial t} - \frac{(\sigma-1)}{\sigma}\frac{(1+KT_b)}{Kp}\frac{dp}{dt} = 0. \tag{3.10a,b}$$

The solutions are of the forms

$$Y_b(\psi,t) = H_b(\psi), \quad T_b(\psi,t) = \frac{1}{K}\left[F_b(\psi)\{p(t)\}^{(\sigma-1)/\sigma} - 1\right]. \tag{3.11a,b}$$

At the flame,

$$Y_b(\psi_f(t),t) = H_b(\psi_f(t)): \quad Y_{bf}(t) = H_{bf}(t), \tag{3.12a}$$

$$T_b(\psi_f(t),t) = \frac{1}{K}\left[F_b(\psi_f(t))\{p(t)\}^{(\sigma-1)/\sigma} - 1\right]:$$

$$T_{bf}(t) = \frac{1}{K}\left[F_{bf}(t)\{p(t)\}^{(\sigma-1)/\sigma} - 1\right]. \tag{3.12b}$$

For $t < t_i$ and/or $\psi_f(t) < \psi_i$, the jump conditions at the flame (between the unburned and the burned bulk gas) yield

$$Y_{bf}(t) = 0 \quad (\text{such that } Y_b(\psi,t) = 0), \tag{3.13a}$$

$$T_{bf}(t) = \{1 + T_u(t)\} \Rightarrow F_{bf}(t) = \frac{\{p(t)\}^{(K-1)/K} + K}{\{p(t)\}^{(\sigma-1)/\sigma}}. \tag{3.13b}$$

For the burned-end-gas region, it follows that

$$\frac{\partial X_b}{\partial t} = 0, \quad \frac{\partial \Theta_b}{\partial t} - \frac{(\chi-1)}{\chi}\frac{(1+K\Theta_b)}{Kp}\frac{dp}{dt} = 0. \tag{3.14a,b}$$

The forms of the solutions are, thus,

$$X_b(\psi,t) = N_b(\psi), \quad \Theta_b(\psi,t) = \frac{1}{K}\left[G_b(\psi)\{p(t)\}^{(\chi-1)/\chi} - 1\right]. \tag{3.15a,b}$$

At the flame,

$$X_b(\psi_f(t),t) = N_b(\psi_f(t)): \quad X_{bf}(t) = N_{bf}(t), \tag{3.16a}$$

$$\Theta_b(\psi_f(t),t) = \frac{1}{K}\left[G_b(\psi_f(t))\{p(t)\}^{(\chi-1)/\chi} - 1\right]:$$

$$\Theta_{bf}(t) = \frac{1}{K}\left[G_{bf}(t)\{p(t)\}^{(\chi-1)/\chi} - 1\right]. \tag{3.16b}$$

For $t > t_i$ and/or $\psi_f(t) > \psi_i$, the flame jump conditions (between the unburned and the burned end gas) yield

$$X_{bf}(t) = 0 \quad (\text{such that } X_b(\psi,t) = 0) \tag{3.17a}$$

$$\Theta_{bf}(t) = \{1 + \Theta_u(t)\} \Rightarrow G_{bf}(t) = \frac{\{p(t)\}^{(\mu-1)/\mu} + K}{\{p(t)\}^{(\chi-1)/\chi}}. \tag{3.17b}$$

3.1. Flame in the Bulk Gas ($\psi_f(t) < \psi_i$)

For the case of $t < t_i$ and/or $\psi_f(t) < \psi_i$, integration of the mapping equation, (2.21c), over the domain $0 \le \psi \le 1$ yields

$$\int_0^{L(t)} dx = \int_0^{\psi_f(t)} \frac{\{1+KT_b(\psi,t)\}}{p(t)} d\psi + \int_{\psi_f(t)}^{\psi_i} \frac{\{1+KT_u(\psi,t)\}}{p(t)} d\psi + \int_{\psi_i}^{1} \frac{\{1+K\Theta_u(\psi,t)\}}{p(t)} d\psi :$$

$$L(t) = \frac{1}{\{p(t)\}^{1/\sigma}} \int_0^{\psi_f(t)} F_b(\psi)\, d\psi + \frac{\{\psi_i - \psi_f(t)\}}{\{p(t)\}^{1/\kappa}} + \frac{(1-\psi_i)}{\{p(t)\}^{1/\mu}}. \tag{3.18}$$

The time derivative of $[\{p(t)\}^{1/\sigma}L(t)]$, as determined from (3.18), with rearrangement, produces

$$\frac{1}{\sigma}\frac{p'(t)}{p(t)}\left[1 - \frac{(\kappa-\sigma)}{\kappa}\frac{\{\psi_i-\psi_f(t)\}}{L(t)\{p(t)\}^{1/\kappa}} - \frac{(\mu-\sigma)}{\mu}\frac{(1-\psi_i)}{L(t)\{p(t)\}^{1/\mu}}\right] + \frac{L'(t)}{L(t)} - \frac{K\psi_f'(t)}{L(t)p(t)} = 0. \tag{3.19}$$

Integration of the mapping equation over the unburned domain $\psi_f(t) < \psi < \psi_i < 1$ yields

$$\int_{x_u(\psi,t)}^{L(t)} dx = \int_{\psi}^{\psi_i} \frac{\{1+KT_u(v,t)\}}{p(t)} dv + \int_{\psi_i}^{1} \frac{\{1+K\Theta_u(v,t)\}}{p(t)} dv :$$

$$x_u(\psi,t) = L(t) - \frac{(\psi_i-\psi)}{\{p(t)\}^{1/\kappa}} - \frac{(1-\psi_i)}{\{p(t)\}^{1/\mu}}. \tag{3.20}$$

Thus, the locations of the flame front and the contact surface are determined to be

$$x_u(\psi_f(t),t) = x_f(t) = L(t) - \frac{\{\psi_i-\psi_f(t)\}}{\{p(t)\}^{1/\kappa}} - \frac{(1-\psi_i)}{\{p(t)\}^{1/\mu}}; \tag{3.21a}$$

$$x_u(\psi_i,t) = x_i(t) = L(t) - \frac{(1-\psi_i)}{\{p(t)\}^{1/\mu}}. \tag{3.21b}$$

The speed of the flame front is

$$\frac{dx_f}{dt}(t) = x_f'(t) = L'(t) + \frac{p'(t)}{p(t)}\left[\frac{1}{\kappa}\frac{\{\psi_i-\psi_f(t)\}}{\{p(t)\}^{1/\kappa}} + \frac{1}{\mu}\frac{(1-\psi_i)}{\{p(t)\}^{1/\mu}}\right] + \frac{\psi_f'(t)}{\{p(t)\}^{1/\kappa}}. \tag{3.22}$$

In turn, the speed of the gas is

$$\frac{\partial x_u}{\partial t}(\psi,t) = u_u(\psi,t) = L'(t) + \frac{p'(t)}{p(t)}\left[\frac{1}{\kappa}\frac{(\psi_i-\psi)}{\{p(t)\}^{1/\kappa}} + \frac{1}{\mu}\frac{(1-\psi_i)}{\{p(t)\}^{1/\mu}}\right]. \tag{3.23}$$

Thus, the speed of the gas at the flame front and that at the contact surface are

$$u_u(\psi_f(t),t) = u_f(t) = L'(t) + \frac{p'(t)}{p(t)}\left[\frac{1}{\kappa}\frac{\{\psi_i-\psi_f(t)\}}{\{p(t)\}^{1/\kappa}} + \frac{1}{\mu}\frac{(1-\psi_i)}{\{p(t)\}^{1/\mu}}\right]; \tag{3.24a}$$

$$u_u(\psi_i,t) = u_i(t) = x_i'(t) = L'(t) + \frac{p'(t)}{p(t)}\left[\frac{1}{\mu}\frac{(1-\psi_i)}{\{p(t)\}^{1/\mu}}\right]. \tag{3.24b}$$

[The notation of (3.24b) reinforces the (known) result that the speed of the gas is continuous across the contact surface at $\psi(t) = \psi_i$.]

Thus, the flame speed, $S(t)$, the difference between the speed of the flame front, $x_f'(t)$, and that of the unburned gas at the flame front, $u_f(t)$, is, from (3.22) and (3.24a),

$$S(t) = \{x_f'(t) - u_f(t)\} = \frac{\psi_f'(t)}{\{p(t)\}^{1/\kappa}}. \tag{3.25}$$

In Appendix A, it is determined that, for α, $n = 1$ and ν_F, $\nu_O = 1$, the flame speed is

$$S(t) \approx \{1 + KT_u(t)\}\{1+T_u(t)\}^2 \exp\left[\frac{\beta}{2}\frac{T_u(t)}{\{1+T_u(t)\}}\right]. \tag{3.26}$$

Thus, with the specification of the motion of the moving wall, i.e., $L(t)$ and $L'(t)$, subject to $L(0) = 1$, and the interface location, ψ_i, the pertinent initial-value problem for the determination of $p(t)$, $\psi_f(t)$, $T_u(t)$, $\Theta_u(t)$, $x_f(t)$, and $x_i(t)$, from (3.19), (3.25) and (3.26), (3.6b) and (3.9b), (3.22), and (3.24b), is

$$\frac{1}{\sigma}\frac{p'(t)}{p(t)}\left[1 - \frac{(\kappa-\sigma)}{\kappa}\frac{\{\psi_i-\psi_f(t)\}}{L(t)\{p(t)\}^{1/\kappa}} - \frac{(\mu-\sigma)}{\mu}\frac{(1-\psi_i)}{L(t)\{p(t)\}^{1/\mu}}\right] = -\frac{L'(t)}{L(t)} + \frac{K\psi_f'(t)}{L(t)p(t)}, \tag{3.27a}$$

$$\frac{\psi_f'(t)}{\{p(t)\}^{1/\kappa}} = \{1+KT_u(t)\}\{1+T_u(t)\}^2\exp\left[\frac{\beta}{2}\frac{T_u(t)}{\{1+T_u(t)\}}\right], \tag{3.27b}$$

$$\frac{KT_u'(t)}{\{1+KT_u(t)\}} = \frac{(K-1)}{K}\frac{p'(t)}{p(t)}, \qquad \frac{K\Theta_u'(t)}{\{1+K\Theta_u(t)\}} = \frac{(\mu-1)}{\mu}\frac{p'(t)}{p(t)}, \tag{3.27c,d}$$

$$x_f'(t) = L'(t) + \frac{p'(t)}{p(t)}\left[\frac{1}{K}\frac{\{\Psi_i-\Psi_f(t)\}}{\{p(t)\}^{1/K}} + \frac{1}{\mu}\frac{(1-\Psi_i)}{\{p(t)\}^{1/\mu}}\right] + \frac{\Psi_f'(t)}{\{p(t)\}^{1/K}}, \tag{3.27e}$$

$$x_i'(t) = L'(t) + \frac{p'(t)}{p(t)}\left[\frac{1}{\mu}\frac{(1-\Psi_i)}{\{p(t)\}^{1/\mu}}\right]; \tag{3.27f}$$

$$p(0)=1,\ \Psi_f(0)=0,\ T_u(0)=0,\ \Theta_u(0)=0,\ x_f(0)=0,\ x_i(0)=\Psi_i. \tag{3.28a-f}$$

This particular problem terminates when $t = t_i$, such that $L(t_i) = L_i$ and

$$\Psi_f(t_i)=\Psi_i,\ p(t_i)=p_i,\ T_u(t_i)=T_{ui},\ \Theta_u(t_i)=\Theta_{ui},\ x_f(t_i)=x_i(t_i)=x_{fi}, \tag{3.29}$$

where p_i, T_{ui}, Θ_{ui}, and x_{fi} are found in the course of solution. Here, the motion of the moving wall is approximated as

$$L(t) = \left[1 - \left\{\frac{(CR)-1}{2(CR)}\right\}(1-\cos\omega t)\right],\quad L'(t) = -\left\{\frac{(CR)-1}{2(CR)}\right\}(\omega\sin\omega t), \tag{3.30a,b}$$

where (CR) [= $(L_{max}/L_{min}) = (1/L_{min}) > 1$] is the compression ratio, and ω [= $2\pi\Omega = 2\pi\{\Omega^*/(u^*_{u0}/L^*_0)\}$, with Ω^* the number of revolutions per unit time] is the (nondimensional) frequency.

3.2 Flame in the End Gas ($\Psi_f(t) > \Psi_i$)

For $t > t_i$ and/or $\Psi_f(t) > \Psi_i$, integration of the mapping equation, (2.21c), over the domain $0 \le \Psi \le 1$ yields

$$L(t) = \frac{1}{\{p(t)\}^{1/\sigma}}\int_0^{\Psi_i} F_b(\Psi)\,d\Psi + \frac{1}{\{p(t)\}^{1/\chi}}\int_{\Psi_i}^{\Psi_f(t)} G_b(\Psi)\,d\Psi + \frac{\{1-\Psi_f(t)\}}{\{p(t)\}^{1/\mu}}. \tag{3.31}$$

Thus, for this case,

$$\frac{1}{\sigma}\frac{p'(t)}{p(t)}\left[1 - \frac{(\chi-\sigma)}{\chi}\frac{\int_{\Psi_i}^{\Psi_f(t)} G_b(\Psi)\,d\Psi}{L(t)\{p(t)\}^{1/\chi}} - \frac{(\mu-\sigma)}{\mu}\frac{\{1-\Psi_f(t)\}}{L(t)\{p(t)\}^{1/\mu}}\right] + \frac{L'(t)}{L(t)} - \frac{K\Psi_f'(t)}{L(t)p(t)} = 0. \tag{3.32}$$

Here, it is approximated that

$$\int_{\Psi_i}^{\Psi_f(t)} G_b(\Psi)\,d\Psi \approx \frac{1}{2}\{\Psi_f(t)-\Psi_i\}\{G_{bf}(t)+G_{bi}\}, \tag{3.33a}$$

where

$$G_{bf}(t) = G_b(\psi_f(t)) = \frac{\{p(t)\}^{(\mu-1)/\mu} + K}{\{p(t)\}^{(\chi-1)/\chi}},$$

$$G_{bi} = G_b(\psi_i) = \frac{p_i^{(\mu-1)/\mu} + K}{p_i^{(\chi-1)/\chi}}. \tag{3.33b}$$

Integration of the mapping equation over the (end-gas) unburned domain $\psi_i < \psi_f(t) < \psi < 1$ yields

$$x_u(\psi,t) = L(t) - \frac{(1-\psi)}{\{p(t)\}^{1/\mu}}. \tag{3.34}$$

Thus, the location and speed of the flame front are

$$x_u(\psi_f(t),t) = x_f(t) = L(t) - \frac{\{1-\psi_f(t)\}}{\{p(t)\}^{1/\mu}}; \tag{3.35a}$$

$$\frac{dx_f}{dt}(t) = x_f'(t) = L'(t) + \frac{p'(t)}{p(t)}\left[\frac{1}{\mu}\frac{\{1-\psi_f(t)\}}{\{p(t)\}^{1/\mu}}\right] + \frac{\psi_f'(t)}{\{p(t)\}^{1/\mu}}. \tag{3.35b}$$

In turn, the speed of the gas is

$$\frac{\partial x_u}{\partial t}(\psi,t) = u_u(\psi,t) = L'(t) + \frac{p'(t)}{p(t)}\left[\frac{1}{\mu}\frac{(1-\psi)}{\{p(t)\}^{1/\mu}}\right]. \tag{3.36}$$

In particular, the speed of the gas at the flame front is

$$u_u(\psi_f(t),t) = u_f(t) = L'(t) + \frac{p'(t)}{p(t)}\left[\frac{1}{\mu}\frac{\{1-\psi_f(t)\}}{\{p(t)\}^{1/\mu}}\right]. \tag{3.37}$$

Thus, the flame speed is

$$S(t) = \{x_f'(t) - u_f(t)\} = \frac{\psi_f'(t)}{\{p(t)\}^{1/\mu}}, \tag{3.38}$$

where, from Appendix A, for α, $n = 1$ and ν_F, $\nu_O = 1$,

$$S(t) \approx \{1 + K\Theta_u(t)\}\{1 + \Theta_u(t)\}^2 \exp\left[\frac{\beta}{2}\frac{\Theta_u(t)}{\{1+\Theta_u(t)\}}\right]. \tag{3.39}$$

Thus, for this domain, the pertinent initial-value problem for the determination of $p(t)$, $\psi_f(t)$, $\Theta_u(t)$, and $x_f(t)$, from (3.32) and (3.33), (3.38) and (3.39), (3.9b), and (3.35), is

$$\frac{1}{\sigma}\frac{p'(t)}{p(t)}\left[1-\frac{(\chi-\sigma)}{\chi}\frac{\{\psi_f(t)-\psi_i\}\{G_{bf}(t)+G_{bi}\}}{2L(t)\{p(t)\}^{1/\chi}}-\frac{(\mu-\sigma)}{\mu}\frac{\{1-\psi_f(t)\}}{L(t)\{p(t)\}^{1/\mu}}\right]=-\frac{L'(t)}{L(t)}+\frac{K\psi_f'(t)}{L(t)p(t)}, \tag{3.40a}$$

$$\frac{\psi_f'(t)}{\{p(t)\}^{1/\mu}}=\{1+K\Theta_u(t)\}\{1+\Theta_u(t)\}^2\exp\left[\frac{\beta}{2}\frac{\Theta_u(t)}{\{1+\Theta_u(t)\}}\right], \tag{3.40b}$$

$$\frac{K\Theta_u'(t)}{\{1+K\Theta_u(t)\}}=\frac{(\mu-1)}{\mu}\frac{p'(t)}{p(t)}, \tag{3.40c}$$

$$x_f'(t)=L'(t)+\frac{p'(t)}{p(t)}\left[\frac{1}{\mu}\frac{\{1-\psi_f(t)\}}{\{p(t)\}^{1/\mu}}\right]+\frac{\psi_f'(t)}{\{p(t)\}^{1/\mu}}; \tag{3.40d}$$

$$p(t_i)=p_i,\ \psi_f(t_i)=\psi_i,\ \Theta_u(t_i)=\Theta_{ui},\ x_f(t_i)=x_{fi}. \tag{3.41a-d}$$

The problem nominally terminates when $t = t_1$, such that $L(t_1) = L_1$ and

$$p(t_1)=p_1,\ \psi_f(t_1)=1,\ \Theta_u(t_1)=\Theta_{u1},\ x_f(t_1)=L_1. \tag{3.42}$$

The above generalization of earlier variable-volume-enclosu. work [7] entails a flame speed that is derived from the adopted reaction rate (cf. Senachin and Babkin [8]) and reduces to the laminar isobaric value as $t \to 0$ (cf. Sivashinsky [9]).

4. DISCUSSION. First, in Appendix B, it is shown that, for the circumstances of interest, heat transfer (to slippery noncatalytic side walls) may be simulated conveniently by assignment of values to the polytropic constants κ, μ, σ, and χ, i.e., the polytropic constants are effectively equivalent to the introduction of heat-transfer coefficients: the smaller the value assigned to the polytropic constant, the greater the heat transfer in the associated domain.

Second, in singular-perturbation terms,end-gas knock is the physical manifestation of the possible nonuniform validity in time of the asymptotic approximation (based on $\beta \gg 1$) that the reaction-rate contribution is negligible in the unburned gas.

However, the asymptotic approximation (based on $\varepsilon \ll 1$) that the diffusion contribution is negligible remains uniformly valid. Thus, upon retaining the chemical-reaction terms in (2.16) and (2.17) that were discarded in writing (3.1) and (3.2) because these terms were taken to be exponentially small, one has

$$\frac{\partial Y}{\partial t} = -\frac{1}{\varepsilon}\Lambda p Y(1+\Phi Y)\exp\left\{-\beta\frac{(1-T)}{T}\right\}, \tag{4.1}$$

$$\frac{\partial T}{\partial t} = \frac{1}{\varepsilon}\Lambda p Y(1+\Phi Y)\exp\left\{-\beta\frac{(1-T)}{T}\right\} + \frac{(\gamma-1)}{\gamma}\frac{(1+KT)}{Kp}\frac{dp}{dt}, \tag{4.2}$$

where $[Y(\psi,t),\ T(\psi,t),\ \gamma] \to [Y_u(t),\ T_u(t),\ \kappa]$ in the unburned-bulk-gas region, and $[Y(\psi,t),\ T(\psi,t),\ \gamma] \to [X_u(t),\ \Theta_u(t),\ \mu]$ in the unburned-end-gas region. It is recalled that adjustment of the polytropic constant γ simulates the effect of heat loss. The above problem, (4.1) and (4.2), subject to the initial conditions $Y(0) = 1$ and $T(0) = 0$, is a variant of the classical spatially homogeneous nonstationary ignition problem in which the criterion for an explosion self-sustained against heat loss is sought. Here, compressional heating, as reflected in the pressure, $p(t)$, raises the temperature of the premixture. Clearly, as T rises in time, the reaction-rate term may become comparable to the other terms of the energy equation. If one substitutes the pressure, $p(t)$, from the solution of the problem posed in Sec. 3, then one can obtain an indication from (4.1) and (4.2) whether significant autoconversion occurs in the gas prior to the completion of flame propagation. If modest autoconversion causes a pressure rise in time during the flame transit beyond that accounted for in the theory of Sec. 3 [9], that development is not easily generalized to include it [4]. Of more concern is the very significant autoconversion of the gas on a time scale comparable to, or less than, the time scale for the completion of flame propagation. Especially if the time scale is less than that for the completion of flame propagation (and less than that for acoustic adjustment), there occurs effectively a constant-volume explosion of the gas. Such a relatively instantaneous reaction of the gas results in a pressure in the end-gas region higher than that in the rest of the gas. The consequence is nonlinear pressure waves and an audible sound from the sympathetic vibration of the engine components. However, there is no incentive to pursue the solution in this spatially nonuniform-pressure domain because the above formulation is inadequate.

The standard asymptotic solution for $\beta \gg 1$ of (4.1) and (4.2), or even of (4.2), with reactant consumption ignored (i.e., $Y \equiv 1$), suffices to establish that, for plausible parameter assignments for hydrocarbon-air premixtures of automotive interest, autoconversion becomes effectively instantaneous (relative to the flame-propagation speed over, say, a half centimeter) for unburned-gas temperatures on the order of 1025 K [4]. Practical experimental data for nonisobaric combustion also give this value [1]. Mathematical treatment beyond that in Appendix C seems to be of limited engineering utility.

Thus, the following computational procedure is evolved to answer the questions posed in Sec. 1. First, with κ and σ taken as given and ψ_i set equal to unity (so that there is no end gas), the first initial-value problem posed in Sec. 3.1, (3.27) and (3.28), is carried out for the parametric assignments of interest. If it holds uniformly in time that the temperature of the gas, T_u, is less than Θ_ε, the (nondimensional) temperature criterion for the onset of autoconversion (see Appendix C), then knock is taken not to occur. If, however, T_u reaches Θ_ε for $\psi_f(t) < 1$, then that value of ψ_f is taken as ψ_i, i.e., ψ_i is the fraction of the charge identified as bulk gas and $(1-\psi_i)$ as end gas. Then, with χ taken as given, the largest value of μ for which $\Theta_u \to \Theta_\varepsilon$ as $\psi_f \to 1$ is sought by the trial-and-error solution of the second initial-value problem of Sec. 3.2, (3.40) and (3.41). That is, the minimal amount of heat transfer from the end gas is sought such that significant end-gas autoconversion is barely precluded. In this manner, the fraction, if any, of the charge that would undergo rapid autoconversion is identified, and the relative heat-transfer provision to preclude such autoconversion is determined.

5. COMPUTATIONAL EXAMPLES. A limited number of examples are presented of the procedure discussed in the last paragraph of Sec. 4. For brevity of reference, the following assignment of (mainly) dimensional parameters is termed nominal:

$$T^*_{u0} = 300\text{ K},\quad T^*_{b0} = 1800\text{ K},\quad T^*_a = 15{,}000\text{ K},\quad \phi = 0.8,$$
$$D^*_{u0} = 0.1\text{ cm}^2/\text{s},\quad u^*_{u0} = 40\text{ cm/s},\quad L^*_0 = 10\text{ cm},\quad \text{Le} = 1,$$
$$\Omega^* = 20\text{ rev/s}. \tag{5.1}$$

It is consistent with these assignments to adopt the following "nominal" values for the dimensionless parameters:

$$\beta = 10,\quad \Lambda = 50,\quad K = 5,\quad \Phi = 4,\quad \varepsilon = 2(10^{-4}),\quad \omega = 31.4\text{ rad},$$
$$\kappa = 1.4,\quad \mu = 1.4,\quad \sigma = 1.4,\quad \chi = 1.4,\quad \text{CR} = 6,\quad \theta_{ig} = 135\text{ deg}. \tag{5.2}$$

The last assignment implicitly gives the time of sparking, or, more precisely, the time at which the flame propagation commences. That is, it is defined that $\theta = \omega t$, so that, at the beginning of compression, $t = 0$, it holds that $\theta = 0$, which corresponds to "bottom dead center" [$L = 1$, the maximum value of L, from (3.30)]. At $\theta = \theta_{ig}$, assigned, i.e., at $t_{ig} = \theta_{ig}/\omega$, the variables $\psi_f(t)$ and $x_f(t)$, held at zero for $0 < t < t_{ig}$, take on values consistent with (3.26) and (3.21a). Sparking and compression can commence simultaneously, so that t_{ig}, $\theta_{ig} = 0$; in general, sparking occurs after a finite interval of compression, so that $t_{ig} > 0$. In the nominal case, $\theta_{ig} = 135$ deg, or flame propagation commences when $L = L_{ig} = 0.289$, i.e., after about five-sevenths of the (compressional) stroke has been completed. All of the parameters are held at their nominal values throughout the computations unless it is explicitly stated otherwise.

The onset of knock is taken to occur at T_u and/or $\Theta_u = \Theta_\varepsilon = 0.4$, or, from (5.1), T^*_u and/or $\Theta^*_u = \Theta^*_\varepsilon = 900$ K--a value somewhat less than the 1025 K cited elsewhere.

The nominal case is found (by design) to be just barely knock-free. Computation indicates that combustion is completed at $t = 0.088$. Since the basis of nondimensionalization is $(L^*_0/u^*_{u0}) = 0.25$ s, combustion lasts 22 ms, or about three-to-four times longer than it does in a typical Otto-cycle-type automotive engine.

At the completion of burning, $T_u = 0.399$, $p = 46.6$, $\theta = 157$ deg, and $L = 0.200$, whereas, at the start of burning, $t = 0.075$, $T_u = 0.129$, $p = 5.69$, and $L = 0.289$. As with all of the cases to be examined, combustion is completed during the compression stroke, i.e., at a crank angle $\theta < 180$ deg. Of course, in an automotive context, the combustion interval is very likely to be (very roughly) symmetric with respect to the time of the piston position at top center.

If the burned-gas polytropic constant, σ, is reduced in value from 1.4 to 1.1, all other parameters being held fixed, then, at the completion of burning, $t = 0.089$, $T_u = 0.378$, $p = 40.945$, $\theta = 159.8$ deg, and $L = 0.192$, whereas, at the start of burning, the corresponding values are (of course) those given in the preceding paragraph. Since modifying the heat transfer from the burned gas is seen to have but modest effect on the temperature of the unburned gas, this parametric variation is not pursued further here.

Table 1 presents results for four parametric variations, all selected to incur knock. The heat-transfer requirement from the end gas that precludes knock is identified in the form of the maximum value of μ, denoted μ_{crit}, that permits $\Theta_u < 0.4$ at all times during the combustion event. These cases entail between 13 and 57% of the initial combustible mass being involved in autoconversion (in the absence of augmented end-gas heat transfer), and, in this sense, represent a spectrum of cases. It may be noted that the larger the fraction of end gas (i.e., the smaller the value of ψ_i), the larger the heat-transfer requirement to preclude knock (i.e., the smaller the value of μ_{crit}). However, it is noteworthy that reduction of K (from its nominal value of 5) to 3, while β is held constant, implies that T_u^* is altered altered from 300 to 450 K. Since $T_u = 0.4$ is retained as the knock criterion, $T_u^* = 990$ K becomes the temperature for explosive conversion of the reactant to product gas.

6. CONCLUDING REMARK. The present work has concentrated on enhancing heat transfer as a countermeasure to end-gas knock. Another (perhaps complementary) strategy is to seek fuel additives that inhibit end-gas autoconversion without impeding flame propagation.

Appendix A.
Flame-Zone Structure

As in the unbounded (isobaric) case, the structure of the flame zone for the bounded (nonisobaric, enclosed) case for $\varepsilon \ll 1$, with $\beta \gg 1$, consists of a convective-diffusive region and a diffusive-reactive region. The results presented are for α, $n = 1$ and ν_F, $\nu_0 = 1$.

For the convective-diffusive region, the appropriate transformation of variables is $(\psi, t) \to (\xi, t)$, where

$$\xi(\psi, t; \beta, \ldots) = M(t; \ldots) E(t; \beta, \ldots)[\{\psi_f(t; \ldots) - \psi\}/\varepsilon]. \tag{A.1}$$

The domain of this region is $-\infty < \xi < 0$, $t > 0$, such that the unburned exterior region is approached as $\xi \to -\infty$. Further, the flame speed is taken to be

$$S(t; \beta, \ldots) = V(t; \ldots) E(t; \beta, \ldots)[1 + O(\beta^{-1})]. \tag{A.2}$$

Here, recalling that $T_{bf}(t; \ldots) = \{1 + T_u(t; \ldots)\}$,

$$E(t; \beta, \ldots) = \exp\left[\frac{\beta}{2} \frac{T_u(t; \ldots)}{\{1 + T_u(t; \ldots)\}}\right]; \tag{A.3}$$

$$M(t; \ldots) = \{1 + T_u(t; \ldots)\}^2, \quad V(t; \ldots) = \{1 + K T_u(t; \ldots)\}\{1 + T_u(t; \ldots)\}^2. \tag{A.4a,b}$$

Note that $\xi \to (-\psi)/\varepsilon$ and $S \to 1$ as $t \to 0$, such that the unbounded case is the initial condition of the bounded case. Through the matching of the solutions of this convective-diffusive region with those of the diffusive-reactive region, it is demonstrated that (A.2) is the correct asymptotic representation for the flame speed.

For this convective-diffusive region, the dependent variables are of the forms

$$Y(\psi, t; \ldots) = X(\xi, t; \ldots)[1 + O(\beta^{-1})], \tag{A.5a}$$

$$T(\psi, t; \ldots) = F(\xi, t; \ldots)[1 + O(\beta^{-1})]. \tag{A.5b}$$

Thus, for these scalings, the specific time-derivative and the reaction-rate terms can be neglected, such that, to leading order of approximation, (2.16) and (2.17) become

$$\frac{\partial^2 X}{\partial \xi^2} - \frac{\partial X}{\partial \xi} = 0, \qquad Le \frac{\partial^2 F}{\partial \xi^2} - \frac{\partial F}{\partial \xi} = 0. \tag{A.6,7}$$

The boundary conditions for these equations are

$$X \to 1, \quad F \to T_u \quad \text{as } \xi \to -\infty; \tag{A.8a}$$

$$X \to 0, \quad F \to T_{bf} = (1 + T_u) \quad \text{as } \xi \to 0_- . \tag{A.8b}$$

The solutions of these convective-diffusive-region boundary-value problems are

$$X(\xi, t) = [1 - \exp(\xi)], \tag{A.9}$$

$$F(\xi, t) = [T_u(t) + \exp(\tfrac{1}{Le}\xi)] = T_{bf}(t) - [1 - \exp(\tfrac{1}{Le}\xi)]. \tag{A.10}$$

It is seen that

$$X \to 1, \quad F \to T_u \quad (\text{exponentially}) \quad \text{as } \xi \to -\infty, \tag{A.11}$$

and that

$$X \sim (-\xi)(1 + \cdots) \to 0_+, \quad F \sim T_{bf} - \tfrac{1}{Le}(-\xi)(1 + \cdots) \to T_{bf-} \quad \text{as } \xi \to 0_- . \tag{A.12}$$

For the diffusive-reactive region, the transformation of variables is $(\psi, t) \to (\eta, t)$, where

$$\eta(\psi, t; \varepsilon, \beta, \ldots) = E(t; \beta, \ldots)[\{\psi_f(t; \ldots) - \psi\}/(\beta^{-1}\varepsilon)], \tag{A.13}$$

with $E(t;\beta,\ldots)$ given in (A.3). The domain of this region is $-\infty < \eta < \infty$, $t > 0$, such that the burned exterior region is approached as $\eta \to \infty$. The flame speed is still that given in (A.2) *et seq* . Again, note that, as $t \to 0$, $\eta \to (-\psi)/(\beta^{-1}\varepsilon)$ and $S \to 1$.

For this region, the dependent variables are taken to be

$$Y(\psi, t; \ldots) = \beta^{-1}\{1 + T_u(t; \ldots)\}^2 Z(\eta, t; \ldots)[1 + O(\beta^{-1})], \tag{A.14a}$$

$$T(\psi, t; \ldots) = \{1 + T_u(t; \ldots)\} - \beta^{-1}\{1 + T_u(t; \ldots)\}^2 G(\eta, t; \ldots)[1 + O(\beta^{-1})]. \tag{A.14b}$$

For $\varepsilon \ll 1$ and $\beta \gg 1$, under these scalings, to leading order of approximation, for this region, the convective terms can be neglected, and (2.16) and (2.17) become

$$\frac{\partial^2 Z}{\partial \eta^2} = \frac{1}{2Le^2} Z \exp(-G), \qquad Le \frac{\partial^2 G}{\partial \eta^2} = \frac{1}{2Le^2} Z \exp(-G). \tag{A.15,16}$$

The boundary conditions are

$$Z, G \to 0 \quad \text{as } \eta \to \infty; \tag{A.17a}$$

$$Z, G \to \infty \quad \text{as } \eta \to -\infty. \tag{A.17b}$$

By subtraction of (A.16) from (A.15) and by use of (A.17), it is found that $Z =$ Le G. Then, (A.15)-(A.17) reduces to

$$\frac{\partial^2 G}{\partial \eta^2} = \frac{1}{2Le^2} G \exp(-G); \tag{A.18a}$$

$$G \to 0 \quad \text{as } \eta \to \infty, \qquad G \to \infty \quad \text{as } \eta \to -\infty. \tag{A.18b}$$

The first integral of this equation is

$$\frac{\partial G}{\partial \eta} = -\frac{1}{Le}\left[1-(1+G)\exp(-G)\right]^{1/2}. \tag{A.19}$$

Thus, it is determined that

$$G \sim \exp\left(-\frac{1}{\sqrt{2}\,Le}\eta\right)(1+\cdots) \to 0 \quad \text{as } \eta \to \infty; \tag{A.20a}$$

$$G \sim \frac{1}{Le}(-\eta)(1+\cdots) \to \infty \quad \text{as } \eta \to -\infty. \tag{A.20b}$$

Returning to the original dependent variables, it is seen that, for the solutions of the convective-diffusive and diffusive-reactive regions,

$$Y \sim \left[(-\xi)+\cdots\right]+\cdots \quad \text{as } \xi \to 0-,$$

$$Y \sim \beta^{-1}\left[(1+T_u)^2\{(-\eta)+\cdots\}\right]+\cdots \quad \text{as } \eta \to -\infty; \tag{A.21a}$$

$$T \sim \left[(1+T_u)-\frac{1}{Le}(-\xi)+\cdots\right]+\cdots \quad \text{as } \xi \to 0-,$$

$$T \sim (1+T_u)-\beta^{-1}\left[(1+T_u)^2\left\{\frac{1}{Le}(-\eta)+\cdots\right\}\right]+\cdots \quad \text{as } \eta \to -\infty. \tag{A.21b}$$

Since $\eta = \beta\xi/(1+T_u)^2$, the solutions (which involve the specification of the leading-order approximation for the flame speed) for these two regions match. In the text, the following expression for this flame speed is employed:

$$S(t;\ldots) \approx S(T_u(t;\ldots);\beta,K) = \{1+KT_u(t;\ldots)\}\{1+T_u(t;\ldots)\}^2 \exp\left[\frac{\beta}{2}\frac{T_u(t;\ldots)}{\{1+T_u(t;\ldots)\}}\right]. \tag{A.22}$$

Appendix B.
Polytropic Law to Simulate Heat Transfer During the Combustion Event in an Enclosure

In conventional notation, from thermodynamics,

$$c_v^* \frac{dT^*}{dt^*} + p^* \frac{d(1/\rho^*)}{dt^*} = q^*, \tag{B.1}$$

where q^* denotes the rate of heat addition per unit mass in the enclosure. Hence, for an ideal gas,

$$\frac{d(\log T^*)}{dt^*} - (\gamma-1)\frac{d(\log \rho^*)}{dt^*} = \frac{d(\log p^*)}{dt^*} - \gamma\frac{d(\log \rho^*)}{dt^*} = \frac{q^*}{c_v^* T^*}. \tag{B.2}$$

For tractability, let $q^* = -E_1^* c_v^* T^*$, with $E_1^* = \text{const} > 0$; the product $E_1^* c_v^*$ plays the role of a heat-transfer coefficient. Then, if the subscript *0* denotes the initial state,

$$\frac{d}{dt^*}\left(\log\left\{\frac{p^*}{\rho^{*\gamma}}\right\}\right) = -E_1^*: \quad \frac{(p^*/p_0^*)}{(\rho^*/\rho_0^*)^\gamma} = \exp(-E_1^* t^*). \tag{B.3}$$

During combustion, as a rough approximation,

$$(p^*/p_0^*) \approx \exp(E_2^* t^*), \tag{B.4}$$

with $E_2^* = \text{const} > 0$. Elimination of t^* between (B.3) and (B.4) produces

$$(p^*/p_0^*)^{\{1+(E_1^*/E_2^*)\}} = (\rho^*/\rho_0^*)^{\gamma}:$$

$$(\rho^*/\rho_0^*)^{\gamma_*} = (p^*/p_0^*), \quad \text{with } \gamma_* = \{1 + (E_1^*/E_2^*)\}^{-1}\gamma. \tag{B.5}$$

Since $(E_1^*/E_2^*) > 0$, it follows that $\gamma_* < \gamma$. That is, the effect of heat transfer from a flow with *rising pressure* may be simulated by adoption of the polytropic constant reduced from the adiabatic value. Furthermore, via (B.5), the polytropic constant may be related to the heat-transfer coefficient (and a characterization of the rate of pressure rise).

Appendix C.
Autoconversion Criterion

In (2.16)-(2.22), with $\varepsilon \ll 1$ and $\beta \gg 1$, it is anticipated that a physically interesting, autoconversion-related case entails the following quantitative relation between the values of the (more generally, independent) parameters ε and β:

$$\varepsilon = \beta^2 \exp(-\beta T_\varepsilon): \quad T_\varepsilon = \beta^{-1} \log(\varepsilon^{-1}\beta^2) \sim O(1), \tag{C.1}$$

such that, when ε and β are specified, T_ε is determined. It is convenient to introduce the notation

$$\Theta_\varepsilon = (1+T_\varepsilon)^{-1} < 1, \quad \text{such that } T_\varepsilon = \Theta_\varepsilon^{-1}(1-\Theta_\varepsilon); \tag{C.2a}$$

$$\beta_\varepsilon = (1+T_\varepsilon)\beta > \beta. \tag{C.2b}$$

For the typical values $\varepsilon = 2(10^{-4})$ and $\beta = 10$, (C.1) and (C.2) give $T_\varepsilon = 1.31$, $\Theta_\varepsilon = 0.432$, and $\beta_\varepsilon = 23.1$; for $\varepsilon = 2(10^{-4})$, but $\beta = 5$ (a somewhat small value), $T^\varepsilon_\varepsilon = 2.35$, $\Theta_\varepsilon = 0.299$, and $\beta_\varepsilon = 16.7$. Thus, for values of ε and β in the range of physical interest, it is confirmed that $T_\varepsilon = \beta^{-1}\log(\varepsilon^{-1}\beta^2) \sim O(1)$. Further, from (C.1) and (C.2), it is seen that

$$\varepsilon = \beta_\varepsilon^2 \Theta_\varepsilon^2 \exp\{-\beta_\varepsilon(1-\Theta_\varepsilon)\}. \tag{C.3}$$

The subscript ε denotes conditions at which it is anticipated that the chemical conversion may no longer be neglected in the gas, although the diffusive transfer remains negligible. This statement may be more readily noted by rewriting (2.16) and (2.17), under (C.1)-(C.3), as

$$\frac{\partial Y}{\partial t} \approx -\frac{1}{2\,Le^2}\, pY(1+\Phi Y)\exp\left\{-\beta_\varepsilon \frac{(\Theta_\varepsilon - T)}{T}\right\}, \tag{C.4}$$

$$\frac{\partial T}{\partial t} \approx \frac{1}{2\,Le^2}\, pY(1+\Phi Y)\exp\left\{-\beta_\varepsilon \frac{(\Theta_\varepsilon - T)}{T}\right\} + \frac{(\gamma-1)}{\gamma}\frac{(1+KT)}{Kp}\frac{dp}{dt}. \tag{C.5}$$

It is seen that, with respect to the temperature, the reaction-rate terms are negligible for $0 \le T < \Theta_\varepsilon$, and they are not neglible for $T \ge \Theta_\varepsilon$; the transition takes place when $\beta_\varepsilon(\Theta_\varepsilon - T) \sim O(1)$.

Thus, the interval in which both the diffusion and reaction-rate terms are negligible with respect to the convection and pressure-gradient terms is

$$0 < t < \tau_\varepsilon: \quad 0 < T < \Theta_\varepsilon, \quad 1 < p < \Pi_\varepsilon; \quad Y = 1. \tag{C.6}$$

This interval terminates when $T = \Theta_\varepsilon$, with $t \to \tau_\varepsilon$ and $p \to \Pi_\varepsilon$ as $T \to \Theta_\varepsilon$. Within this interval, the solution is given by (3.6a,b) in the unburned bulk gas, and by (3.9a,b) in the unburned end gas. From (2.13b), for $T^*_{u0} = 300$ K, $T^*_{b0} = 1800$ K, the condition $T = \Theta_\varepsilon$ implies that $T^* = \Theta^*_\varepsilon$ $(= T^*_{u0}(1+K\Theta_\varepsilon)) = 948$ K for $\beta = 10$, and $T^* = 749$ K for $\beta = 5$. The value $T^* = \Theta^*_\varepsilon = 900$ K is used in the text; this value corresponds to $T = \Theta_\varepsilon = 0.4$.

BIBLIOGRAPHY

1. C. F. Taylor, The Internal-Combustion Engine in Theory and Practice, Vol. 2, MIT Press, Cambridge, MA, 1968, pp. 34-85.

2. E. F. Obert, Internal Combustion Engines and Air Pollution, Intext Educational Publishers, New York, 1973, pp. 105-110, 291-341.

3. S. L. Hirst, L. J. Kirsch, "The Application of a Hydrocarbon Autoignition in Simulating Knock and Other Engine Phenomena," Combustion Modeling in Reciprocating Engines, Plenum Press, New York, 1980, pp. 193-229.

4. G. F. Carrier, F. E. Fendell, S. F. Fink, P. S. Feldman, "Heat Transfer as a Determent of End-Gas Knock," Combust. Sci. Technol. 38(1984), 1-48.

5. G. F. Carrier, F. E. Fendell, W. B. Bush, P. S. Feldman, "Nonisenthalpic Interaction of a Planar Premixed Laminar Flame with a Parallel End Wall," SAE Paper 790245, 1979.

6. S. F. Fink, F. E. Fendell, W. B. Bush, "Nonadiabatic Propagation of a Planar Premixed Flame: Constant-Volume Enclosure," AIAA J. 23(1985), 424-431.

7. G. F. Carrier, F. E. Fendell, P. S. Feldman, "Nonisobaric Flame Propagation," Dynamics and Modeling of Reactive Systems, Academic Press, New York, 1980, pp. 333-351.

8. P. K. Senachin, V. S. Babkin, "Self-Ignition of Gas in Front of the Flame Front in a Closed Vessel," Fizika Goreniya i Vzryve 18:1(1982), 3-8.

9. G. I. Sivashinsky, "Hydrodynamic Theory of Flame Propagation in an Enclosed Volume," Acta Astronaut. 6(1979), 631-645.

WBB: KING, BUCK & ASSOCIATES, 2384 SAN DIEGO AVE., SAN DIEGO, CA

FEF: ENGINEERING SCIENCES LAB., TRW, REDONDO BEACH, CA

SFF: HARDNESS AND SURVIVABILITY LAB., TRW, REDONDO BEACH, CA

Table 1. Parametric variations on the nominal (threshold-knock) case[a]

Case	Flame enters end gas ($\psi_f = \psi_i$)							Flame propagation complete ($\psi_f = 1$)				
	t	θ^0	L	x_f	ψ_f	p	Θ_u	t	θ^0	L	p	μ_{crit}
CR = 12	0.082	147.6	0.16	0.13	0.71	45.6	0.35	0.083	149.1	0.15	63.0	1.3604
$K = 3$	0.083	150.5	0.22	0.14	0.41	15.7	0.30	0.088	157.8	0.20	30.3	1.3002
$\omega = 51.6$ rad	0.056	166.4	0.18	0.17	0.87	45.6	0.38	0.057	167.7	0.18	52.6	1.3836
$\theta_{ig} = 165^0$	0.098	177.0	0.17	0.16	0.83	46.2	0.37	0.099	177.9	0.17	54.3	1.3795

[a]For the case $K = 3$, if $p_0^* = 1$ atm, since $T_{u0}^* = 450$ K, then the initial density ρ_{u0}^* is reduced to two-thirds that of the other cases.

Lectures in Applied Mathematics
Volume 24, 1986

NONLINEAR STABILITY OF SOLID PROPELLANT COMBUSTION

F.J. Higuera and A. Linan

1. INTRODUCTION. The stability and transient response of solid propellant combustion has received considerable attention in the literature, using, after Zeldovich (1942 and 1964) and Denison and Baum (1961), the quasisteady approximation for the gas phase. This approximation results from the small value of the ratio of gas-to-solid densities.

The analysis of the stability of a simple model of solid propellant combustion is given here. The model includes a surface pyrolisis reaction and a gas phase reaction, assuming the gas phase combustion process to be quasisteady and quasiplanar. The quasiplanar assumption results from the moderately small value of the ratio of thickness of the gas-to-solid transport zones, or from large values of the effective nondimensional activation energy of the gas phase reaction.

Two parameters, ε_g^{-1} and ε_s^{-1}, characterizing the sensitivity of the gas phase and pyrolisis reaction with temperature, enter into the linear stability analysis and a third parameter appears in the nonlinear anaysis. Travelling waves of non-zero wavenumber are found on the stability boundary $R(\varepsilon_g, \varepsilon_s) = 0$. These results are quoted in Section 2.

A nonlienar stability analysis, close to the stability boundary, is carried out in Section 3 for the case of practical importance corresponding to large values of the gas phase activation energy ε_g^{-1}. In this limit the characteristic transverse length and response time become large compared to the thickness of the transport zone in the solid and the residence time in this zone, respectively. Three different time scales are found from the linear stability analysis. A system of nondispersive waves is found in the shorter time scale; these waves are weakly dispersive and then dispersive effects can be counterbalanced by small nonlinear effects in a second time scale. The resulting problem coincides with that associated with the propagation of gravity wave in shallow water, and the Korteweg-deVries equation describes the evolution of waves travelling in any direction. Under very general conditions the asymptotic state of a wave train can be predicted by the inverse scattering method and consists in a series of solitons and a residual wave train. The analysis corresponds to an infinite system; results for an interface of finite size have been obtained by Margolis (1985) for another two different models of combustion of solids.

Finally, the growth or decay effects predicted by the linear analysis appear in the longest time scale, resulting in a small perturbation to the Korteweg-deVries equation that includes new nonlinear terms in addition to those coming from the linear theory. The perturbation changes slowly the amplitude and velocity of each of the solitons toward a defined value or zero, depending on the initial size of the soliton and the values of the parameters. The analysis for a single soliton is carried out in Section 4.

In the last section of equations describing the slow evolution of a nonlinear modulated wavetrain is obtained from a pseudovariational formulation. The hyperbolic character of these equations is not altered by the perturbations due to the real part of the growth rate.

2. FORMULATION, STEADY STATE AND RESULTS FROM THE LINEAR STABILITY ANALYSIS. We anlayze the nonlinear stability of the combustion of a solid propellants subject to nonplanar perturbations, for the simple model of Denison and Baum (1961), which includes a surface heterogeneous reaction and a gas phase reaction. Both are supposed to be global one-step reactions following Arrhenius laws, the second one with large activation energy. No further reactions occur inside the homogeneous solid and the transport coefficients and specific heats are constant in each phase.

A relation of the form

$$m=m_s(T_s) = B \exp(-E_s/RT_s) \tag{1}$$

for the mass flux m crossing the interface, as a function of the interface temperature T_s, is assumed to represent the surface reaction. E_s is the activation energy for this reaction, and B is a preexponential constant.

The reaction zone in the gas phase is a thin layer behind a transport zone; from the analysis of this layer and its matching to the transport zone a second relation

$$m_f(T_f) = C \exp(-E/2RT_f) \tag{2}$$

between the mass flux and the flame temeprature T_f is obtained. E is the activation energy for the gas phase reaction, and E/RT_f is assumed to be large. If the gas phase transport zone is assumed to be planar and steady, $m_f(T_f) = m$.

The thickness of the transport zone in the gas and the conduction zone in the solid are given by $\delta_g \sim \lambda_g/mc_p$ and $\delta_s \sim \lambda_s/mc_s$, respectively, as results from the balance between convective transport due to the recession of the interface and conduction. It turns out that $\delta_g << \delta_s$, due to the fact that the thermal conductivity of the gas, λ_g, is much smaller than that of the solid, λ_s. The residence times in these two zones are $\tau_g \sim \rho_g \lambda_g / m^2 c_p$ and $\tau_s \sim \rho_s \lambda_s / m^2 c_s$, and $\tau_g << \tau_s$, because of the small value of the gas-to-solid density ratio. The anaysis that follows is addressed to the response of the solid phase, and the times involved are of the order of τ_s; in addition the characteristic tranverse length of the perturbations that can grow is of the order of δ_s. Both facts taken together allow a quasisteady and locally one-dimensional treatment of the gas phase; so that in particular, the relations (1-2) are still valid in the transient analysis. The problem is reduced to solve the heat conduction equation in the interior of the solid, and to determine the shape of the perturbed interface. Most of the analysis can be directly applied to other combustion models. There is a steady state solution in which the variables depend only on the distance ζ to the planar interface, $\zeta=0$; with the solid lying at $\zeta<0$. The steady flame temperature T_{fo} is given by

$$c_p T_{fo} = c_s T_\infty + Q \tag{3}$$

Where Q is the heat released per unit mass by both reactions and T_∞ is the temperature of the solid far from the interface. Here and in the following the subscript o is used for the magnitudes in the steady state. Eq. (3) is an overall energy balance. The steady mass flux m_o results then from (2) and the steady interface temperature T_{so} from (1). The final step would be to solve the energy conservation equation in the solid to find the steady temperature profile.

In a reference frame moving relative to the solid with the steady recession velocity the general unsteady problem can be formulated as follows

$$\frac{\partial\Theta}{\partial t} + \frac{\partial\Theta}{\partial\zeta} = \nabla^2\Theta \tag{4}$$

$$\zeta \to -\infty : \quad \Theta = 0 \tag{5}$$

$$\zeta = X_s : \begin{cases} \Theta = [1+ \frac{\omega}{\gamma-1} \ln \mu]/(1-\omega\ln\mu) & (6) \\ \frac{\partial\Theta}{\partial n} - \mu\Theta = - \varepsilon_g \mu \ln \mu/(1-\omega_g \ln \mu) & (7) \end{cases}$$

where $\Theta = (T-T_\infty)/(T_{so}-T_\infty)$ is the nondimensional temperature and $\mu = m/m_0$ is the nondimensional mass flux. The distances have been referred to $\lambda_s/m_o c_s$ and the time to $\rho_s \lambda_s/m_o^2 c_s$. The parameters that appear are

$$\omega = RT_{so}/E_s, \ \gamma = T_{so}/T_\infty, \ \omega_g = 2RT_{fo}/E,$$
$$\varepsilon_g = \frac{c_p}{c_s} \frac{2RT_{fo}^2}{E(T_{so}-T_\infty)} \quad \text{and} \quad \varepsilon_s = \frac{RT_{so}^2}{E_s(T_{so}-T_\infty)} \tag{8}$$

The shape of the interface is given by $\zeta = X_s\ (y, z, t)$, where y and z are transverse coordinates, and the nondimensional mass flux can be written as

$$\mu = (1-X_{s_t}) \sqrt{1 + X_{sy} + X_{sz}^{\ 2}} \tag{9}$$

Finally n is normal to the interface. The first condition at the interface comes from (1), and the second one results from an energy balance between the solid side of the interface and the gas behind the flame, after (2) is used to eliminate the flame temperature.

In order to analyze the linear stability of the steady state, small perturbations are considered with a dependence on time and the transversal coordinates through a factor of the form $\exp(\Omega t+iky)$. (Nothing new comes from the dependence on on the other transverse coordinate in the linear analysis.) The results, after the whole problem is linearized, lead to the compatibility condition, or dispersion relation, of the type

$$\varepsilon_s^2\Omega^3 + \{\varepsilon_s^2 k^2 + (1+\varepsilon_g)\,\varepsilon_s - (1-\varepsilon_g)^2\,\Omega^2$$

$$+ (2\,\varepsilon_s\,k^2 + \varepsilon_g)\,\Omega + k^2 = 0 \qquad (10)$$

which has been written as a cubic equation for Ω. It is worth noticing that only two parameters, ε_s and ε_g, appear in (10); their inverses characterize the temperature sensitivity of the interface and gas phase reactions (see Higuera and Linan, 1982).

Due to the large value of the activation energy for the gas phase reaction, the parameter ε_g can take very small values, and this is the case considered in what follows. The stability limit for $\varepsilon_g=0$ is given by $\varepsilon_s=2$ and the steady state is unstable when $\varepsilon_s<2$. For small values of ε_g the perturbations that can grow correspond to large wavelengths, $k=\delta K$ with $\delta=\varepsilon_g^{\frac{1}{4}} << 1$ and $K\sim 1$. The dispersion relations close to the stability limit can be written in the form (with $\varepsilon_s = (1+\delta^2\alpha)$ and $\alpha \sim 1$)

$$\Omega = \pm\, i\delta K \mp i\delta^3 K(\alpha+2k^2) - \delta^4\{1/2 + 2k^2(\alpha+4k^2)\} + \ldots$$

and is sketched in Fig. 1.

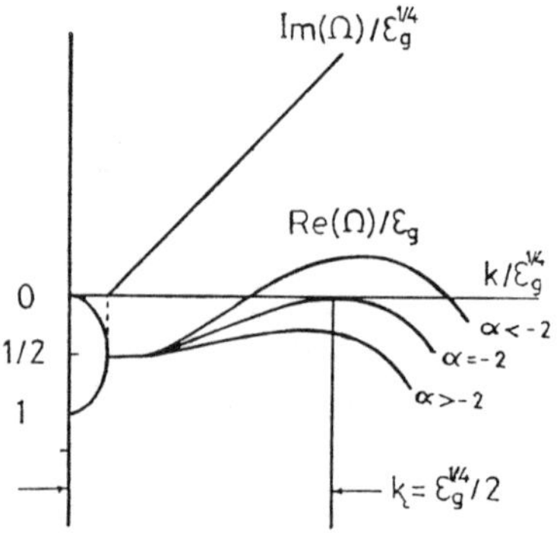

Fig. 1

The role of the imaginary part, which is larger than the real part, can be interpreted as follows: Every heated band of the solid is almost independent of its neighbours, because the wavelength of the perturbations is large. In first approximation the changes in the temperature distribution inside the solid when the burning rate increases are due to two effects (Fig. 2). On one hand the whole temperature profile is squeezed against the interface and on the other the temperature rises because, at the interface, it is an increasing function of the burning rate. When ε_s is close to 2, the second effect dominates over the first one and therefore the temperature in most of the band increases with the burning rate, $-X_{st}$, and so does the heat content. Now the origin of these changes in the heat content and the burning rate is the heat conduction tangential to the interface, coupling different heated bands of the solid, which is due to temperature differences at a fixed depth when the interface is deformed, see Fig. 2. The heat flux is proportional to the slope X_{sy} and the net flux received by a given band goes like $-X_{sy}$; this being the origin of the temperature changes, one obtains the relation

$$(-X_{st})_t \sim X_{syy}$$

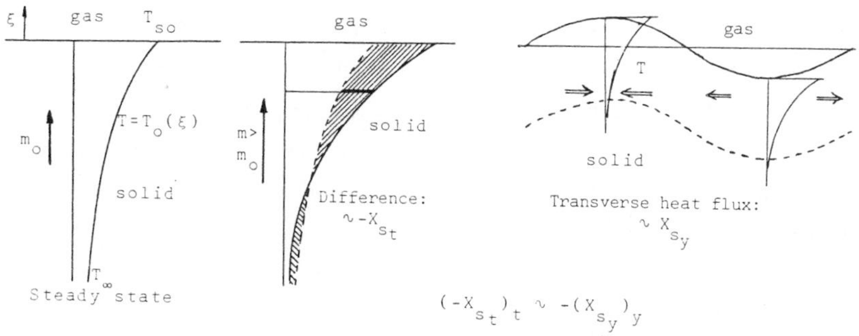

Fig. 2

accounting for the leading order of the expansion (11). The next term in (11) involves a correction proportional to K^3 representing dispersion. This comes from the fact that the temperature distributions in the last sketch in Fig. 2, responsible for the heat conduction, are not exactly equal to those of the steady state profile, but include a small change proportional to the second derivative of the front position and, in addition, the time derivatives in the heat conduction equation provide other small term proportional to $X_{s_{ttt}}$ in the heat content of the band. Both effects lead to a small correction, proportional to the fourth derivative of the front position, in the previous estimate, accounting, after the square root is taken, for the K^3 term in (11).

3. DERIVATION OF THE NONLINEAR EVOLUTION EQUATIONS.

The three different time scales appearing in the dispersion relation are explicitly introduced in the nonlinear analysis through a multiscale method in which the functions depend on the variables

$$\zeta, \quad \eta = \delta y, \quad \tau = \delta t, \quad T = \delta^3 t \quad \text{and} \quad T' = \delta^4 t \qquad (12)$$

where only one variable is introduced in the plane of the interface because, for the moment, the analysis is restricted to onedimensional waves.

The only new piece of information to be introduced is the order of magnitude of the (finite) perturbations. The clue for this choice comes from the fact that nonlinearity must counterbalance dispersion, because otherwise the amplitude of any initially localized perturbation would decay as $1/\sqrt{E}$, before the growth predicted by the linear analysis can have any effect. In order to get this balance one must use expansions of the type

$$X_s = \delta X + \delta^2 X^{(2)} + \dots$$

$$\Theta - \Theta_o = \delta^2 \Theta^{(2)} + \delta^3 \Theta^{(3)} + \dots \qquad (13)$$

into (4-7).

The results of the multiscale analysis are simply quoted, skipping over the intermediate steps.

In the faster time scale the waves on the interface satisfy the wave equation

$$X_{\tau\tau} = X_{\eta\eta} \tag{14}$$

and from here on only waves travelling to the left will be considered, so that

$$X = X\ (\eta+\tau,\ T,\ T') \tag{15}$$

Waves travelling to the right can be accounted for in a similar way and, for perturbations that are initially of bounded support, the final results to be obtained are valid in spite of the nonlinear interaction between both families of waves.

By proceeding with the multiscale analysis the following equation results for the evolution in the intermediate scale T.

$$X_{\eta T} = -\alpha X_{\bar{\eta}\bar{\eta}} + 2 X_{\bar{\eta}\bar{\eta}\bar{\eta}\bar{\eta}} + 2\ (\omega-2)\ X_{\bar{\eta}}\ X_{\bar{\eta}\bar{\eta}} \tag{16}$$

where $\bar{\eta} = \eta + \tau$ and ω is a third parameter, defined before, that does not appear in the linear stability analysis, and takes values between 0 and 2, when ε_s is close to 2. Eq. (16) can be written in the rescaled variables

$$x = -2^{-1/3}(\bar{\eta}-\alpha T)\ ,\ V = 2^{1/3}(2-\omega)X/6\ ,\ W = V_x \tag{17}$$

in the form

$$W_T + 6WW_x + W_{xxx} = 0 \tag{18}$$

which is a standard form of the Korteneg-deVries equation, so that up to here the resulting problem coincides with that associated with the propagation of gravity waves in shallow water. It can be seen that the nonlinear term provides the typical hyperbolic steepening of the profiles, counter-balancing the dispersive behaviour coming from the third derivative.

Finally the growth or decay effects predicted by the linear analysis appear in the slow time scale. The results can be summarized in the form of a perturbation to the K-dV equation, which takes the form

$$W_T + WW_x + W_{xxx} = \delta\Gamma(W) \tag{19}$$

with

$$\Gamma(W) = -2^{5/3}W_{xxxx} + 2^{1/3}\alpha W_{xx}$$

$$- W/2-3\cdot 2^{2/3}\frac{1-\omega}{2-\omega}(W^2)_{xx} \tag{20}$$

4. PERTURBATION OF A SOLITON. Let us begin by considering the unperturbed equation (18). This is a prototype for the long wavelength behaviour of many systems; it was first found in the analysis of shallow water waves (Korteweg and deVries, 1895). It is well known that (18) can be exactly solved by the inverse scattering transform, under fairly general initial conditions.

The idea of the inverse scattering transform is to associate the linear eigenvalue problem

$$\phi_{xx} + [\lambda^2+W(x,T)]\,\phi=0\ , \quad -\infty<x<\infty \tag{21}$$

with the nonlinear equation (18). The solution $W(x_1T)$ that we are looking for appears here as a potential in the Schrodinger equation (21) and T plays the role of parameter. It is assumed that W tends to zero sufficiently fast when $x \to \pm\infty$ and that $\int_{\infty-}^{\infty}(1+|x|)|W|dx$ is finite for all T. In these conditions all the real values of λ constitute spectrum of (21), and it is possible to find eigenfunctions $\phi(x, \lambda, T)$ normalized according to the conditions

$$\phi(x, \lambda, T) \to \begin{cases} e^{-i\lambda x}, & x \to -\infty \\ a(\lambda, T)\, e^{-i\lambda x} + b(\lambda, T)\, e^{i\lambda x}, & x \to \infty \end{cases} \tag{22}$$

which represent a wave travelling from the right; on arriving to the region where W is different from zero this wave is partially transmitted and partially reflected backward; b/a is known as the transmission coefficient. The function $a(\lambda)$ admits an analytic extension into the upper half complex λ-plane and its zeros on the imaginary axis $(\lambda_k = iV_k, V_k > 0)$ are the discrete eigenvalues of (21). The corresopnding eigenfunctions ϕ_k tend to zero as $\exp(V_k x)$ when $x \to -\infty$ and as $b_k \exp(V_k x)$ when $x \to \infty$, where b_k is the analytic extension of $b(\lambda)$ to λ_k. As is well known, (Gelfand and Levitan, 1951), the set of scattering data

$$s = b/a(\lambda), \ \lambda \text{ real}; \ \lambda_k = iV_k, \ \gamma_k = b_k/a_k' \tag{23}$$

where $a_k' = (\partial a/\partial\lambda)_{\lambda_k}$, contains sufficient information to reconstruct the potential W. The second point is that when $W(x,T)$ is a solution of the K-dV equation it is possible to find ordinary differential equations for the evolution of the various spectral components, separated from each other (see, p.e., Newell, 1978). The resolution of (18) proceeds therefore along the following lines: given the initial profile $W(x, 0)$, one can state (21) and find the set of initial scattering data. These provide the initial conditions for the evolution equations for the scattering data and, once these equations have been solved, the solution $W(x,T)$ can be constructed from them.

There is a soliton associated with each eigenvalue of the discrete spectrum. One of these solitons taken alone is a solution of (18) of the form

$$W = 2V^2 \operatorname{sech}^2[V(x-4V^2T)] \tag{24}$$

where iV is the only discrete eigenvalue of (21) for this $W(x,T)$. In the original variables the solution (24) takes the form

$$\bar{\underline{X}} = 2^{2/3} \cdot 6 \frac{V}{2-W} \operatorname{Th} [V(\frac{\eta+z-\alpha T}{2^{1/3}})] \tag{25}$$

which represent a step on the interface travelling toward the left with velocity $1-\delta^2 (\alpha-2^{1/3}\cdot 4V^2)$.

Asymptotically, when $T\to\infty$, the solution of (18) tends to a series of solitons plus a residual wave train. The number and sizes of the solitons depend on the initial conditions and, as can be seen, the bigger the soliton is the faster it travels, so that, after some time, they become ordered and separate away from each other.

When $W(x,T)$ evolves according to the perturbed K-dV equation (19), it is no longer possible to separate the evolution equations for the various components of the scattering set. Nevertheless, if $\delta << 1$, the change of scattering data can be obtained through some perturbation method applied to the coupled system, which can be written as, (newell, 1978).

$$\frac{d}{dT} (\frac{b}{a}) = 8i\lambda^3 \frac{b}{a} - \frac{\delta}{2i\lambda a^2}\int_{-\infty}^{\infty} \Gamma(W)\phi^2 \, dx \tag{26}$$

$$\frac{d}{dT} (\lambda_k) = \frac{\delta}{2i\lambda_k\gamma_k a_k'^2}\int_{-\infty}^{\infty} \Gamma(W)\phi_k^2 \, dx \tag{27}$$

$$\frac{d}{dT}(\gamma_k)+\left(\frac{a_k''}{a_k'}+\frac{1}{\lambda_k}\right)\gamma_k\frac{d\lambda k}{dT} = 8i\lambda_k^3\gamma_k \frac{\delta}{2i\lambda_k a_k'^2}\int_{-\infty}^{\infty}\Gamma(W)\left(\frac{d\phi^2}{d\lambda}\right)_{\lambda_k} dx \qquad (28)$$

It is not clear how the general solution of these equations would look like. Here we are going to consider only an initial perturbation consisting in a single soliton and are going to look for its slow evolution taking the solution in the form

$$W = 2V^2 \operatorname{sech}^2 [V(x-\sigma)] \qquad (29)$$

where V is a slowly changing variable and $\sigma = \int^T 4V^2 dT$ is a phase. This should be an important case, according to what was said before, and even in this amplified approach there appear complicated features, which are not going to be addressed here. This concern, in particular the possible birth of secondary solitons from the continuous spectrum, which always produces an oscillatory tail behind the main step. The eigenfunction associated with $\lambda_1 = iV$ and the scattering data for (29) can be written as

$$\phi_1=\phi(\lambda= iV) = \frac{b_1}{2} e^{-V\sigma}\operatorname{sech} [V(x-\sigma)] \qquad (30)$$

$$a(\lambda) = \frac{\lambda-iV}{\lambda+iV}, \quad b_1 = b_1(0)e^{V\sigma}, \quad \gamma_1 = 2iV e^{2V\sigma}$$

Using now these and similar expressions to evaluate the right hand sides of (26-28) a closed system would result. In particular

$$\frac{dV}{dT} = - \frac{\delta V}{4}\left\{\frac{4}{3} + \frac{32}{15} 2^{1/3}\alpha V^2 + \frac{16^2}{21}\left(1 - \frac{6}{5}\frac{1-\omega}{2-\omega}\right)2^{2/3} V^4\right\} \qquad (31)$$

is obtained from the central equation (27), which becomes uncoupled from the others. In so far as this perturbation approach is valid Eq. (31) describes the slow evolution of the soliton. There is a critical value α_c, depending on ω, such that when $\alpha > \alpha_c$ the initial amplitude of the soliton decreases, continuously until it disappears. Whereas if $\alpha < \alpha_c$ the asymptotic state will depend on the initial size of the soliton; which will disappear if it is smaller than the value V_L in the sketch in Fig. 3, and will tend to V_∞ otherwise. The critical value of α is

$$\alpha_c(\omega) = -\frac{10}{\sqrt{7}}\left[1 - \frac{6}{5}\,\frac{1-\omega}{2-\omega}\right]^{1/2} \tag{32}$$

and is plotted in Fig. 4 together with $V_c(\omega)$

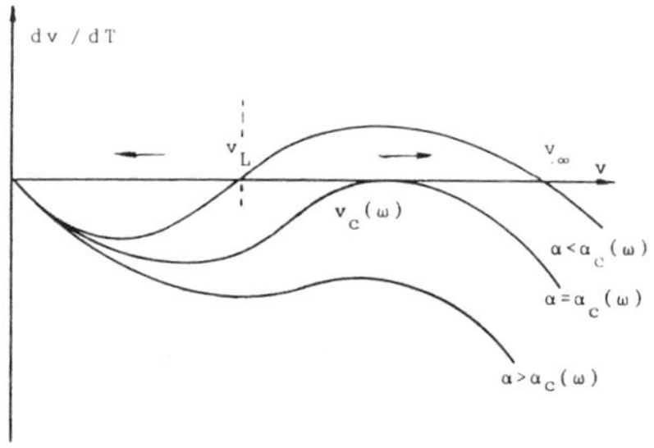

Fig. 3

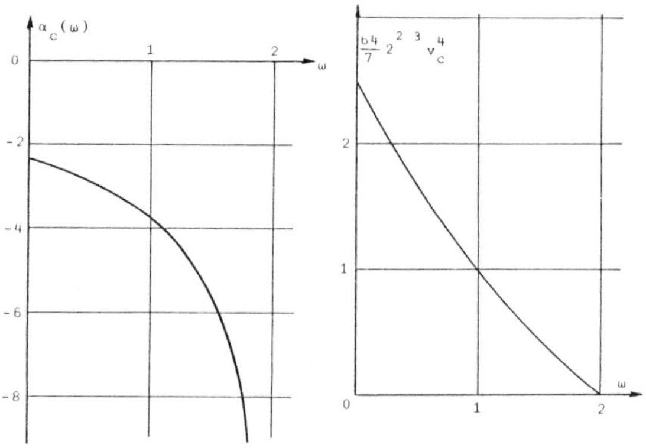

Fig. 4

Up to this point the waves are restricted to travel in a fixed direction. We are now going to comment briefly on the effects coming from the geometry in the propagation of a solitary wave on the interface, when these effects are of the same order as those coming from the real part of Ω in the disperson relation (leading to the right hand side in (19)). This accounts for sufficiently small curvatures of the front;

when the curvature is larger the geometrical effects dominate and determine the evolution of the wave without any significant influence from the dispersion or the growth predicted by the linear stability analysis. Let the successive positions of the front be given by $f_1(\eta,\zeta) - \tau = 0$ and let $f_2(\eta,\zeta) = \text{const.}$ be the orthogonal trajectories or rays. Both families taken together define an orthogonal curvilinear coordinate set (as sketched in Fig. 5). The length of an elementary arc can be written as $ds^2 = S^2 df_1^2 + B^2 df_2^2$, where the metric coefficients satisfy the geometrical condition

$$\frac{\partial}{\partial f_1}\left(\frac{1}{S}\frac{\partial B}{\partial f_1}\right) + \frac{\partial}{\partial f_2}\left(\frac{1}{B}\frac{\partial S}{\partial f_2}\right) = 0 \qquad (33)$$

In order to determine the functions f_1 and f_2 the best procedure is to write down equations for the angle Θ between the rays and a fixed direction in the plane of the interface. As can be seen from Fig. 5, Θ is the solution of

$$\frac{\partial\Theta}{\partial f_2} = \frac{1}{S}\frac{\partial B}{\partial f_1}$$
$$\qquad (34)$$
$$\frac{\partial\Theta}{\partial f_1} = -\frac{1}{B}\frac{\partial S}{\partial f_2}$$

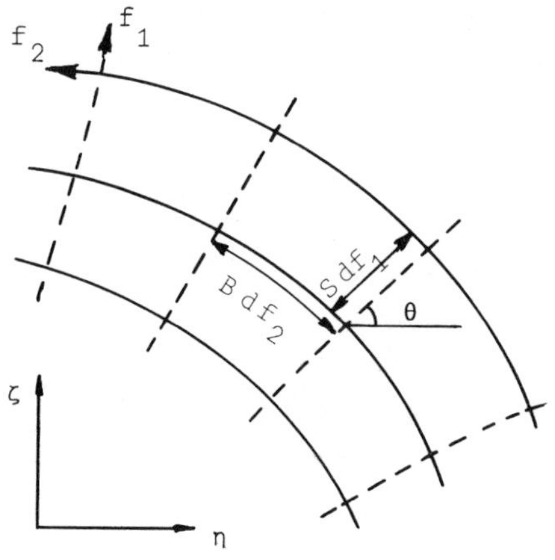

Fig. 5

To close the problem the following two relations must be taken into account.

$$S = 1-\delta^2 \alpha + 4\cdot 2^{1/3} \delta^3 V^2 \tag{35}$$

$$\frac{\partial V/\partial f_1}{S^2} = - \frac{\delta^3 V}{4} \{\frac{4}{3} + \frac{32}{15} 2^{1/3} \alpha V^2 + \frac{16^2}{21} (1 - \frac{6}{5} \frac{1-\omega}{2-\omega})^{2/3} V^4\} - \frac{2}{3}\Gamma_1 V \tag{36}$$

where

$$\Gamma_1 = \frac{\partial B/\partial f_1}{2SB} \tag{37}$$

The first of these is simply the expression of the normal wave velocity in terms of its amplitude, and the second is a generalization of (32), accounting also for the slow variation of the breadth B along the "channel" between two rays, which leads to the final term. A term of this form appears in the theory of shallow water waves in a slowly varying channel (see p.e. Miles, 1979). A definite approximation is involved here because of course the rays are not confining walls.

The system (33-37) is a closed problem whose solution would give Θ and the local wave amplitude as functions of f_1 and f_2. Finally, f_1 and f_2 can be tied to the fixed directions η and ζ on the interface through the solution of

$$\partial f_1/\partial\eta = \cos\Theta/S \qquad \partial f_1/\partial\zeta = \sin\Theta/S$$
$$\partial f_2/\partial\eta = -\sin\Theta/B \qquad \partial f_2/\partial\zeta = \cos\Theta/B \tag{38}$$

as can be seen from Fig. 5.

As a specific example let us consider the case of a symmetrical wave travelling away from a point. In this case the final term in (36) takes the form $-V/3r$, where r is the distance to the source of the wave. The analysis applies when $r \sim \delta^3$ and, as can be seen, the geometrical divergence tends to decrease the amplitude of the wave at a rate that decreases as the wave evolves and r increases. So the divergence changes the limiting value V_L found before for planar waves and some waves initially larger than this limiting value can end up decaying and disappearing.

5. MODULATION APPROACH. There is a second type of travelling wave solutions of the K-dV equation, cn wave, which are periodic travelling waves of the form $W=W(\Theta)$ with $\Theta = kx-\Omega T$. Here Ω and k are constants and $W(\Theta)$ is a 2π periodic function. (In fact the solitary wave can be obtained as the infinite wavelength limit from the cn wave). It is possible to obtain evolution equations for the slowly varying amplitude and frequency of a cn wave under the effect of the perturbation in the right hand side of (19), by taking advantage from one (appropriate) conservation law of the infinite number that there exist for the unperturbed K-dV equation. A somewhat more general approach comes from modulation theory, which accounts for the evolution of slowly changing waves including their slow dependence on spatial corodinates in addition to their temporal changes. The solution is taken in the form

$$V = \Psi + \phi\ (\Theta) \qquad (39)$$

whose $\Psi = \Psi(x',T')/\delta$ is the slowly varying mean level of the interface and $\Theta = \theta(x',T')/\delta$ is a phase variable. $\phi(\Theta) = \phi(\Theta+2\pi)$ represents the oscillatory part of the wave, and $x' = \delta x$ and $T' = \delta T$ are the slow variables. Now the local values

$$\beta = \Psi_x\ , \quad \gamma = -\Psi_T\ , \quad k = \Theta_x \ \text{ and } \ \Omega = -\Theta_T \qquad (40)$$

still depend slowly on space and time. In particular, $W = \beta + k\,\phi_\Theta$. Carrying (40) into (18), the following equations results for the leading order in an asymptotic expansion in powers of δ

$$k^2 W_{\Theta\Theta\Theta} + 6WW_\Theta - UW_\Theta = 0 \qquad (41)$$

where $U = \Omega/k$. After two integrations the expression

$$k^2 W_\Theta^2 + 2W^3 - UW^2 + 2BW - 2A = 0 \qquad (42)$$

is obtained. Here A and B are slowly varying functions. A further integration allows to calculate $W(\Theta)$ in terms of A, B, U and k. The equations describing the slow evolution of these variables, as well as Ψ, would result from the appropriate resolubility conditions on the higher order terms in the multiscale analysis. Instead of going directly into these higher order approximations on the equation itself, it proves to be more efficient, for the unperturbed K-dV equation, to take into account the existence of a variational formulation, and to carry the expressions (39-42) into the lagrangian density, taking variations only after the fast variables have been suppressed by averaging the lagrangian over a cycle of the oscillation. The perturbation in the right hand side of (19) destroys the variational formulation, however, it is still useful to apply the averaging method on a pseudo-variational principle, in which the lagrangian depends on the functions to be varied and the solution of the problem. This can be used to describe irreversible processes and was introduced by Jimenez and Whitham (1976) as an extension of the modulation approach for lagrangian systems (see Whitham, 1974). The perturbation in the right hand side of (19) can be written as

$$\delta\Gamma(V) = \delta \frac{\partial}{\partial x} \{\Gamma'(V)\} \qquad (43)$$

where

$$\Gamma'(V) = -2^{5/3} V_{xxxx} + 2^{1/3}\alpha\, V_{xx} - V/2 - 3\cdot 2^{2/3}\, \frac{1-\omega}{2-\omega}(V_x^2)_x \qquad (44)$$

and one possible pseudo-variational principle leading to (19) results from the lagrangian density

$$L = -\frac{1}{2} V_T V_x - V_x^3 + \frac{1}{2} V_{xx}^2 + \delta V_x \tilde{\Gamma}' \tag{45}$$

where the tilde means that Γ' is kept constant when variations are taken. Now if (39) is taken into (45) and (41-42) are used, the averaged lagrangian

$$\mathcal{L} = \langle L \rangle = kZ(A,B,U) = \beta B + \frac{1}{2}\beta\gamma - \frac{1}{2}U\beta^2 - A - \frac{1}{2}\beta\tilde{\Psi} - \delta k\,(\Theta-\tilde{\Theta})\,\langle \phi_{\Theta\Theta}\tilde{\Gamma}' \tag{46}$$

results, where

$$Z = \frac{1}{2\pi} \oint [2A-2BW + UW^2 - 2W^3]^{1/2}\, dW \tag{47}$$

and

$$\langle\cdot\rangle = \frac{1}{2\pi} \int_0 (\cdot)\, d\Theta \tag{48}$$

The averaged variational equations coming from (46) are

$$\mathcal{L}_B = 0, \quad -\frac{\partial \mathcal{L}_\gamma}{\partial T'} + \frac{\partial \mathcal{L}_\beta}{\partial x'} - \mathcal{L}_\Psi = 0, \quad \mathcal{L}_A = 0 \quad \text{and}$$

$$\frac{\partial \mathcal{L}_\Omega}{\partial T'} \quad \frac{\partial \mathcal{L}_k}{\partial x'} - \mathcal{L}_\Theta = 0 \tag{49a}$$

together with the consistency relations

$$\frac{\partial \beta}{\partial T'} + \frac{\partial \gamma}{\partial x'} = 0 \quad \text{and} \quad \frac{\partial k}{\partial T'} + \frac{\partial \Omega}{\partial x'} = 0 \tag{49b}$$

which result from the definitions (40).

After a certain amount of algebra these expressions can be transformed into the system

$$\frac{\partial Z_B}{\partial T'} + U \frac{\partial Z_B}{\partial x'} + Z_A \frac{\partial B}{\partial x'} = -Z_B/2 \tag{50a}$$

$$\frac{\partial Z_u}{\partial T'} + U \frac{\partial Z_u}{\partial x'} - Z_A \frac{\partial A}{\partial x'} = k <\phi_{\Theta\Theta}\ \Gamma'> - Z_B^2/2\ Z_A \tag{50b}$$

$$\frac{\partial Z_A}{\partial T'} + U \frac{\partial Z_A}{\partial x'} - Z_A \frac{\partial U}{\partial x'} = 0 \tag{50c}$$

describing the slow evolution of the wave train, whereas the other variables are given by

$$k = 1/W_A \ , \ \Omega = U/W_A \ , \ \beta = -\ W_B/W_A \quad \text{and}$$

$$\gamma = -\ UW_B/W_A - B \tag{51}$$

in terms of A, B and U.

A few general comments may be made about (50). First it can be seen that it is a hyperbolic system, and can be written in characteristic form if the roots p, q and r of the cubic in the integrand of (47) are introduced as new variables in place of A, B and U. The hyperbolic character of (50) means that a modulated wavetrain is stable in the Whitham sense (Whitham 1974). The second fact to notice is that the perturbation in the right hand side of (19) is reflected only in the right hand sides of (50), and these terms do not involve derivatives of the variables, so that they do not modify the local slope of the characteristic curves in the (x',T') plane. Their main effect is to destroy the Riemann invariants that, otherwise, would be the combinations p+q , q+r and r+p. Finally, it must be remarked the coupling between the slow variations in wave amplitude and phase and the slow variations in the mean level of the interface.

ACKNOWLEDGEMENT

We acknowledge the partial support for this research by the Spanish CAICYT under Project 2291-83.

REFERENCES

Denison, R. and Baum, E. (1961). A simplified model of unstable burning of solid propellants. ARS J. 31 (8), pp. 1112-1122.

Gelfand, I.M. and Levitan, B.M. (1951). On the determination of a differential equation from its spectral function. Am. Math. Transl. (2) 1, pp. 253-304.

Higuera, F.J. and Linan, A. (1982). Stability of solid propellant combustion subject to nonplanar perturbations. In Dyanmics of Flames and Reactive Systems, Progress in Astronautics and Aeronautics, Vol. 95, edited by J. R. Bowen et al., pp. 248-258.

Jimenez, J. and Whitham, G.B. (1976). An averaged Lagrangian method for dissipative wavetrains. Proc. R. Soc. London. A. 349, pp. 277-287.

Korteweg, D.J. and deVries, G. (1895). On the change of form of long stationary waves. Phil. Mag. (5) 39, pp. 422-443.

Margolis, S.B. and Armstrong, R.C. 1985. Two asymptotic models for solid propellant combustion. Submitted to Combustion Science and Technology.

Miles, J.W. 91979). On the Korteweg-deVries equation for a gradually varying channel. J. Fluid Mech. 91, pp. 181-190.

Newell, A.C. (1978). Near-integrable systems, nonlinear tunnelling and solitons in slowly changing media. In Nonlinear Evolution Equations Soluable by the Spectral Transform, Research Notes in Mathematic, F. Cabgero (Ed.), pp. 127-179.

Whitham, G.B. (1974). Linear and Nonlinear Waves. Wiley, New York.

Zel'dovich, Ya.B. (1942). Theory of the combustion of gunpowders and explosive substances. Zh. Eksp. Tear. Fiz. 12 (11), pp. 498-524.

Zel'dovich, Ya.B. (1964). Rate of combustion of gunpowders under varying pressure. Zh. Prikl. Mekh. Tekhn. Fiz. 5 (3), pp. 126-130.

E.T.S. Ingenieros Aeronauticos. Universidad Politecnica de Madrid. Pza Cardenal Cianeros 3, 28040 Madrid. Spain.

Lectures in Applied Mathematics
Volume 24, 1986

TWO ASYMPTOTIC MODELS FOR SOLID PROPELLANT COMBUSTION

Stephen B. Margolis and Robert C. Armstrong

In the present work we summarize the results of a recent study (Margolis and Armstrong, 1985) in which we derive two different asymptotic models that describe the nonsteady, nonplanar burning of certain types of homogeneous solid propellants. Motivated in part by recent work on ammonium perchlorate deflagration (Guirao and Williams, 1971), we assume, in the first model, that a fraction of the propellant is pyrolyzed directly to product gases at a solid/gas interface, while the remainder sublimes and burns in the gas phase. In the second model, there is a thin liquid layer between the solid and gas, with combustion occurring in both the liquid and gas phases. Here, we confine our attention to the first model, which is illustrated in Fig. 1.

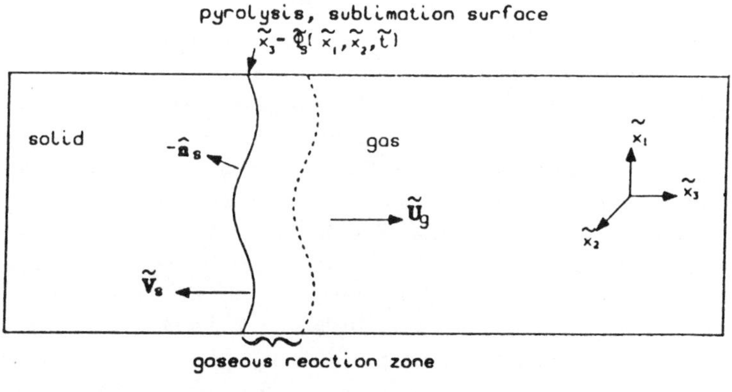

Figure 1. Geometry of propellant deflagration (Model I).

In our analysis, we consider a regime in which the gaseous flame lies near the propellant surface. Then, exploiting the fact that the activation energies of the pyrolysis and gas phase reactions are large, we employ the method of matched asymptotic expansions to derive a flame sheet model analogous to those derived for strictly gaseous (cf. Matkowsky and Sivashinsky, 1979) and strictly condensed (Margolis, 1983, 1985) combustion. Solutions on either side of the sheet, which is actually a thin gaseous reaction zone adjacent to the solid/gas interface, are then connected by nonlinear jump conditions which depend on local conditions there. We also are able to derive an analytical expression for the burning rate eigenvalue for the special case of steady, planar burning (cf. Buckmaster, Kapila and Ludford, 1976). In particular, we find that the burning rate approaches a constant value in the limit of complete conversion to products by pyrolysis at the solid/gas interface or in the limit that the pyrolysis reaction is rate-limiting. Similarly, the burning rate becomes small in the limit that the gas-phase reaction is rate-limiting or in the limit that the gas-to-solid density ratio ρ approaches zero, in which case the magnitude of the velocity field in the gas phase becomes large.

Aside from determining an expression for the burning rate, the main advantage of the asymptotic model over the original equations with a distributed reaction rate is that the former allows us to write the solution for the special case of steady, planar burning explicitly, thereby facilitating investigations of stability and bifurcation phenomena. Indeed, linearizing about this basic solution, we obtain a linear stability problem which admits complex perturbations ϕ in the burning surface of the form

$$\phi \sim \exp(i\omega t \pm imx_1 \pm inx_2)\,,$$

where t is the time variable, x_1 and x_2 are the transverse spatial coordinates, m and n are the transverse wavenumbers of the disturbance, and $\omega(k, \Xi, \rho)$ is the complex frequency. The latter is determined by a dispersion relation as a function of the wavenumber $k = (m^2 + n^2)^{1/2}$, a modified activation energy parameter Ξ, and the gas-to-solid density ratio ρ. Setting the real part of $i\omega$ equal to zero then determines the pulsating neutral stability boundaries shown in Fig. 2. We observe that a decreasing density ratio is destabilizing. In the limiting case $\rho = 1$, the neutral stability boundary is identical to that which is obtained for strictly condensed phase combustion (Sivashinsky, 1981; Margolis, 1983, 1985), although the definition of Ξ is in general different for the two types of problems. Finally, we mention that intrinsic pulsating instabilities in solid propellant combustion have also been predicted using other models (cf. Denison and Baum, 1961; Krier et al., 1968; De Luca, 1976; Higuera and Liñán, 1984). However, these other analyses assume a quasi-steady and, in the multidimensional case considered in the last study, a quasi-planar gas phase, whereas no such assumptions are employed here.

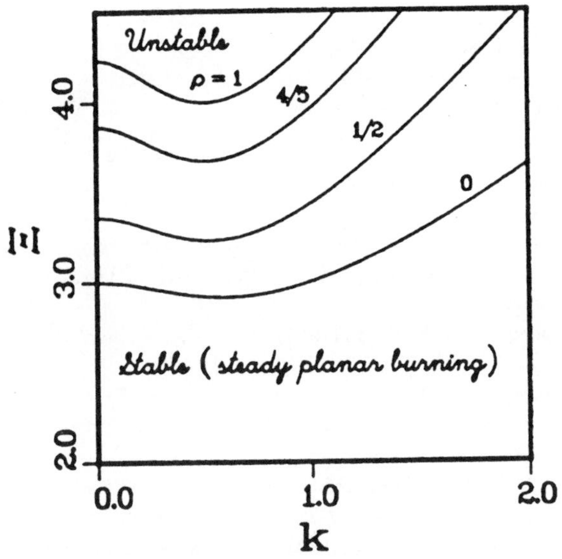

Figure 2. Neutral stability boundaries for various ρ.

The pulsating neutral stability boundaries depicted in Fig. 2, which are thus very similar to those which describe the reaction-diffusion instability in condensed phase combustion and the pulsating diffusional-thermal instability in gaseous combustion (cf. Margolis and Matkowsky, 1983), suggest a transition to a nonsteady, nonplanar mode of burning as the activation energy parameter Ξ is increased above a critical value. Indeed, bifurcation analyses in a cylindrical geometry for the condensed problem (Matkowsky and Sivashinsky, 1979; Margolis et al., 1985; Booty, Margolis and Matkowsky, 1985) predict a rich variety of pulsating and spinning phenomena, some of which are well-documented experimentally (cf. Merzhanov et al., 1973; Dvoryankin et al., 1982; Strunina et al., 1983). In particular, the last theoretical study, as well as a similar analysis for gaseous combustion (Margolis and Matkowsky, 1985), consider bifurcation near a double eigenvalue Ξ_0 of the linearized problem to obtain bifurcation diagrams which exhibit various secondary branching and mode-jumping phenomena, and which predict the existence of stable quasi-periodic combustion waves.

Acknowledgment: This work was supported by the Office of Energy Research, U. S. Department of Energy, and by a Memorandum of Understanding between the Department of the Army and the Department of Energy.

REFERENCES

1. M. R. Booty, S. B. Margolis and B. J. Matkowsky (1985). *Interaction of Pulsating and Spinning Waves in Condensed Phase Combustion*, **SIAM J. Appl. Math.**, to appear.
2. J. D. Buckmaster, A. K. Kapila and G. S. S. Ludford (1976). *Linear condensate deflagration for large activation energy*, **Acta Astronautica 3**, pp. 593 - 614.
3. L. De Luca (1981). *Nonlinear burning stability theory of heterogeneous thin flames*, **Eighteenth Symposium (International) on Combustion**, pp. 1439 - 1450.
4. M. R. Denison and E. Baum (1961). *A simplified model of unstable burning in solid propellants*, **ARS J. 31**, pp. 1112 - 1122.
5. A. V. Dvoryankin, A. G. Strunina and A. G. Merzhanov (1982). *Trends in the spin combustion of thermites*, **Combustion, Explosion and Shock Waves 18**, pp. 134 - 139.
6. C. Guirao and F. A. Williams (1971). *A model for ammonium perchlorate deflagration between 20 and 100 atm*, **AIAA J. 9**, pp. 1345 - 1356.
7. F. J. Higuera and A. Liñán (1984). *Stability of solid propellant combustion subject to nonplanar perturbations*, in **Dynamics of Flames and Reactive Systems - Progress in Astronautics and Aeronautics 95**, pp. 248 - 256.
8. H. Krier, J. S. T'ien, W. A. Sirignano and M. Summerfield (1968). *Nonsteady burning phenomena of solid propellants: theory and experiments*, **AIAA J. 6**, pp. 278 - 285.
9. S. B. Margolis (1983). *An asymptotic theory of condensed two-phase flame propagation*, **SIAM J. Appl. Math. 43**, pp. 351 - 369.
10. S. B. Margolis (1985). *An asymptotic theory of heterogeneous condensed combustion*, **Combust. Sci. Technol. 43**, pp. 197 - 215.
11. S. B. Margolis and R. C. Armstrong (1985). *Two Asymptotic Models for Solid Propellant Combustion*, **Combust. Sci. Technol.**, to appear.
12. S. B. Margolis and B. J. Matkowsky (1983). *Nonlinear stability and bifurcation in the transition from laminar to turbulent flame propagation*, **Combust. Sci. Technol. 34**, pp. 45 - 77.
13. S. B. Margolis and B. J. Matkowsky (1985). *Flame propagation in channels: secondary bifurcation to quasi-periodic pulsations*, **SIAM J. Appl. Math. 45**, pp. 93 - 129.
14. S. B. Margolis, H. G. Kaper, G. K. Leaf and B. J. Matkowsky (1985). *Bifurcation of pulsating and spinning reaction fronts in condensed two-phase combustion*, **Combust. Sci. Technol. 43**, pp. 127 - 165.
15. B. J. Matkowsky and G. I. Sivashinsky (1978). *Propagation of a pulsating reaction front in solid fuel combustion*, **SIAM J. Appl. Math. 35**, pp. 465 - 478.

16. A. G. Merzhanov, A. K. Filonenko and I. P. Borovinskaya (1973). *New phenomena in combustion of condensed systems*, **Dokl. Phys. Chem. 208**, pp. 122 - 125.
17. G. I. Sivashinsky (1981). *On spinning propagation of combustion waves*, **SIAM J. Appl. Math. 40**, pp. 432 - 438.
18. A. G. Strunina, A. V. Dvoryankin and A. G. Merzhanov (1983). *Unstable regimes of thermite system combustion*, **Combustion, Explosion and Shock Waves 19**, pp. 158 - 163.

Combustion Research Facility
Sandia National Laboratories
Livermore, California 94550

Lectures in Applied Mathematics
Volume 24, 1986

Plane shock initiation of homogenous and heterogenous condensed phase explosives with a sensitive rate.

D. S. Stewart

Abstract. This lecture shows how classic thermal explosion theory can be generalized to treat the induction phase of the shock initiation to detonation in homogeneous and heterogeneous condensed phase explosives. The forward reaction rate is assumed to be sensitive; small changes in the thermodynamic state lead to moderate to large changes in the rate. A large parameter, such as a dimensionless activation energy, characterizes the rate. We show that the initiation behaviour and the location of the ignition point within the explosive is a strong function of the initiating shock pressure, the chosen form of the sensitive rate and the character of the heterogeneities.

1. INTRODUCTION. The initial value problem of plane shock initiation is that of an unreacted explosive at rest that is suddenly shocked by the collision of the explosive with a piston (or flyer plate). The shock then serves to start the reaction, which under appropriate circumstances, leads to the development of a detonation wave that propagates through the material.

Certain classes of condensed phase explosives, liquid nitromethane is an example, as well as explosive gases exhibit what is refered to as thermal explosive behaviour. In a typical experiment, the first phase of the initiation transient involves the passage of the initiating shock through the explosive, producing little reaction that releases heat. However, as the reaction proceeds, even by small amounts, the thermodynamic state is changed. When reaction rate is extremely sensitive to small changes in the state the reaction rate dramatically accelerates after a characteristic time from the initiating event. The characteristic time is called the induc-

0075-8485/86 $1.00 + $.25 per page

tion time and the dramatic acceleration of reaction rate and the violent processes accompanying it is known as a thermal explosion.

The standard theory, applied to this class of explosives, assumes that after the passage of the shock, each explosive particle undergoes a spatially uniform thermal explosion (in the particle coordinate) starting from the shocked, unreacted state. The wave motion between the shock and piston is completely neglected.

Neglecting the wave motion simplifies the analysis of the problem enormously and a description of the induction phase is obtained by assuming small reaction depletion and that the state is close to the unreacted, shocked state. Then an explicit time can be derived that describes when a perturbation of the shocked pressure (say) becomes infinite. This time is identified as the induction time.

One of the main points of the lecture is to illustrate that the effects of wave propagation, for a model that is complex enough to describe condensed phase explosives, can dramatically influence the induction phase behavior and the observed induction time. Also, we will show that the solution of the wave propagation problem actually determines the point of ignition within the shocked explosive.

2. FORMULATION. The explosive, in the shocked state, is assumed to be a reacting Euler fluid. The complete formulation of the initiation problem consists of the equations of motion for a given rate law, the boundary condition that the piston velocity is attained at the piston surface and the shock conditions at the leading shock. We adopt the notation and convention of Fickett and Davis [1]. The equations and variables that appear are usually dimensionless. Dimensional variables are indicated by a tilde. The pressure and density reference values are those of the unreacted, shocked state corresponding to the initiating shock. The velocity scale is the speed of sound of the initial unreacted, shocked explosive. The time scale is left arbitrary for now with the length scale chosen as the product of the velocity and time scales.

Thus we introduce the dimensionless variables

$$P = \tilde{P}/\tilde{P}_s, \quad \rho = \tilde{\rho}/\tilde{\rho}_s, \quad c = \tilde{c}/\tilde{c}_s, \quad u = \tilde{u}/\tilde{c}_s, \quad D = \tilde{D}/\tilde{c}_s, \qquad t = \tilde{t}/\tilde{t}_c, \quad x = \tilde{x}/(\tilde{c}_s\tilde{t}_c). \tag{1}$$

The "s" subscript refers to the reference unreacted, shocked state. P, ρ, λ, c, u, D, t and x represent pressure, density, the reaction progress variable of a forward exothermic reaction, the sound velocity, particle velocity, the shock velocity and time and distance in the lab frame.

We will assume that the equation of state of the material is that of a polytropic fluid for simplicity. The polytropic equation, for the specific internal energy $\tilde{E}$, is given by

$$\tilde{E} = \tilde{P}/[(\gamma - 1)\tilde{\rho}] - \tilde{q}\lambda, \tag{2}$$

where γ is the polytropic parameter. The speed of sound c is then given as

$$c = \sqrt{\gamma P/\rho}. \tag{3}$$

In addition $\tilde{q}$ appears in (2) and represents the heat of reaction. $\tilde{q}$ appears only in the dimensionless heat release parameter called the thermicity defined here as

$$\sigma = \alpha(\gamma + 1)\tilde{\rho}_0\tilde{q}/(\gamma\tilde{P}_s), \qquad \alpha \equiv \left[(\gamma - 1)\tilde{\rho}_s/((\gamma + 1)\tilde{\rho}_0)\right]. \tag{4}$$

Note that $\tilde{\rho}_0$ is the unshocked density of the explosive. Later, we will employ the strong shock approximation which sets the parameter $\alpha = 1$. Thus σ is inversely dependent on the pressure $\tilde{P}_s$.

Finally we prescribe the rate. The rate is assumed to be sensitive to changes in the state reflected by a large parameter. Again for simplicity we restrict our discussion to the Arrhenius form. An indication of how the theory can be generalized to a wider class of rate laws is given later in the lecture. Thus we assume the rate to be of the form

$$r = \tilde{t}_c\tilde{k}(1 - \lambda)e^{-\theta/T}. \tag{5}$$

In the above $\tilde{k}$ is the characteristic inverse reaction time. Also the variable T appears, which can loosely be thought of as the temperature. The appropriate form of rate laws for condensed phase explosives are marginally known at best so that a particular form is not sacred and is only accepted as a representative model. Thus (5) represents a model and is not necessarily justified from kinetic theory as for gases. T as a function of P, ρ and λ ultimately prescribes the form of the rate.

Examples where T takes on a precise form usually follow from defining a thermal equation of state for the explosive. For example, consider the forward reaction with a mole change

$$A \rightarrow (1+\delta)B, \quad \delta > 0. \tag{6}$$

For a mixture of polytropic fluids, the thermal equation of state, modeled after a gas, is

$$P/(\rho T) = (1+\delta\lambda). \tag{7}$$

The large parameter θ can now be properly identified as

$$\theta = \tilde{E}_+/(\tilde{R}\tilde{T}_s), \tag{8}$$

where $\tilde{E}_+$ is the dimensional activation energy and $\tilde{R}$ is the gas constant.

The complete formulation of the initiation problem can now be posed. The equations of motion are written in characteristic form as

$$P_{,t} + (u+c)P_{,x} + \gamma\rho c\big[u_{,t} + (u+c)u_{,x}\big] = \gamma\sigma\rho c^2 r, \tag{9}$$

$$P_{,t} + (u-c)P_{,x} - \gamma\rho c\big[u_{,t} + (u-c)u_{,x}\big] = \gamma\sigma\rho c^2 r, \tag{10}$$

$$P_{,t} + uP_{,x} - \gamma[\rho_{,t} + u\rho_{,x}] = \gamma\sigma\rho c^2 r, \tag{11}$$

$$\lambda_{,t} + u\lambda_{,x} = r. \tag{12}$$

The piston condition is simply

$$u = u_{\mathrm{p}} \quad \text{at } x = u_{\mathrm{p}} t. \tag{13}$$

Finally we supply the shock conditions. The shock conditions that are presented incorporate two assumptions, the form of the equation of state and the strong shock approximation. The latter is a standard assumption for condensed phase explosives and assumes that

$$\tilde{P}_o/\tilde{P}_s \ll 1, \tag{14}$$

where $\tilde{P}_o$ is the pressure of the unshocked explosive. The strong shock relations for a polytropic fluid are given by

$$\rho = 1, \quad u = 2D/(\gamma+1), \quad P = 2\gamma(\gamma-1)D^2/(\gamma+1), \quad \lambda = 0 \tag{15}$$

at

$$x = x_s \equiv \int_0^t D\,dt. \tag{16}$$

3. CLASSICAL THERMAL EXPLOSION THEORY. It is appropriate to review classic thermal explosion theory. The results derived here will be used to compare with our later calculations.

The assumptions of classic thermal explosion theory neglect spatial derivatives and the boundary conditions associated with the confinement of the piston and its interaction with the leading shock. The other assumptions are that the state is close to the shocked state and that only a small amount of reaction has occurred. Thus ignoring spatial derivatives in the particle coordinate equation, (9)–(12) become

$$P_{,\tilde{t}} + \gamma\rho c u_{,\tilde{t}} = \gamma\sigma\rho c^2 \tilde{r}, \tag{17}$$

$$P_{,\tilde{t}} - \gamma\rho c u_{,\tilde{t}} = \gamma\sigma\rho c^2 \tilde{r}, \tag{18}$$

$$P_{,\tilde{t}} - \gamma\rho_{,\tilde{t}} = \gamma\sigma\rho c^2 \tilde{r}, \tag{19}$$

$$\lambda_{,\tilde{t}} = \tilde{r}. \tag{20}$$

We have reintroduced the dimensional time for convienence. From the above equations we immediately conclude that

$$\rho_{,\tilde{t}} = u_{,\tilde{t}} = 0, \quad \rho = 1, \quad u = u_{\mathrm{p}}, \tag{21}$$

and

$$P_{,\tilde{t}} = \gamma\sigma\rho c^2 \tilde{r}, \quad \lambda_{,\tilde{t}} = \tilde{r}. \tag{22}$$

Taking the state close to the shocked state we write

$$P = 1 + P', \quad \lambda = \lambda', \quad T \equiv P/[\rho(1+\delta\lambda)] = 1 + (P' - \delta\lambda'), \tag{23}$$

where

$$|P'|, |\lambda'| \ll 1. \tag{24}$$

These approximations, substituted into (22), lead to the simple problem for P' and λ'

$$\begin{gathered} P'_{,\tilde{t}} = \gamma\sigma\tilde{r}', \qquad \lambda'_{,\tilde{t}} = \tilde{r}', \qquad \tilde{r}' \equiv \tilde{k}e^{-\theta}e^{\theta(P'-\delta\lambda')}, \\ \lambda' = P' = 0 \quad \text{at } \tilde{t} = 0, \end{gathered} \tag{25}$$

with the solution

$$\begin{gathered} P' = -\theta^{-1}[\gamma\sigma/(\gamma\sigma - \delta)]\ln[1 - (\gamma\sigma - \delta)\tilde{k}\theta e^{-\theta}\tilde{t}], \\ \lambda' = P'/(\gamma\sigma). \end{gathered} \tag{26}$$

The result (26) clearly identifies a characteristic time for thermal runaway, namely,

$$\tilde{t}_{\mathrm{i}} = t_{\mathrm{i}}^{\mathrm{u}}(\tilde{k}\theta)^{-1}e^{\theta}, \qquad t_{\mathrm{i}}^{\mathrm{u}} \equiv (\gamma\sigma - \delta)^{-1}. \tag{27}$$

The "i" subscript refers to induction while the "u" superscript refers to spatially uniform. Note that (26) properly identifies the asymptotic dependence of the induction time on the large parameter and the size of the perturbation of the shocked state to be $O(\theta^{-1})$.

4. THE INDUCTION PHASE WITH WAVE PROPAGATION.

Based on the previous section we choose the time scale

$$\tilde{t}_{\mathrm{i}} = (\tilde{k}\theta)^{-1}e^{\theta}. \tag{28}$$

Also it is convenient to introduce the piston coordinate

$$\xi = x - u_{\mathrm{p}}t, \tag{29}$$

where $\xi = 0$ is the piston location. In the induction phase, we perturb the solution about the shocked state, so we expand

$$\begin{gathered} \rho = 1 + \rho_1/\theta + \cdots, \quad u = u_{\mathrm{p}} + u_1/\theta + \cdots, \quad P = 1 + P_1/\theta + \cdots, \\ c = 1 + \cdots, \quad T = 1 + T_1/\theta + \cdots, \quad \lambda = \lambda_1/\theta + \cdots, \\ D = D_{\mathrm{s}} + D_1/\theta + \cdots. \end{gathered} \tag{30}$$

With these expansions, equations (9)–(12) lead to

$$P_{1,t} + P_{1,\xi} + \gamma[u_{1,t} + u_{1,\xi}] = \gamma\sigma h, \tag{31}$$

$$P_{1,t} - P_{1,\xi} - \gamma[u_{1,t} - u_{1,\xi}] = \gamma\sigma h, \tag{32}$$

$$P_{1,t} - \gamma\rho_{1,t} = \gamma\sigma h, \tag{33}$$

$$\lambda_{1,t} = h, \tag{34}$$

where

$$h \equiv e^{T_1}, \quad T_1 = P_1 - \rho_1 - \delta\lambda_1. \tag{35}$$

The piston condition simply requires that

$$u_1 = 0 \quad \text{at } \xi = 0. \tag{36}$$

The shock relations require that

$$\rho_1 = 0, \quad u_1 = \left[2/(\gamma+1)\right] D_1,$$
$$P_1 = \left[2\sqrt{\gamma(\gamma-1)}\,/(\gamma+1)\right] D_1, \quad \lambda_1 = 0, \tag{37}$$

at

$$\xi = [D_s - u_p]t = \sqrt{(\gamma-1)/2\gamma}\; t, \tag{38}$$

where

$$D_s \equiv (\gamma+1)\Big/\sqrt{2\gamma(\gamma-1)}, \quad u_p \equiv \sqrt{2/\gamma(\gamma-1)}. \tag{39}$$

The above problem is solved numerically, by the method of characteristics, until a thermal runaway (a singularity in the state variables) is observed at a location between the piston and the leading shock. The results of these calculations, for the prediction of the induction time, can be summarized by the following formula

$$\tilde{t}_i = t_i(\gamma, \sigma, \delta)(\tilde{k}\theta)^{-1} e^{\theta}. \tag{40}$$

The factor t_i is found from the numerical computations while the spatially uniform thermal explosion time is given by (27). Indeed, one of the main points of this lecture is to observe that generally

$$t_i \neq t_i^u. \tag{44}$$

The ratio t_i/t_i^u can now be computed explicitly as the parameters γ, σ, and δ are varied. In particular, varying σ corresponds to changing the initial shock strength, δ corresponds to changing the rate. Other effects that include sensitivity of the rate to changes in the density or pressure and other effects due to changes in the kinetic scheme can be included, with additional parameters that may influence t_i.

5. NUMERICAL RESULTS FOR A HOMOGENEOUS EXPLOSIVE. This section illustrate the results of studying the solution to (31)–(39) as δ is varied. A representative value for the polytropic exponent for a homogenous explosive, such as nitromethane, is $\gamma = 2.86$. For initiating pressures in the range of 5 to 12 GPa, σ varies between 1.6 and .7.

Figures 1. show the result of a typical calculation for a homogeneous explosive with no mole change, $\delta = 0$. The figures show the space time evolution of the perturbations of pressure, temperature and progress variable up to the time of thermal runaway. The results show the development of a localized explosion (a singularity in space and time) first occuring at the piston face. This calculation is in qualitative agreement with Jackson's and Kapila's [2] result for the initiation of an ideal gas.

The current result shows that a local hot spot (ignition point) develops at the piston boundary. However, for modeling condensed phase explosives, the ideal equation of state must be regarded as a special case. Numerical solutions with other choices for the material and thermal equation of state, that are defined by γ, σ and δ, show that the ignition point does not always occur at the piston face.

The results shown in Figures 2. correspond to a positive mole change, $\delta = 2$, and show dramatic qualitative differences from the previous case. In particular the ignition point develops at the leading shock instead of at the piston face. The calculations show clearly that the reaction proceeds most rapidly at the leading shock and much more slowly, by comparison, in the interior.

Figures 3. show the results of a parameter study for the ratio of the induction time to the spatially uniform induction time when δ is changed holding γ and σ fixed. This graph demonstrates the general conclusion that the proper inclusion of the wave mechanics for the initiation of a thermal explosion by a shock can lead to dramatically different predictions depending on the exact form of the reaction rate and its sensitivity to the rate.

Indeed, one can conclude that there are instances when the standard theory may work well, but given the complexity of the heat release processes in condensed phase explosives, it seems unlikely that all explosives, identified as having thermal explosive behavior, have properties such that $t_i = t_i^u$.

6. GENERALIZING THE RATE LAW. In this section we show how more general rate laws can be accommodated. For the induction phase, the state is a perturbation of the unreacted shocked state. If ε is the inverse of the large rate parameter, then we represent the perturbed state as

$$P = 1 + \varepsilon P_1 + \cdots, \quad \rho = 1 + \varepsilon \rho_1 + \cdots, \quad \lambda = \varepsilon \lambda_1 + \cdots. \tag{42}$$

If the dimensionless rate, $r(P, \rho, \lambda)$, when perturbed about the shocked state, can be represented such that it has the form

$$r = \varepsilon \tilde{t}_c \tilde{g}(\varepsilon)\left[h(P_1, \rho_1, \lambda_1) + \cdots\right], \tag{43}$$

then $\tilde{t}_c$ must be chosen such that it has the form

$$\tilde{t}_c \tilde{g}(\varepsilon) = 1. \tag{44}$$

This assumption simply reflects that we are considering materials such that the perturbations of the shocked state and those due to the reaction influence each other. This assumption is physically motivated and is not very restrictive for real explosives.

An example of a non-Arrhenius form is a commonly used pressure dependent rate law such as given by

$$r = \tilde{t}_c \tilde{k}(1 - \lambda)(P/P_o)^n, \tag{45}$$

where P_o is a reference pressure. If n is large, the rate is sensitive to a small change in the pressure and we would choose

$$\varepsilon = n^{-1}, \quad \tilde{t}_c \equiv \tilde{g}^{-1} = (\tilde{k}n)^{-1} P_o^n, \quad h = e^{P_1}. \tag{46}$$

As mentioned previously, there are many uncertainties in posing the appropriate form of the rate law for condensed phase explosives. Thus it is of physical interest to consider the mathematical questions that arise when $h(P_1, \rho_1, \lambda_1)$ is replaced by a general functional form with only modest constraints. We can then pose a host of interesting questions to the analysts, the answers to which have physical implications and may constrain the form of the rate law. One obvious question is, given the problem posed by equations (31)–(34) and boundary and shock conditions (36) and (37), what functional form of $h(P_1, \rho_1, \lambda_1)$ leads to thermal runaway?

7. MODELING OF HETEROGENEOUS EXPLOSIVES. In this section we model the induction phase behaviour and the induction time dependence on the properties of a heterogenous explosive. We will consider a material whose thermodynamic description is, to first order, that of a base explosive. We will consider variations in state properties that are of $O(\varepsilon)$ or smaller, where ε is inverse large rate parameter. For an activation energy $\theta = 10$ this represents an allowable variation of 10 percent.

We now formulate the induction problem allowing for variation of the local thermodynamic state on the identity of the material particle. Thus the state generally depends explicitly on the Lagrangian coordinate. The governing equations remain those of (31)–(34). However c, γ, σ and r generally depend on the local thermodynamics. In other words, the equation of state and the form of the reaction term depend on the identity of the material particle. For example, we suppose that the internal energy function (modeled after a polytropic fluid) is given as

$$\tilde{E}(\tilde{P}, \tilde{\rho}, \lambda; \varsigma) = \tilde{P} / \left[(\hat{\gamma} - 1)\tilde{\rho}\right] - \tilde{q}(\varsigma)\lambda, \tag{47}$$

where explicit particle dependence is given by $\hat{\gamma}(\varsigma)$ and $\tilde{q}(\varsigma)$. Note that the relation between the Lagrangian and the lab coordinate x is given by

$$x = \varsigma + \int_0^t u\big(x(\varsigma, \bar{t}), \bar{t}\big)\, d\bar{t}. \tag{48}$$

The shock relations also reflect the existence of the heterogeneities through the energy equation. Conservation of energy through the leading shock requires that

$$\begin{aligned} \tilde{E}(\tilde{P}_s, \tilde{\rho}_s, 0; \varsigma) + \tilde{P}_s/\tilde{\rho}_s + (\tilde{D} - \tilde{u}_s)^2/2 = \\ \tilde{E}(\tilde{P}_o, \tilde{\rho}_o, 0; \varsigma) + \tilde{P}_o/\tilde{\rho}_o + \tilde{D}^2/2. \end{aligned} \tag{49}$$

Additionally, we introduce the parameter ν to measure the amount of deviation that the thermodynamic properties have from the mean state of the base explosive. Using the shocked state, with $\nu = 0$ to define the scales, we then formulate the induction problem under the formal assumption that

$$0 < \nu < \varepsilon \ll 1. \tag{50}$$

Equations (31)–(34) still hold for $A \to B$. The effects of the heterogeneities are exhibited by the first order shock relations which replace (37). In the strong shock approximation and for a polytropic equation of state these become

$$\begin{gathered} \rho_1 = -\big[2\gamma(\gamma-1)/(\gamma+1)\big]\Omega, \\ u_1 = \big[2/(\gamma+1)\big]D_1 - \big[(\gamma-1)\sqrt{2\gamma(\gamma-1)}\,/(\gamma+1)\big]\Omega, \\ P_1 = \big[2\sqrt{2\gamma(\gamma-1)}\,/(\gamma+1)\big]D_1 - \big[\gamma(\gamma-1)^2/(\gamma+1)\big]\Omega, \\ \lambda_1 = 0, \end{gathered} \tag{51}$$

where we have assumed that

$$\hat{\gamma} = \gamma + \nu\bar{\gamma}(\varsigma). \tag{52}$$

Ω is a function that is determined by the character of the heterogeneity and is given by

$$\Omega = (\nu/\varepsilon)\bar{\gamma}(\varsigma)/\big[\gamma(\gamma-1)^2\big]. \tag{53}$$

For example, if the heterogeneities have a periodic structure in the unshocked explosive, then Ω is periodic in ς. Finally notice that since $u = u_{\mathrm{p}}$, in the induction phase, (48) shows that to leading order, the piston cooordinate ξ and the Lagrangian coordinate ς are identical;

$$\varsigma = \xi + o(1). \tag{54}$$

Figures 4 show a representative numerical solution when Ω is chosen to be a periodic function of the form

$$\Omega = \alpha\sin(\pi\omega\varsigma), \tag{55}$$

representing the variations caused by $\bar{\gamma}$. Figures 4. are for the values $\alpha = .2$, $\omega = 30$, while γ, σ and δ have the values that correspond to the homogenous explosive of Figure 1; $\gamma = 2.86$, $\sigma = 1$, $\delta = 0$. The solution obtained for the heterogenous explosive has a different induction time than that of the homogeneous explosive and is reduced by about 1/2. Generally the induction time depends significantly on α and δ.

The most striking aspect of these results is that they show that ignition occurs first in the interior of the explosive and not at the piston or the shock. This effect can only be attributed to the presence of the heterogeneities and the exact location of the ignition point can only be calculated by including the effects of wave propagation. These results are preliminary, but serve to point out how sensitive the induction behaviour is to the presence of heterogeneities.

8. CONCLUSIONS. The lecture has presented a formulation for the shock induced ignition of explosives, that have a well identified induction phase. A simple explosive, such as an ideal reacting gas, with no mole change, the case pioneered by Jackson and Kapila [2], has ignition occurring at the piston boundary. However, the lecture has demonstrated that modifying the rate, or the equation of state, in a way that is justified for more complex explosives, leads to a richer description where ignition can occur at the shock and possibly within the interior of the explosive. Finally, we have given a formulation for modeling heterogenous explosives that indicates that the initiation behaviour is strongly dependent on the heterogeneity.

BIBLIOGRAPHY

1. Fickett, W. and Davis, W. C. (1979), *Detonation*, University of California Press.
2. Jackson, T. L. and Kapila, A. K. (1985), "Shock induced thermal runaway," *SIAM Journal on Applied Mathematics*, Vol. 45, No. 1, pp. 130–137.

DEPARTMENT OF THEORETICAL AND APPLIED MECHANICS
UNIVERSITY OF ILLINOIS AT URBANA - CHAMPAIGN
URBANA, ILLINOIS 61801

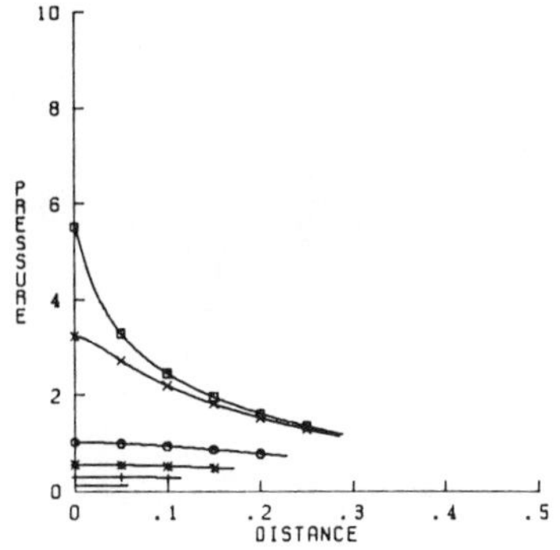

(1a)

KEY

Symbol	Time
—	.1
+	.2
✱	.3
O	.4
×	.5
□	.505

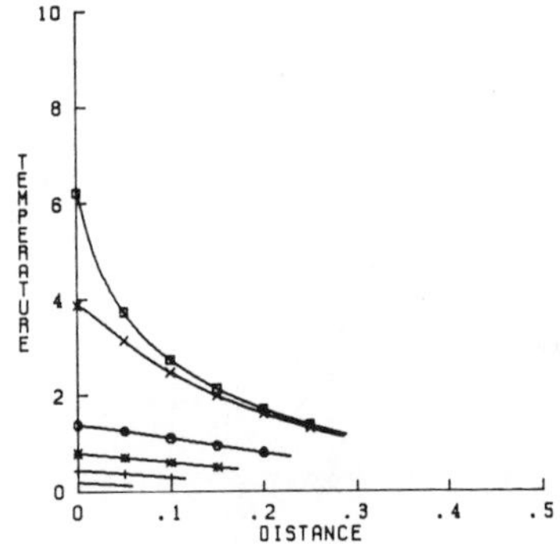

(1b)

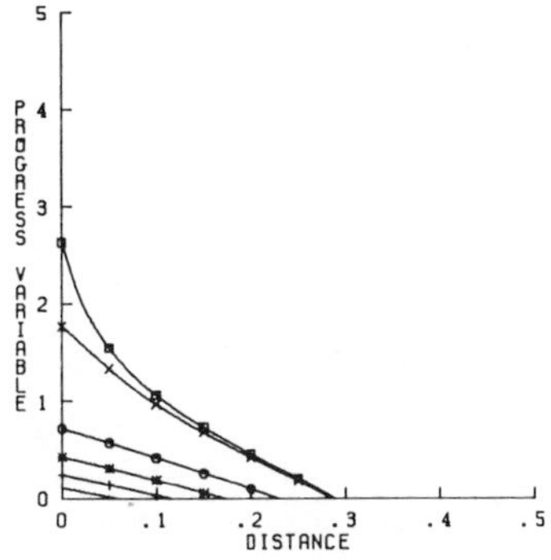

(1c)

Figures (1a)–(1c) show the space-time evolution of the perturbations of pressure, temperature, and progress variable of the shocked state as solutions of (31)–(39). $\gamma = 2.86$, $\sigma = 1$, $\delta = 0$. Distance $= \xi$, time $= t$. A uniform mesh was used with space and time increments .005. The induction time was found to be $t_i = .505$. Thermal runaway occurs at the piston.

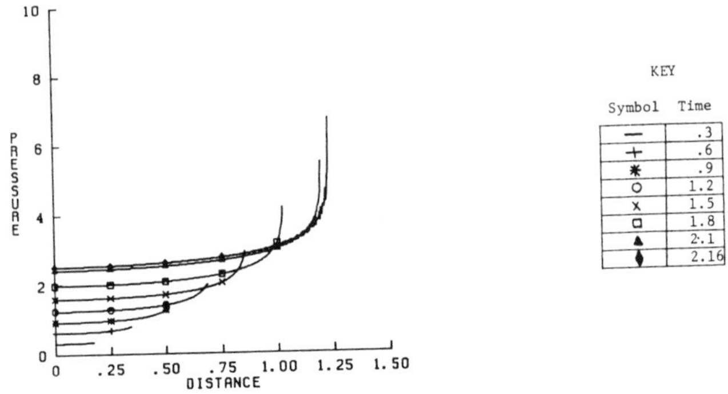

(2a)

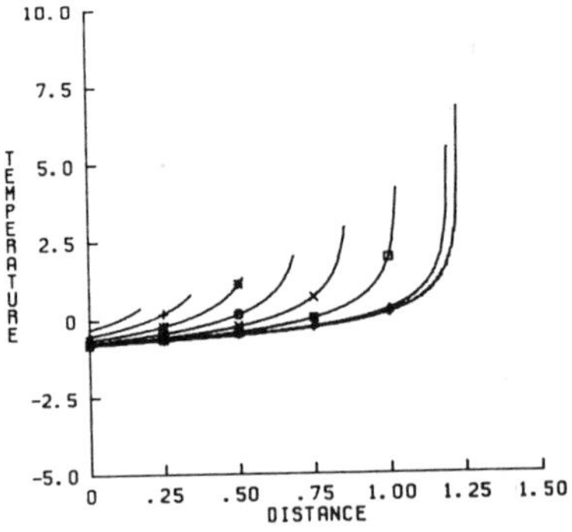

(2b)

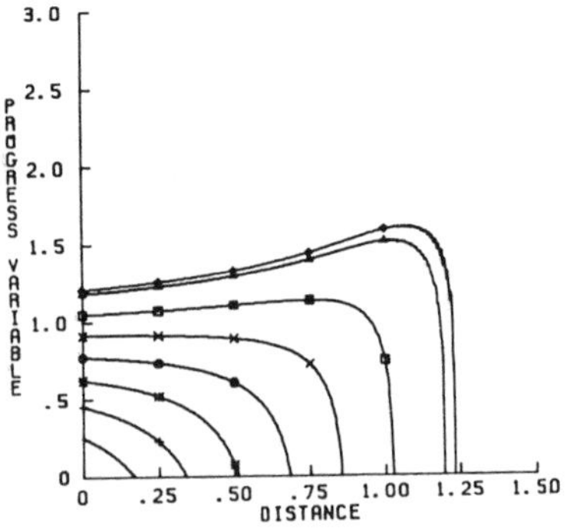

(2c)

Figures (2a)–(2c) show the results of solving (31)–(39) for $\delta = 2$, $\gamma = 2.86$, $\sigma = 1$. The mesh increment used was .0025. Some numerical noise is exhibited by the solution due to the steep gradients that develop near the shock prior to thermal runaway. The induction time calculated was $t_i = 2.16$. Thermal runaway occurs at the leading shock.

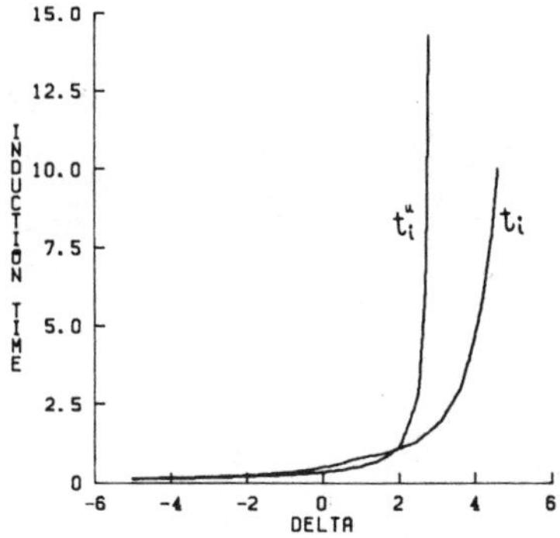

(3a)

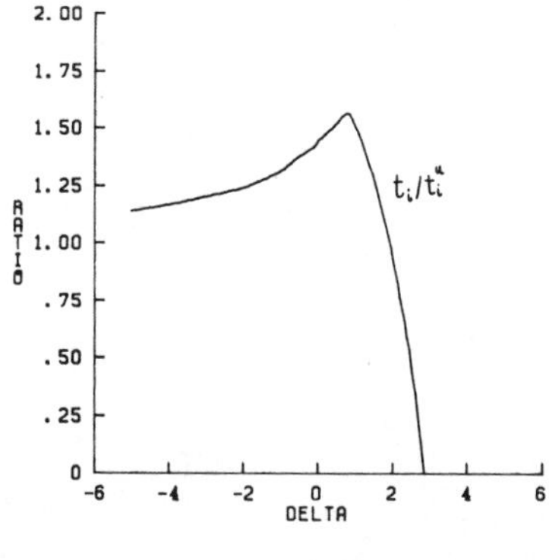

(3b)

Figure 3. Shows the effect of varying δ on the induction time t_i defined by (31)–(39). $\gamma = 2.86$, $\sigma = 1$.

KEY

Symbol	Time
—	.05
+	.1
✳	.15
○	.2
×	.25
□	.3
▲	.323

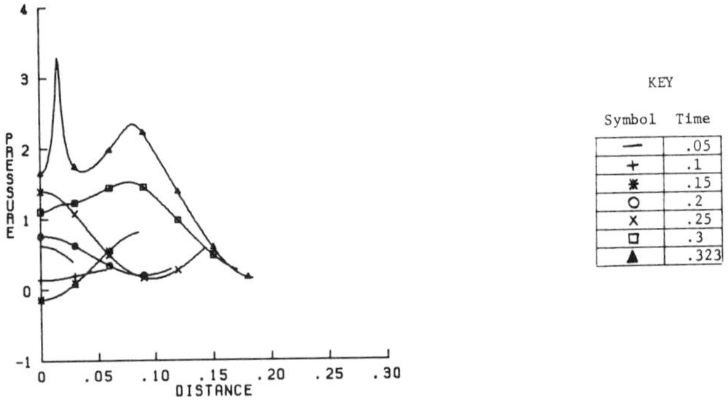

(4a)

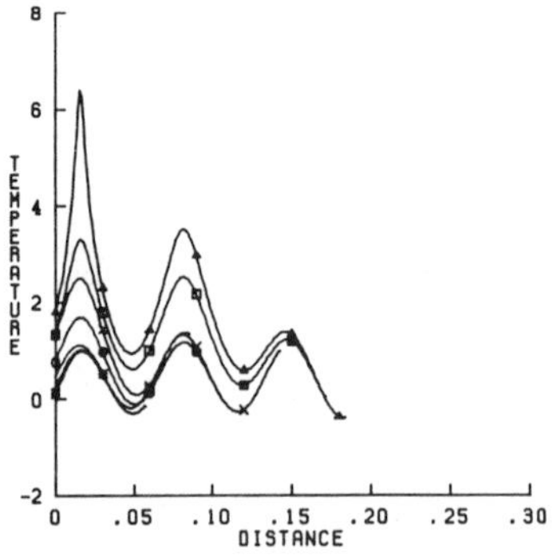

(4b)   (4c)

Figures (4a)–(4c) show the induction behaviour of a hetero-geneous explosive as defined in Section 7. $\Omega = .2\sin(30\pi\varsigma)$. $\gamma = 2.86$, $\sigma = 1$, $\delta = 0$. The induction time was calculated as $t_i = .323$. The uniform mesh size was .001. The thermal runaway occurs in the interior of the explosive.

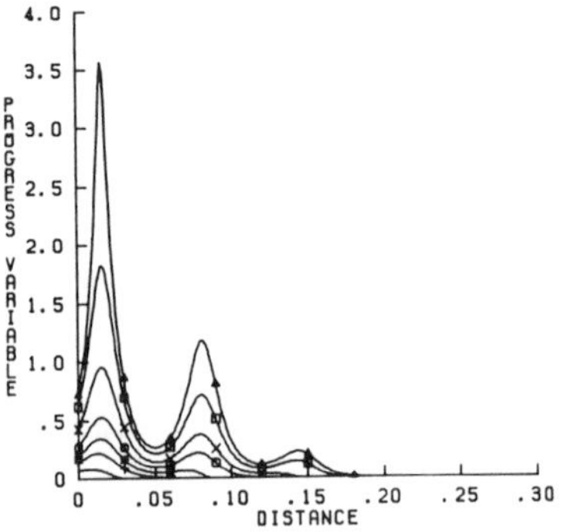

Lectures in Applied Mathematics
Volume 24, 1986

THE STRUCTURE AND STABILITY OF PLANE DETONATION WAVES FOR LARGE ACTIVATION ENERGY

J.D. Buckmaster

ABSTRACT. The linear stability of plane detonations to planar disturbances is considered for one-step Arrhenius kinetics in the limit of infinite activation energy. It is shown that the explosion front, the thin zone where reaction is vigorous, is stiff and resists displacement. As a consequence, stability in the limit is determined solely by the nature of the solution in the induction zone, upstream of the explosion front.

1. INTRODUCTION

It is well known from experiments [1] and numerical calculations [2,3], that detonation waves are often highly unstable. It is therefore important to understand the origins of the instability, the manner in which it evolves, and the nature of the unsteady three-dimensional structure that it can lead to.

Theoretical analyses of the one-dimensional (plane) problem have made clear the origins of the instability [2,4,5,6,7]. Roughly speaking, the steady structure consists of a shock-wave, followed by an induction zone in which chemical reaction occurs at a modest level, followed by a region of intense chemical activity, followed, in turn, by a uniform region of burnt gas. Perturbations within the main reaction zone generate acoustic signals, some of which propagate forwards and perturb the lead shock. This generates temperature perturbations behind the shock which are transmitted downstream (acoustically and advectively) and effect the level of activity within the main reaction zone. This coupling can lead to resonance and divergent oscillations.

In order to carry out a stability analysis it is necessary first to construct the steady solution. This can be done numerically as part of an overall numerical treatment, the path taken by Erpenbeck, [4]. An alternative approach is to provide an analytical description by modelling, 'irrational' approximation [8], or a rational (asymptotic) treatment.

Zaidel's work [6] is an example of the first. In the context of one-step Arrhenius kinetics with a large activation energy, the temperature can be approximated by a piecewise constant function, the 'square wave', and Zaidel uses this as the basis of a stability analysis. Instability can be predicted in this way, but Erpenbeck [5] has shown that the results are not correct.

Zaidel's analysis is rotted in an appeal to activation energy asymptotics, but is not an asymptotic treatment. Our purpose in the present paper is to construct a formal asymptotic description of the stability problem, valid in the limit of infinite activation energy. Whether or not this is of physical value is discussed in the final section.

2. GOVERNING EQUATIONS AND THE STEADY STRUCTURE

The detonation structure downstream of the shock wave is quite thick, so that diffusion can be neglected. For this reason, an appropriate set of nondimensional equations can be written in the form

$$\frac{D\rho}{Dt} + \rho \frac{\partial v}{\partial x} = 0 \ , \quad p = \rho T \ ,$$

$$\rho \frac{Dv}{Dt} = -\frac{1}{\gamma}\frac{\partial p}{\partial x} \ , \quad \rho \frac{DT}{Dt} = \frac{\gamma - 1}{\gamma}\frac{Dp}{Dt} + \beta \Lambda Y e^{-\theta/T} \ , \tag{2.1}$$

$$\rho \frac{DY}{Dt} = -\Lambda Y e^{-\theta/T} \ .$$

These have been nondimensionalized using the unshocked density, sound-speed, temperature, and pressure. The reference length is that of the induction zone which will be defined precisely during the course of the analysis.

The immediate post-shock conditions are defined by the Rankine-Hugoniot conditions. In a shock-fixed frame they are

$$v_s = \frac{2+(\gamma-1)M^2}{(\gamma+1)M},\ \rho_s = \frac{M}{v_s},\ T_s = 1+\frac{\gamma-1}{2}(M^2-v_s^2)\ ,\ p_s = \rho_s T_s,\ Y_s = 1 \qquad (2.2)$$

where M is both the Mach No. and the mass flux in the unshocked gas. We are concerned with piston driven detonations for which the steady value M_s is prescribed.

Consider now the steady structure in the limit $\theta \to \infty$. If ϕ_o denotes a steady variable, ϕ_{os} its value at the shock $(x=0)$, and

$$\delta\phi_o \equiv \phi_o - \phi_{os}\ , \qquad (2.3)$$

we may express all the $\delta\phi_o$'s in terms of δY_o. Indeed, from (2.1),

$$(\gamma + 1)\delta v_o = \frac{(M_o^2 - 1)}{M_o} - \left((\frac{M_o^2 - 1}{M_o})^2 + 2(\gamma + 1)\beta\delta Y_o\right)^{\frac{1}{2}}\ ,$$

$$\delta p_o = -\gamma M_o \delta v_o,\ \delta\rho_o = \frac{-\rho_{os}\delta v_o}{v_{os}+\delta v_o},\ \delta T_o = -\beta\delta Y_o - \frac{(\gamma-1)}{2}\left(2v_{os}\delta v_o + (\delta v_o)^2\right). \qquad (2.4)$$

To describe the structure then, it is only necessary to solve a single equation for Y_o,

$$M_o \frac{dY_o}{dx} = -\Lambda Y_o e^{-\theta/T_o}\ . \qquad (2.5)$$

Following the shock-wave is the induction zone in which all variables change by only $O(1/\theta)$ amounts. For example, changes in temperature may be written as

$$\delta T_o = \frac{T_{os}^2}{\theta} t , \qquad (2.6)$$

and then, (2.4) and (2.5) yield

$$\frac{dt}{dx} = e^{bt} , \quad t(0) = 0 \qquad (2.7)$$

provided

$$\frac{\theta K \Lambda e^{-\theta/T_{os}}}{M_o T_{os}^2} = 1 \qquad (2.8)$$

where

$$K = \frac{-\beta(c_{os}^2 - \gamma u_{os}^2)}{(c_{os}^2 - u_{os}^2)} \qquad (2.9)$$

(c is the sound speed). A choice for Λ corresponds to a choice for the reference length, and (2.8) is equivalent to taking the latter to be the length of the induction zone. This length is well defined, for the solution to (2.7) is

$$t = -\ell n(1 - x) \qquad (2.10)$$

and induction ends with thermal runaway at $x = 1$. The corresponding expression for δY_o is

$$\delta Y_o = \frac{T_{os}^2}{\theta K} \ell n(1 - x) \qquad (2.11)$$

so that, using (2.4), the changes in all of the variables can be described.

The dimensionless length of the induction zone is 1 ($x_i = 1$). Its dimensional length, x_i', satisfies the condition

$$x_i' \propto e^{-\theta/T_{os}} . \tag{2.12}$$

It follows that a perturbation in T_{os} of the form

$$T_{os} = \bar{T}_{os}(1 + \varepsilon) \tag{2.13}$$

leads to a change in x_i' of magnitude

$$\frac{\delta x_i'}{x_i'} = \frac{\varepsilon\theta}{\bar{T}_{os}} . \tag{2.14}$$

In other words, an $0(\varepsilon)$ perturbation of the induction zone structure is naturally associated with an $0(\varepsilon\theta)$ displacement of the explosion front, the exponentially thin region that immediately follows the induction zone.

The final state of the burnt gas (ϕ_{ob}) is obtained by setting $\delta Y_o = -1$. The transition from the description (2.11) to this state (the explosion front) is described in terms of the variable σ , where

$$x = 1 - e^{-\theta\sigma} , \quad \sigma > 0 . \tag{2.15}$$

Such nonlinear scalings are familiar in thermal explosion theory [10]. Equation (2.5) is then

$$\frac{dY_o}{d\sigma} = \frac{-T_{os}^2}{K} Y_o \exp\left(\frac{\theta}{T_{os}} - \frac{\theta}{T_o} - \theta\sigma\right) , \tag{2.16}$$

with asymptotic solution

$$\frac{1}{T_{os}} - \frac{1}{T_o} = \sigma , \quad 0 < \sigma < \frac{1}{T_{os}} - \frac{1}{T_{ob}} . \tag{2.17}$$

3. PERTURBATION ANALYSIS

Linear stability is defined by the behavior of infinitesimal perturbations, i.e.,

$$\phi = \phi_o + \varepsilon\phi_1 \ , \quad \varepsilon << 1 \ . \tag{3.1}$$

Substituting into the continuity equation (2.1a), and retaining only those terms that are linear in ε , yields (in a frame attached to the moving shock, $x = h(t)$)

$$\frac{\partial \rho_1}{\partial t} + \frac{\partial}{\partial x}(\rho_o v_1 + v_o \rho_1) - \frac{\partial h}{\partial t}\frac{\partial \rho_o}{\partial x} = 0 \tag{3.2}$$

Linear versions of the other equations (2.1) can be obtained in the same manner.

Boundary conditions are provided by the perturbed shock conditions, which we shall not write down, and a statement about downstream conditions. We are concerned with overdriven detonations, driven by a piston for an infinite time, so that a radiation condition holds: there are no upstream propagating waves in the burnt gas.

The coefficient functions in the perturbation equations, ρ_o , v_o , etc., are described by the analysis of §2. Within the induction zone they are constant, to leading order, and the complete set of perturbation equations simplies to

$$\frac{D\rho_1}{Dt} + \rho_{os}\frac{\partial v_1}{\partial x} = 0 \ , \quad p_1 = \rho_{os}T_1 + T_{os}\rho_1 \ ,$$

$$\rho_{os}\frac{Dv_1}{Dt} = -\frac{1}{\gamma}\frac{\partial p_1}{\partial x} \ , \quad \rho_{os}\frac{DY_1}{Dt} = -w_1 \ , \tag{3.3}$$

$$\rho_{os}\frac{DT_1}{Dt} = \frac{\gamma-1}{\gamma}\frac{Dp_1}{Dt} + \beta w_1 \ ,$$

where w_1 is the perturbation reaction term. This alone depends on the two-term induction-zone expansion for the steady state, and is given by the formula

$$w_1 = \frac{\Lambda\theta e^{-\theta/T_{os}}}{T_{os}^2} \cdot \frac{T_1}{1-x} \tag{3.4}$$

The singularity originates with the logarithm in the result (2.10).

The general solution to the system (3.3,4) is singular at $x = 1$, i.e., in the neighborhood of $x = 1$

$$\phi_1 \sim (1 - x)^{\alpha}, \quad \alpha < 0 . \tag{3.5}$$

In view of the much weaker logarithmic singularity in ϕ_o, it follows that the expression (3.1) is not uniformly valid.

This difficulty should be no surprise. As we noted in §2, an $O(\varepsilon)$ perturbation in the induction zone is naturally associated with an $O(\varepsilon\theta)$ displacement of the explosion front, and a perturbation analysis that fails to account for this is unlikely to be uniformly valid in the neighborhood of the front. Thus we modify the expansion (3.1) by introducing a new variable s instead of x where

$$s = x + \varepsilon\theta f(s,t) + O(\varepsilon^2) . \tag{3.6}$$

Now the induction zone is terminated at

$$s = 1, \quad \text{i.e.,} \quad x = 1 - \varepsilon\theta f(1,t) \ldots , \tag{3.7}$$

so that f is an $O(1)$ quantity. An example of the application of this coordinate perturbation to a steady problem is given in [9].

The function f must be chosen to eliminate the singularity (3.5), if possible. It is not difficult to show that the best that can be done is to make T_1 regular while the other variables have logarithmic singularities. There is an entire class of functions f that yield this condition.

We turn now to the explosion front which is thin, quasi-steady, and suffers $0(\epsilon\theta)$ displacements. Previous results for the stability problem make it clear that we are concerned with oscillations that have $0(1)$ frequencies, so that the perturbation velocity of the front is $0(\epsilon\theta)$.

The front establishes connection conditions between the induction zone and the region of burnt gas. To leading order these conditions are simply jumps, and the $0(\epsilon\theta)$ perturbation to the front speed (defined by the function f) perturbs these jumps by $0(\epsilon\theta)$ amounts. In other words, the flow field in the burnt gas behind the front experiences $0(\epsilon\theta)$ perturbations.

The picture that we have drawn at this point is quite simple. The shock wave and the induction zone experience $0(\epsilon)$ perturbations, the explosion front and the burnt gas $0(\epsilon\theta)$ perturbations. This is perfectly acceptable provided the perturbations in the burnt gas are consistent with the fact that there can be no upstream propagating waves in that region. Now the perturbation pressure and velocity are related in a simple wave by the formula

$$\gamma c_o \rho_o v_1 = p_1 \, , \tag{3.8}$$

and this condition must be satisfied in the burnt gas. However, it is easy to show that the perturbation jump conditions are not consistent with this relationship.

We conclude that the only way in which the perturbation jump conditions and the relation (3.8) can be accomodated is if $0(\epsilon\theta)$ perturbations are permitted within the induction zone. But this leads to a 'bootstrapping' paradox, since it implies $0(\epsilon\theta^2)$ displacements of the reaction front which, in turn, must generate $0(\epsilon\theta^2)$ perturbations within the induction zone ...

This leads us to the essential conclusion of this section. Order $0(\epsilon\theta)$ perturbation velocities of the reaction front are not permitted. Order $0(\epsilon\theta)$ displacements can be accommodated provided they occur slowly (on an $O(1/\theta)$ time scale), a choice that is not of interest for the stability problem. On the $O(1)$ time scale the displacements must be $O(\varepsilon)$; the reaction front is 'stiff'.

4. FORMULATION AND DISCUSSION OF THE STABILITY PROBLEM

In §3 we were led to the coordinate stretching (3.6) on account of the singular behavior of the induction zone perturbations identified in (3.5). This, however, leads to a paradox when the downstream boundary conditions are invoked (no upstream waves) so that a different resolution of the difficulty must be found. Equations for p_1 and T_1 that are valid within the induction zone can be deduced from (3.3), namely

$$\frac{D^2 p_1}{Dt^2} - c_{os}^2 \frac{\partial^2 p_1}{x^2} = \gamma \Omega \frac{D}{Dt}\left(\frac{T_1}{1-x}\right),$$

$$\rho_{os} \frac{DT_1}{Dt} = \frac{\gamma-1}{\gamma} \frac{Dp_1}{Dt} + \Omega \frac{T_1}{1-x}, \tag{4.1}$$

$$\Omega \equiv \beta \frac{M_o}{K}.$$

With time derivatives replace by $i\omega$, where ω is the frequency (i.e., p_1, $T_1 \sim e^{i\omega t}$) this is a third order system of linear o.d.e.'s and we seek a solution of Frobenius type in the neighborhood of $x = 1$,

$$T_1 = \sum_{n=0}^{\infty} a_n (1-x)^{n+\alpha}, \quad p_1 = \sum_{n=0}^{\infty} b_n (1-x)^{n+\alpha}. \tag{4.2}$$

Of the three solutions constructed in this way, one is singular corresponding to

$$\alpha = -\frac{\Omega}{\rho_{os} v_{os}} \frac{c_{os}^2 - \gamma v_{os}^2}{c_{os}^2 - v_{os}^2}, \tag{4.3}$$

and must be discarded. The other two are regular ($\alpha=0$) and are characterized by the condition

$$T_1 = 0 \quad \text{at} \quad x = 1 \ . \tag{4.4}$$

Now the perturbation shock conditions lead (by elimination of the perturbation mass flux) to a pair of homogeneous boundary conditions for p_1 and T_1 at $x = 0$, so that in this way we have homogeneous boundary conditions for the system (4.1), and the eigenvalue problem is properly posed. It is remarkable for depending only on the steady-state structure within the induction zone.

The contrast between the present formulation and that of Zaidel [6] is striking. Zaidel's model neglects reaction within the induction zone so that only acoustic equations ((4.1) with $\Omega=0$) need to be solved there. Similar equations are solved in the burnt gas and the solutions in the two regions are linked by jump conditions across the reaction front whose changing location must be specified in some fashion. The moving front plays a dominant role.

In the formulation here the reaction front is <u>fixed</u> and it is this condition, together with the effects of reaction within the induction zone, that determine the eigenvalues.

There are certain similarities with the analysis of Abousief and Toong [3]. They adopt various assumptions, the most important of which is equivalent to that of small heat release, although that is not the context in which they justify their work (more precisely, they linearize the flow terms - not the reaction term - about the post-shock state). In essence they apply the system (3.3), here justified only for the induction zone, to the entire combustion field. w_1 is properly calculated as the local perturbation term and is not given the representation (3.4). It is difficult to see how this can be correct and yet the results obtained in [3] agree well with previous numerical calculations. Perhaps this can be understood in the light of the asymptotic results obtained here. If numerical integration through the main reaction zone effectively imposes the stiffness condition (4.4), then the fact that the integration is, in detail, quite inaccurate in that zone is irrelevant.

We conclude this discussion by noting that, although the formulation of the stability problem described here is surely the correct limit problem when $\theta \to \infty$, as it stands it is likely to be of limited physical interest. For since K is proportional to β, we note that Ω is independent of β and equations (4.1) do not contain the heat release parameter. The focus of most discussions of plane detonation stability is the impact of heat release and the speed with which the wave is driven.

It is conceivable that a distinguished limit can be derived of greater physical interest than the simple limit described here, but we propose an alternative method of incorporating β into the picture. The motivation for this stems from the observation that for finite θ the square-wave structure might be a reasonable approximation because of the exponential dependence of reaction on temperature, whilst the uncritical neglect of all $0(1/\theta)$ terms might not be. Thus consider the perturbation reaction term which has the unsimplified form

$$w_1 = \Lambda e^{-\theta/T_o} (Y_i + \frac{\theta T_i Y_o}{T_o^2}) . \tag{4.5}$$

The approximation (3.4), valid within the reaction zone, neglects the Y_1 term. If we retain it, then (3.4) is

$$w_1 \simeq \frac{\Lambda\theta e^{-\theta/T_{os}}}{T_{os}^2} \frac{\left(T_1 + \frac{T_{os}^2}{\theta} Y_1\right)}{(1 - x)} , \tag{4.6}$$

the stiffness condition (4.4) is replaced by

$$T_1 + \frac{T_{os}^2}{\theta} Y_1 = 0 \quad \text{at} \quad x = 1 \; , \tag{4.7}$$

and the equation for Y_1 must be added to the system (4.1). This system is exactly that treated in [3] with the exception that we apply it only to the induction zone and replace integration through the main reaction zone by the condition (4.7). More detailed discussion, together with numerical results, will be presented in [9].

ACKNOWLEDGEMENT

This work was supported by the National Science Foundation and the Army Research Office. I am grateful to the Humboldt Foundation for a US Senior Scientist Award which made it possible to do this work in the congenial environment of the RWTH, Aachen. N. Peters' hospitality during this period is gratefully acknowledged. In addition, I am thankful to D.S. Stewart for helpful discussions on detonation theory.

REFERENCES

1. R.A. Strehlow, Combustion Fundamentals, McGraw-Hill, NY 1984.
2. W. Fickett and W. Wood, Physics of Fluids, 9, 903-916, 1966.
3. G.E. Abousief and T.Y. Toong, Combustion and Flame, 45, 67-94, 1982.
4. J.J. Erpenbeck, 9th Symposium (International) on Combustion, The Combustion Institute, Pittsburgh, p. 442-453, 1963.
5. J.J. Erpenbeck, Physics of Fluids, 7, 684-696, 1964.
6. R.M. Zaidel, Dokl. Akad. Nauk, SSSR (Phys. Chem. Sect.) 1142-1145, 1961.
7. W. Fickett, 'Stability of the square-wave detonation in a model system', Physica D, in press.
8. M. Van Dyke, Perturbation Methods in Fluid Mechanics, Parabolic Press, Stanford, 1975.
9. J. Buckmaster, 'On the Stability of Planar Detonations,' to be published.
10. J.D. Buckmaster and G.S.S. Ludford, Theory of Laminar Flames, Cambridge University Press, NY, Chapter 12, 1982.

Aero & Astro Engineering
University of Illinois
Urbana, IL 61801

Lectures in Applied Mathematics
Volume 24, 1986

STABILITY THEORY FOR SHOCKS IN REACTING MEDIA: MACH STEMS IN DETONATION WAVES.

Rodolfo R. Rosales[†]

ABSTRACT. We review some of the mathematical theories that attempt to deal with the complex transversal wave structures (Mach stems) observed experimentally in shock waves in reacting media (detonation waves). In particular, we review the weakly nonlinear theory of A. Majda and the author in the context of a simplified physical situation, where the algebraic complications of this theory are reduced to a minimum.

1. INTRODUCTION.

In this paper we review some of the mathematical theories that attempt to deal with the complex transversal wave structures observed experimentally in detonation waves. No claim of completeness is made in this, in fact we are heavily biased towards the results of references [37] and [38]. The plan of the paper is as follows:

In this section we first review, briefly, the theory of shocks in nonre-

1980 Mathematics Subject Classification. 76–06,76L05.
[†]Partially supported by NSF grant number DMS–8402757. This research was performed in part while the author was an Alfred P. Sloan Research Fellow.

acting gaseous media (for it is the starting point for most of the theories for detonation waves). Then we introduce, again briefly, the three main models for detonation waves, namely: the simplest, Chapman-Jouguet (CJ) theory; the Zeldovich-von Neumann-Doering (ZND) theory; and the square wave (SW) model. We do not consider situations where transport effects are important. Next, we review some of the observed experimental facts and the difficulties these theories have when confronted with them. Finally, we review the results of the linear hydrodynamic stability analysis for the models above.[1] We concentrate here on the ZND and SW theories. These two models offer the most difficulties to the analysis, and have been (so far) the less useful in providing a bridge to the next order of approximation (see the next paragraph).

In (§ 2) we concentrate on the CJ theory. Because of its simplicity, a complete (explicit) analytic treatment of the linear stability analysis is possible in this case. One can then gain enough knowledge and understanding of the situation to go on to the next order of approximation and incorporate nonlinear (albeit weak) effects into the stability analysis. A theory can then be produced that begins to explain some of the experimentally observed phenomena. Of course, this is only a modest beginning and much remains to be done. The results of this theory are summarized in this second section. In the appendix, (§ 6), polytropic gas equations of state are considered.

The remainder of the paper is dedicated to a detailed presentation of the fundamental elements of the nonlinear theory introduced in (§ 2). However, rather than present the derivation in its original complexity, we choose to simplify the physical situation by assuming isentropic flow. There is no pretense here that the physical situation ever satisfies this assumption; rather, this is a mathematical device to simplify the algebra. No essential *qualitative* feature is lost, at least in the present case.[2] In

[1] Quite clearly the first step in attempting to understand what is wrong.

[2] In fact this theory applies whenever the following general mathematical situation occurs: Consider a system of m hyperbolic conservation laws in several space dimensions and, in this system, shocks satisfying Lax's κ-shock "entropy" inequalities for $\kappa = m$ (see [34]). If such shocks are linearly unstable *only* to spontaneous emission of radiation, then the theory applies. The model of detonation waves in isentropic flow is the simplest case of this situation. It seems possible to generalize the theory to cases where $\kappa \neq m$. This may be of use for potential applications in M.H.D., say, where generally $\kappa \neq m$.

(§ 3) the equations of the simplified model are introduced. In (§ 4) the results of the linear stability analysis are presented and, finally, (§ 5) introduces the details of the weakly nonlinear stability analysis.

1.1. SHOCKS IN NONREACTING MEDIA. The basic elements in the theoretical understanding of shocks in nonreacting gaseous media are some rather simple solutions of the mathematical equations of compressible gasdynamics: the plane, steady, traveling waves. When transport[3] and relaxation[4] effects are neglected,[5] these solutions are indeed simple: the fluid dynamical variables are constant everywhere, except across a moving, plane, infinitely thin surface (the shock), across which they have jump discontinuities. We now describe these solutions (see reference [11] for an at length discussion):

We use subscripts 0 and 1 to indicate quantities in the unshocked and shocked gases, respectively. Let p be the pressure, ρ the density, $\tau = \rho^{-1}$ the specific volume, v the gas velocity normal to the shock (we always use the convention that the normal to the shock points into the unshocked gas), e the specific internal energy and D the velocity of the shock normal to itself. Then the equations of compressible gas dynamics reduce to the following Rankine-Hugoniot algebraic relationships (obtained from the conservation of mass, energy and momentum, see [11], [28]):

$$p_1 - p_0 = -\left[\rho_0(D - v_0)\right]^2 (\tau_1 - \tau_0), \tag{1.1}$$

$$e_1 - e_0 + \frac{1}{2}(p_1 + p_0)(\tau_1 - \tau_0) = 0, \tag{1.2}$$

$$(v_1 - v_0) = -\rho_0(D - v_0)(\tau_1 - \tau_0), \tag{1.3}$$

with the tangential component of the flow velocity continuous across the shock. In addition one has the flow condition:

$$D - v_0 > 0. \tag{1.4}$$

The assumption of local thermodynamic equilibrium implies that:

$$e = E(p, \tau), \tag{1.5}$$

[3]Heat conduction, viscosity, diffusion and radiation.

[4]So that the gas is in local thermodynamic equilibrium at every point.

[5]In fact, we will also neglect body forces.

where E is the equation of state. Finally, the thermodynamic requirement that the entropy S should increase across the shock, i.e.: $S_1 > S_0$, yields:

$$p_1 > p_0, \quad \rho_1 > \rho_0, \quad v_1 - v_0 > 0. \tag{1.6}$$

That is, shocks are compressive and accelerate the fluid in their direction of motion. For future reference we introduce here the compression ratio:

$$\mu = \rho_1/\rho_0 = \tau_0/\tau_1 = (D - v_0)/(D - v_1) > 1.$$

Let now $U = (D - v_0)$ be the speed of the shock normal to itself, relative to the unshocked gas. Consider in the (p, τ) plane both the

$$\text{Rayleigh line}: \quad R = (p - p_0) + (\rho_0 U)^2(\tau - \tau_0) = 0 \tag{1.7}$$

and the

$$\text{Hugoniot curve}: \quad H = E(p,\tau) - e_0 + \frac{1}{2}(p + p_0)(\tau - \tau_0) = 0. \tag{1.8}$$

Quite clearly (1.1) and (1.2) indicate that both the initial and final states of the shock are on the intersections of these two curves, see Fig. 1.

Normally, the equation of state is such that along the Hugoniot curve[6]:

$$(dp/d\tau)_H < 0 \text{ and } (d^2p/d\tau^2)_H > 0. \tag{1.9}$$

Thus, the Hugoniot and Rayleigh lines intersect in at most two points: the initial (p_0, τ_0) and final (p_1, τ_1) states of the shock transition.

Remark 1.1 (1.9) is not always true, and it may fail when materials undergo shock-induced phase transformations or yield plastically at the elastic limit (see [6], [12] and [14] for discussions of general equations of state), but for the gaseous shocks of interest to us here, it is adequate.

An assumption related to (1.9), which we will also make, is:

$$E_p > 0 \text{ and } p + E_\tau > 0. \tag{1.10}$$

The following inequalities, for the Mach numbers M_0 and M_1, apply:

$$M_0 = (D - v_0)/c_0 > 1 \text{ and } 0 < M_1 = (D - v_1)/c_1 < 1, \tag{1.11}$$

[6]We use here the thermodynamic convention: variables displayed as subscripts are to be held constant during derivation.

where $c > 0$, $c^2 = (dp/d\rho)_S = \tau^2(p + E_\tau)/E_p$, is the speed of sound. This finishes our description of the plane, steady, shocks. Of course, shocks are generally neither plane, nor steady. The theory, however, holds that we can represent general shocks as infinitely thin, smooth surfaces and that the above equations apply locally at each point along the surface. Elsewhere, away from shocks, and where the flow is continuous, the Euler equations of compressible gasdynamics apply, i.e.:

$$\dot{\rho} + \rho \; div \; \mathbf{u} = 0, \tag{1.12}$$

$$\dot{\mathbf{u}} + \tau \; grad \; p = 0, \tag{1.13}$$

$$\dot{e} + p\dot{\tau} = 0, \tag{1.14}$$

where $\mathbf{u}$ is the flow velocity and a dot indicates time derivative along particle paths, i.e.: $\dot{f} = f_t + (\mathbf{u} \cdot \; grad \;)f$. An alternative form of (1.14) is:

$$\dot{S} = 0, \tag{1.15}$$

as follows from the thermodynamic identity $de = TdS - pd\tau$.

This theory, as outlined here, agrees rather well with experimental evidence (with some exceptions which are not the subject of this paper).

Remark 1.2 Where different shock waves interact, the above picture must be modified. For this to occur, however, one must force (by some external, large, noninfinitesimal perturbation) different shocks to interact. In other words, these situations do not arise spontaneously.

Remark 1.3 Where several shocks interact, generally one needs to introduce contact discontinuities (slip lines). Across them there is no flow and the pressure is continuous, but the tangential component of the fluid velocity may be discontinuous, as well as the density. These discontinuities, generated by shock interactions in more than one dimension, are unstable and a source of turbulence in real fluids.

Remark 1.4 Finally, we recall that shocks are not self supporting. An external source of energy is required to maintain them. Otherwise they decay, their mechanical energy transformed into thermal energy. This behavior is related to the fact that $M_1 < 1$ in (1.11), so that rarefaction waves on the back of the shock can overtake it.

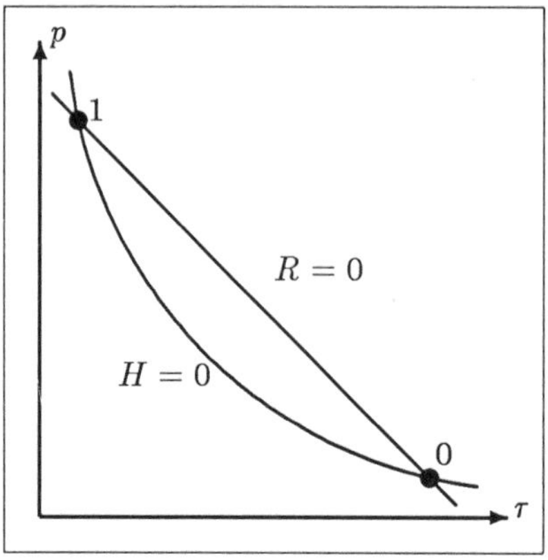

Figure 1: Hugoniot diagram for a shock wave.

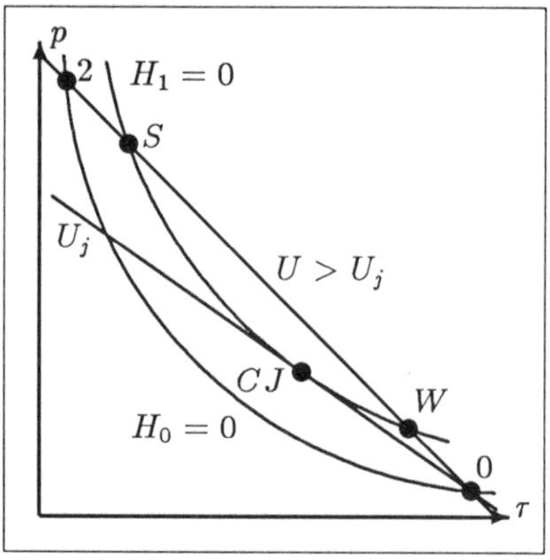

Figure 2: Hugoniot diagram for a detonation wave.

1.2. CJ THEORY OF DETONATIONS. The elements of this simplest theory were formulated around the turn of the century by Chapman [7] and Jouguet [32] (see also [11] and [28]). It is assumed that the detonation front is infinitely thin and that the gas particles crossing the front instantaneously adjust from thermodynamic equilibrium in the unburnt gas ahead of the front to thermodynamic equilibrium in the burnt gas behind the front: the chemical reaction occurs instantaneously. Again, transport effects are neglected. The same equations of (§ 1.1) apply, but with two equations of state, one for the unburnt gas and another one for the burnt gas i.e.:

$$e_0 = E_0(p_0, \tau_0) \text{ and } e_1 = E_1(p_1, \tau_1). \tag{1.16}$$

Thus, we also have two Hugoniot curves, $H_0 = 0$ and $H_1 = 0$. For an exothermic reaction the Hugoniot curve for the products is above the Hugoniot curve for the unburnt gas. Fig. 2 summarizes the situation.

Note there is a minimum speed of propagation, corresponding to a *finite amplitude wave*, i.e.: U_j in Fig. 2, with final state CJ. The Chapman-Jouguet hypothesis holds that these are self supporting detonations (the energy to maintain them being supplied now by the chemical reaction, see remark 1.4). For $U > U_j$ two final states are possible, S (for strong or overcompressed detonation) or W (for weak detonation). We have:

$$S \text{ detonations}: \quad 0 < M_1 < 1 < M_0, \tag{1.17}$$

$$W \text{ detonations}: \quad 1 < M_1 \text{ and } 1 < M_0, \tag{1.18}$$

$$CJ \text{ detonations}: \quad M_1 = 1 < M_0. \tag{1.19}$$

In this inviscid, hyperbolic, theory *weak detonations must be rejected*, for they do not satisfy Lax's κ−shock "entropy" inequalities [34][7].

Remark 1.5 Because of the change in the equation of state across the plane of discontinuity, we can no longer exclude solutions for which $p_1 < p_0$, as in the case of shocks in nonreacting media (see (1.6)). As a matter of fact some of these transitions do occur in reality (flames), but in the context of this inviscid theory they must, again, be rejected, for they do not satisfy Lax's κ−shock "entropy" inequalities [34]. One needs transport and finite reaction rate effects to determine their speeds and avoid nonuniqueness in the solution of the initial value problem. We mention these transitions (deflagrations) here, for we will encounter one of them in the context of the square wave model.

1.3. THE ZND THEORY OF DETONATIONS. The basic elements of this model where developed by Zeldovich [55], von Neumann [51] and Doering [13] (see also [28]). Transport effects are neglected, but the reaction rate is taken finite. The reaction is started by a non reacting shock (the precursor shock) that raises the temperature of the gas violently. To this shock the equations of (§ 1.1) apply. The theory can then be summarized as follows:

(i) In the unreacted gas, ahead of the wave, the flow is described by equations (1.12), (1.13) and (1.14). No reaction is presumed to occur here. The situation is then the same as in (§ 1.1).

(ii) There is then a purely fluid dynamical shock (the precursor shock), again described by the theory in (§ 1.1). When the gas particles go across this shock, the reaction begins.

[7] Basically: there are not enough equations to determine all the unknowns. In particular one can chose the speed of the wave more or less arbitrarily. This produces an ill posed problem, with nonuniqueness of the solution. For a discussion of this problem see [11]. When finite reaction rates and/or transport effects are considered, weak detonations may arise for certain special values of the speed U, which is now determined by the reaction and the transport effects (eliminating the nonuniqueness).

(iii) Behind the precursor shock, in the reacting shocked gas, the flow is again described by equations (1.12) through (1.14), supplemented by the chemical reaction equations:

$$\dot{\boldsymbol{\lambda}} = \mathbf{W}(p, \tau, \boldsymbol{\lambda}), \tag{1.20}$$

where $\mathbf{W}$ is the vector of chemical reaction rates and $\boldsymbol{\lambda}$ is the vector of progress variables. (Thus $\boldsymbol{\lambda}$ describes chemical composition). In addition we must know the complete equation of state:

$$e = E(p, \tau, \boldsymbol{\lambda}). \tag{1.21}$$

Thus we have a different Hugoniot curve for each possible $\boldsymbol{\lambda}$.

Remark 1.6 All thermodynamic variables other than chemical composition are assumed to be in local thermodynamic equilibrium everywhere.

Remark 1.7 Clearly (1.15) is no longer true in the reacting media.

Let us consider now the plane, steady, solution of these equations. Then, ahead of the precursor shock, which is now plane, all the variables are constant. Behind it, they are functions only of the distance s to the percursor shock,[8] but without any transversal dependence. Clearly, at any point in space, given the local chemical composition $\boldsymbol{\lambda}$, all the jump relations (1.1), (1.2) and (1.3) must apply (with e as in (1.21)). Thus all the variables $p, \rho, \mathbf{u}$ can be written in terms of the (as yet unknown) chemical composition $\boldsymbol{\lambda}$. To determine it we must solve (1.20), which now reduces to the O.D.E.:

$$-\{D - v(\boldsymbol{\lambda})\}\frac{d}{ds}\boldsymbol{\lambda} = \mathbf{W}(p(\boldsymbol{\lambda}), \tau(\boldsymbol{\lambda}), \boldsymbol{\lambda}), \tag{1.22}$$

with initial condition $\boldsymbol{\lambda}|_{s=0} = \boldsymbol{\lambda}_0$ (where $\boldsymbol{\lambda}_0$ is the composition of the unburnt gas). As $s \downarrow -\infty$ we expect $\boldsymbol{\lambda} \to \boldsymbol{\lambda}_1$, where $\boldsymbol{\lambda}_1$ is the equilibrium composition of the burnt gas.

Clearly these objects are hardly trivial: one can rarely solve (1.22) explicitly, and must resort to numerical computations. For a detailed theoretical study of these waves see [52]. Fig. 2 summarizes this discussion: the precursor shock takes the state point from 0 to 2. Then the

[8] If y is the coordinate normal to the precursor shock then $s = y - Dt - y_0$, where y_0 is the position of the shock at time $t = 0$.

state point moves (following the Rayleigh line), under the influence of the chemical reaction, to S or W. In Fig. 2: $H_0 = 0$ and $H_1 = 0$ are the Hugoniot curves corresponding to $\boldsymbol{\lambda} = \boldsymbol{\lambda}_0$ and $\boldsymbol{\lambda} = \boldsymbol{\lambda}_1$, respectively. Intermediate Hugoniot curves are not displayed.

Remark 1.8 The motion of the state point along the Raleigh line is not monotone. In general, this theory does not exclude weak detonations, although they are exceptional objects.

Remark 1.9 This theory eliminates the uncertainties regarding which states are thermodynamically achievable (see the discussion following (1.19)). But the extra complications are nontrivial.

1.4. THE SQUARE WAVE MODEL. For many reactions hardly any heat is released initially, until a certain stage is reached, when large quantities of heat are released rather quickly. For a reaction of this type, a ZND steady wave would have a pressure versus distance profile as shown in Fig. 3, with the other fluid dynamical variables behaving similarly.

Based on this observation Shchelkin [42] made the following simplification of the ZND model (we follow Zaidel [53] here):

(iv) In (1.20) assume a single progress variable, with $\lambda = 0$ in the unburnt gas. Assume an irreversible reaction, with $W > 0$.

(v) Assume no heat is released by the reaction until a certain critical value $\lambda = \lambda_c > 0$ is reached. Thus, for $0 \leq \lambda < \lambda_c$, E in (1.21) is equal to E_0, the equation of state for the unburnt gas.

(vi) When λ reaches λ_c all the heat is released instantaneously and the reaction terminates, with E in (1.22) jumping discontinuously to E_1, the equation of state for the burnt gas.

With these approximations Fig. 3 is now replaced by Fig. 4. This is the square wave model. The wave consists of two discontinuous transitions (see Fig. 2): first a shock wave (the precursor shock) from 0 to 2, followed after a while by a weak deflagration (the flame) from 2 to S.

Remark 1.10 As mentioned earlier (see remark 1.5), deflagration waves are rejected in the context of the inviscid theory. However, the weak deflagration of the square wave model is not isolated: its velocity is uniquely determined by its relationship to the precursor shock via the chemical reaction. Therefore it would seem that nonuniqueness and ill posedness are not a problem here.

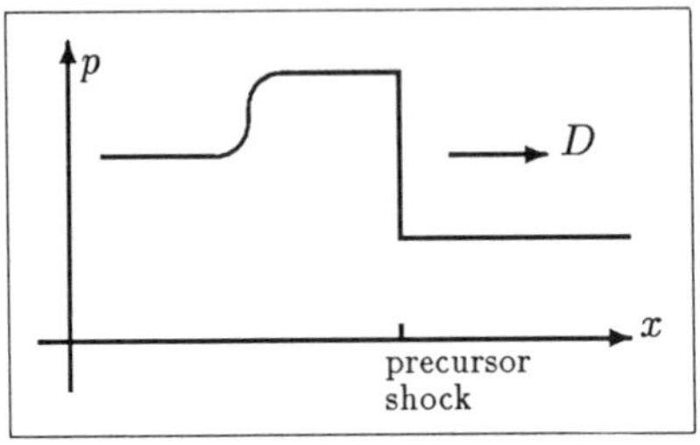

Figure 3: See the text.

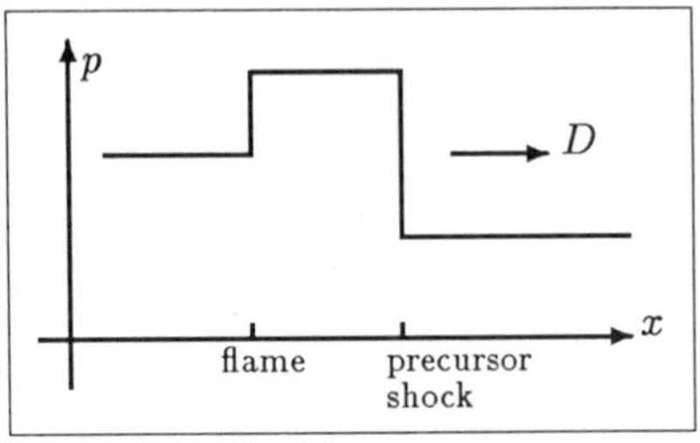

Figure 4: Square wave detonation.

Remark 1.11 This model has strong intuitive appeal. It is simple, yet it gives some longitudinal structure to the waves (unlike the CJ theory). It has been very useful in providing an intuitive understanding of some of the observed phenomena described next.

1.5. TRANSVERSAL WAVE STRUCTURE. All the reviewed theories of detonation waves assume the set up described for shock waves after (1.11) and in remark 1.2 (with the plane steady waves the fundamental elements of the description). Experiments contradict this (see [28], [39], [44] and [46]). The flow behind detonations is observed to be somewhat turbulent, and disturbances occur moving transverse to the main detonation front (Fig. 5 is a schematic representation of this situation).

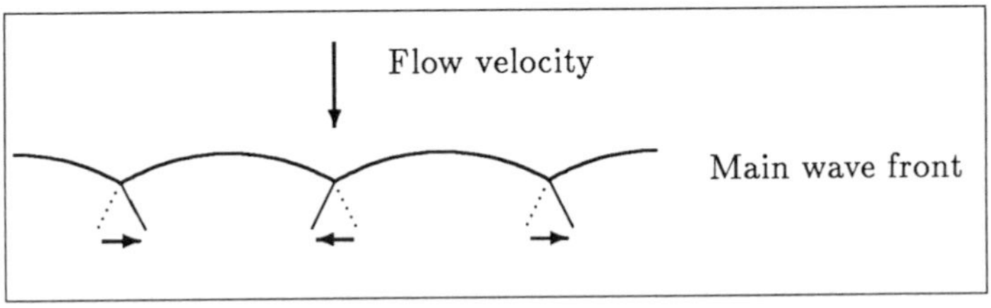

Figure 5: Transversal wave structures in a detonation wave.

Close examination shows that, in fact, each one of the transverse waves is a complex triple-shock, slip-line wave configuration (similar to the ones observed on Mach reflection of purely fluid dynamical shock waves), with the turbulence being generated by the slip-lines (see remark 1.3). A schematic drawing of one of these waves is shown in Fig. 6.

Remark 1.12 In fact, these Mach stems are the most common type of the special situations mentioned in remark 1.2. Here, however, they appear *spontaneously*, without any type of external perturbation.

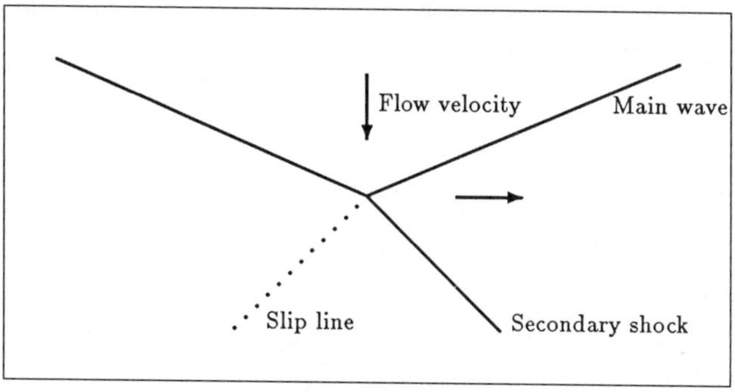

Figure 6: Individual transversal Mach stem configuration.

There is a large body of literature where, partly assuming the structures just described and using a mixture of intuitive, mathematical and phenomenological arguments, various authors have attempted to predict some of the observed features (e.g.: transversal wave length). Many of these efforts have produced results in good qualitative, if not quantitative, agreement with experiments (see [3], [4], [10], [39], [44]). In particular, Strehlow and his co-workers have developed an interesting theory of regular spacings of these Mach stems, partly based on geometrical optics and ray tracing ideas ([5], [46], [47], [48], [49]). However, the *basic issues* of:

(i) Why do transverse Mach stems form in reacting gases?
(ii) What are the mechanisms responsible for spontaneous Mach stem formation? (1.23)
(iii) Why is it that transverse Mach stems never form spontaneously in unidirectional shock fronts in inert polytropic gases?

are not addressed in these papers and remain the subject of investigation. They are the main objective of the work being reviewed here.

In order to study the questions in (1.23), it seems obvious that, one must first investigate the stability of the steady waves in the models just described. If these solutions of the equations are unstable, then a study of the nature of the instability can provide clues as to what may be wrong with the present mathematical picture, and how to correct it. Of course, any study of stability must begin with the simplest type first: stability to infinitesimal perturbations (linear). It cannot stop there, however. The phenomena under consideration are obviously highly nonlinear, and we cannot hope to understand them just with a linear theory.

1.6. STABILITY OF PLANE, STEADY, ZND WAVES. In the linear analysis one considers the plane, steady, wave solution of the equations (as described after remark 1.7). Then one linearizes the full equations of the model near this solution, in the usual fashion of hydrodynamic stability (being careful to consider the free surface in the problem, the position of the precursor shock, properly[9]). One obtains a set of linear equations in the perturbed variables, which must then be analyzed. The bulk of this work is due to Erpenbeck ([16], [19], [20], [21], and [23]) although others have contributed as well (e.g.: [2], [40])[10].

Let now x and z denote the coordinates transversal to the unperturbed plane steady precursor shock and let y denote the normal coordinate, with $y = 0$ at the unperturbed shock. Then, in the linear equations, one can separate the (x, z, t) dependence, $e^{i(kx+mz-\omega t)}$, and look for normal modes (subject to appropriate conditions of causality and boundedness).

Remark 1.13 Rather than separate time $e^{-i\omega t}$ as above, Erpenbeck actually uses Laplace transforms. The approach is, of course, equivalent.

The separated equations yield O.D.E.'s in y, with ω a sort of eigenvalue. The stability question hinges on whether or not there are normal modes with $\operatorname{Im}\omega > 0$ (exponentially unstable modes), $\operatorname{Im}\omega = 0$ (neutrally stable modes) or $\operatorname{Im}\omega < 0$ (exponentially stable modes).

Remark 1.14 ω is not an eigenvalue in the standard fashion, for it also appears in the boundary conditions for the O.D.E. at $y = 0$ (from the linearization of the shock jump conditions).

The main difficulty in this analysis arises from the fact that the O.D.E. in y is not constant coefficients (as the unperturbed wave does

[9]So that the dependent variables at positions where the perturbed flow is shocked, but where the unperturbed flow is not, are not expanded around their unshocked values, and vice versa.

[10]Generally only overcompressed detonations have been considered.

have nontrivial y dependence). Further, in general we know these coefficients only through the results of numerical computations (see the discussion following (1.22)). Thus, the analysis of the linearized normal mode equations is very hard and can generally be done only numerically.

Because of the situation just described, it is very difficult to get an idea of what the general picture for the linear stability analysis looks like. The prospects for a nonlinear theory seem dim at the the moment.

Remark 1.15 In a situation where one is near the "stability boundary" and just two ω's (related by $\omega_1 = \omega_2^*$, since solutions are real) are slightly above the real axis, then an analysis for a Hopf bifurcation (see [45]) is possible. Such a computation has been performed ([22], [24]). Due to the limitation to a single pair of modes, structures such as the ones described in (§ 1.5) are outside the scope of this approach. For numerical calculations and observations related to this type of nonlinear oscillatory behavior ("galloping" detonations) see [2], [25], [26], [27] and [28].

1.7. SQUARE WAVE MODEL STABILITY. In this case there is a series of semi intuitive approximate treatments of the stability question, some of them purely analytical and some including numerical computations ([8], [9], [42], [43] and [54] to name a few). The results of these analyses often contradict each other and, since the basic hypothesis are not laid out clearly, it is not clear what one should conclude.

In [53] Zaidel attempted a more conventional stability analysis, linearizing around the steady state square wave solution of the model (as stated in (§ 1.4)). There are some errors in his analysis, corrected later by Erpenbeck [18]. The result of this analysis is that plane steady state square waves appear to be catastrophically linearly unstable[11], a most undesirable situation. An analysis of the stability of the square wave in a simplified model of combustion [29] confirms this result. It is then clear that the arguments quoted in the first paragraph above are either wrong, or contain unstated assumptions that make the model better behaved than what this second set of arguments predicts (hopefully the second situation applies).

From the previous discussion it seems clear that the square wave

[11]That is: one can find modes with arbitrarily large growth rates (provided one considers small enough spatial wavelengths).

model has serious mathematical difficulties, that must be resolved before it can be used effectively for a theoretical understanding of the structures discussed in (§ 1.5). In fact, this model is so appealing for its simplicity that this seems a very important problem to investigate (see [41]).

Remark 1.16 We stress again that for semiquantitative treatments this model is very useful, if dealt with care. For example: Alpert and Toong [1] have presented a complete semiquantitative picture of the mechanism of longitudinal oscillation in a square wave detonation, which seems to be in good agreement with some of the experimental observations for blunt projectiles fired into gaseous explosives.[12]

2. STABILITY THEORY FOR THE SIMPLEST (CJ) THEORY OF DETONATIONS.

In this section we consider the stability question, both at the linear and weakly nonlinear level, for the plane steady waves of the theory reviewed in (§ 1.2). We only deal with strong (overcompressed) detonations, and restrict ourselves to a two dimensional set up.[13] First we review the results of the linear analysis and then introduce the weakly nonlinear theory of [37] and [38], including the "steady wave bifurcation" analysis of [38].

2.1. THE LINEAR STABILITY ANALYSIS. The objective is to consider the equations for infinitesimal perturbations of the plane, steady, overcompressed detonation. These are a linear system of P.D.E.'s, with constant coefficients in $y > 0$ and $y < 0$ plus boundary conditions at $y = 0$ relating the solution for $y > 0$, $y < 0$ and the perturbation to the wave front.

The solutions of the equations above can be written explicitly as

[12]Say; Hydrogen-Oxygen mixtures, for which the square model is a fairly good representation of the pressure/time detonation wave profile.

[13]Through out this section we work in the coordinate frame of the unperturbed detonation wave: x is the coordinate along the unperturbed front and y is the coordinate normal to the unperturbed front, with $y = 0$ the unperturbed front and $y > 0$ the unreacted gas. It is further assumed, W.L.O.G., that the component of the fluid velocity in the x direction vanishes for the unperturbed detonation.

sums of exponentials. One then looks for normal mode solutions (i.e.: solutions with time dependence $e^{-i\omega t}$, where ω is the same throughout) that satisfy both boundedness and causality. The calculations are long and cumbersome, but can be carried out explicitly (we will illustrate the process in our (§ 4), in the context of isentropic gas flow). One of the results of these calculations is (see [15], [17], [30] [31], [33] and [50]):

> With the kind of assumptions that seem reasonable for gaseous detonations, i.e.: (1.9) and (1.10), no exponentially growing modes can exist. (2.1)

In fact, it can be seen that the conditions on the equations of state necessary to guarantee existence of exponentially growing modes[14] arise under circumstances where shock splitting can occur[15] (see [30], [31], [50]). These situations are, clearly, not relevant to the questions in (1.23).

One must also consider the possibility of neutrally stable linear modes. These are possible, as D'yakov [15] noticed (although his analysis has errors). These modes exist if the unperturbed wave satisfies (see [37]):

$$(\Gamma + 1)^{-1} < (\mu - 1)M_1^2 < (1 + M_1)\Gamma^{-1}, \tag{2.2}$$

where $\Gamma = \{\rho_1(E_1)_p(p_1, \tau_1)\}^{-1} > 0$ is the Gruneisen coefficient of the reaction products (we use here the notation of (§ 1)). This is a reasonable condition for detonations, but much harder on shocks in nonreacting gases (not satisfied at all by shocks in inert polytropic gases). See the appendix for an analysis of this inequality for polytropic equations of state.

When (2.2) applies the detonation will, spontaneously, emit radiation into the combustion products region $y < 0$. This radiation is emitted in two uniquely determined directions (one the mirror image of the other). We refer now to Fig. 7, which depicts the right moving emissions:

(i) A is a wave front of the emitted acoustic wave, $x - st - \lambda_A y =$ const., where $s > 0$ and λ_A are uniquely determined.[16]

---

[14] In particular $(dp/d\tau)_H > 0$, at least somewhere, is required.

[15] a single shock may split spontaneously into two shocks travelling in the same or opposite directions (see [6], [12], [14]) Experimentally this is observed, for example, when shocks induce phase transformations or force materials beyond the elastic limit.

[16] Since this is an acoustical wave: $s = -v_1\lambda_A + c_1\sqrt{1 + \lambda_A^2}$.

(ii) P is a wave front of the emitted entropy and shear waves,[17] $x - st - \lambda_p y = \text{const}$, where λ_p is uniquely determined.

The perturbed detonation front has the form, for ε infinitesimal,

$$y = \varepsilon\eta(x - st), \tag{2.3}$$

where η is arbitrary. The strength of the emitted waves is proportional to $\varepsilon\eta'(x - st - \lambda_A y)$ and $\varepsilon\eta'(x - st - \lambda_p y)$, respectively.

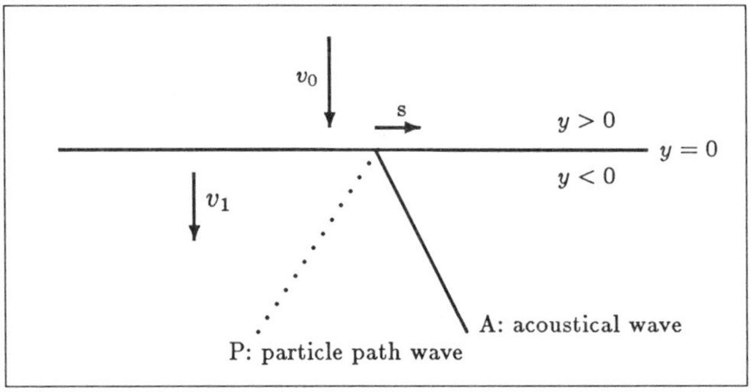

Figure 7: Spontaneously emitted radiating boundary waves.

Remark 2.1 Because the emitted waves in Fig. 7 propagate into a medium with nonzero flow velocity (in the detonation wave frame of reference), energy does not flow normal to the wave fronts. This must be considered when implementing causality (i.e.: energy must be moving away from the detonation wave). D'yakov [15] did not implement this condition correctly, the error was corrected later in [33].

Remark 2.2 Fowles ([30], [31]) showed that spontaneous emission is related to overreflection: a sound wave impinging on the detonation wave from behind (recall $M_1 < 1$) is reflected with a reflection coefficient bigger than 1.[18] This occurs exactly for the same range of parameters (2.2). Further, at a certain angle of incidence the reflection coefficient is ∞: then a reflected wave without an incident one is possible (spontaneous emission). He suggested that spontaneous emission maybe related to formation of transversal Mach stems, but carried no computations to back up the claim.

[17] Carried with the flow, thus $s = -v_1\lambda_p$.

[18] In an enclosed cavity this leads to growth in time of the linearized solution.

Remark 2.3 When $(\mu - 1)M_I^2 < (\Gamma + 1)^{-1}$ all the modes in the linear stability analysis problem decay exponentially and a rigorous proof of fully nonlinear structural stability of the wave front can be made ([35], [36]). Spontaneous Mach stem formation is not possible in this regime, which is the one that applies to shocks in nonreacting polytropic gases.

2.2. NEAR LINEAR THEORY. Where the inequalities (2.2) are satisfied, one can search for weakly nonlinear corrections (i.e.: $O(\varepsilon^2)$, see (2.3)) to the radiating boundary waves discussed in (§ 2.1). [19] Care must be taken to implement causality at the second order so that:

$$\text{Additional incoming radiating waves are never generated at the perturbed detonation front, only outgoing radiating waves are generated there.} \tag{2.4}$$

We remark that incoming waves can be generated through the interactions of outgoing waves in the region $y < 0$, without violating causality. In fact, this is an essential feature of the near linear theory, as it introduces nonlocal interactions of different parts of the perturbed wave front.

One immediate consequence of the analysis is that the perturbed wave front is no longer translational invariant as in (2.3), but presents typical nonlinear distortion in time. In the near linear theory this occurs slowly (as the nonlinear effects are small) and we have:

$$y = \varepsilon\eta(\xi, t^*) + O(\varepsilon^2) \tag{2.5}$$

for the perturbed wave front position, where $\xi = x - st$ (s as in the linear theory) and $t^* = \varepsilon t$. Further, the following equation must be satisfied:

$$0 = q_{t^*} + \{a_1 q^2 + a_2 \int_0^\infty q(\xi + \hat{\beta} y, t^*) q_\xi(\xi + y, t^*) dy\}_\xi, \tag{2.6}$$

where $q = \eta_\xi$ and $\hat{\beta} > 1, a_1, a_2$ are constants (these constants have complicated expressions, which nevertheless can be computed explicitly in terms of the unperturbed detonation parameters). See [37] for this analysis .

Remark 2.4 The integral term in (2.6) arises through the interaction in the region $y < 0$ of the emitted waves, as pointed out following (2.4). See (§ 5.2) for a detailed exposition of this phenomena.

[19] We illustrate this procedure in (§ 5).

Example 2.1. An Exact Solution: Mach Stems. It is not too hard to figure out what the exact traveling wave solutions of (2.6) are. Let

$$q = q_+ \quad \text{for} \quad \xi - \hat{w}t^* > 0 \quad \text{and} \quad q = q_- \quad \text{for} \quad \xi - \hat{w}t_* < 0, \tag{2.7}$$

where[20] $q_+ > q_-$ (see remark 2.5) and $q_+, q_-, \hat{w}$ are constants. Then $q_\xi = (q_+ - q_-)\delta(\xi - \hat{w}t)$, where $\delta(\cdot)$ is Dirac's delta function, so that

$$\int_0^\infty q(\xi+\hat{\beta}y, t^*)q_\xi(\xi+y, t^*)dy = \begin{cases} 0 & \text{for } \xi - \hat{w}t^* > 0, \\ (q_+ - q_-)q_+ & \text{for } \xi - \hat{w}t^* < 0. \end{cases} \tag{2.8}$$

Substituting (2.7) and (2.8) in the conservation form (2.6) yields:

$$0 = -\hat{w}(q_+ - q_-) + a_1(q_+^2 - q_-^2) - a_2 q_+(q_+ - q_-) \Rightarrow$$

$$\hat{w} = (a_1 - a_2)q_+ + a_1 q_-. \tag{2.9}$$

We now note that (2.7) yields, for some constant r:

$$\eta = \begin{cases} r + q_+(\xi - \hat{w}t^*) & \text{for } \xi - \hat{w}t^* > 0, \\ r + q_-(\xi - \hat{w}t^*) & \text{for } \xi - \hat{w}t^* < 0. \end{cases} \tag{2.10}$$

Thus, this solution represents a *corner* in the perturbed wave front, propagating at speed $(dx/dt) = s + \varepsilon\hat{w}$. Since the radiated acoustical and entropy/shear waves have strengths proportional to εq (see after (2.3)), they will have discontinuities proportional to $\varepsilon(q_+ - q_-)$ along the wave fronts $x - (s + \varepsilon\hat{w})t - \lambda_A y = 0$ and $x - (s + \varepsilon\hat{w})t - \lambda_p y = 0$, respectively. But a discontinuous near linear acoustical wave is a *shock* and a discontinuous near linear entropy/shear wave is a *contact discontinuity* (or slip line). Thus this solution represents a *triple shock-slip line configuration*, as in Fig. 6. Exactly the type of structure we were looking for!

Remark 2.5 The shock the acoustical wave has become in the example above must satisfy the entropy condition (1.6). A study of this condition shows that one must have $q_+ > q_-$. Thus this *must* be the admissibility criteria ("entropy condition") for discontinuous solutions of (2.6). This implies that the perturbed detonation front can break only as in Fig. 6. This agrees well with experimental evidence, as breaks with the corner pointing towards the unburnt gas, $y > 0$, are not observed.

[20]One may also consider $q_+ = q_-$, but this is of course a trivial solution.

One can show that in general, even for smooth initial data, solutions of (2.6) will develop discontinuities in q (which look locally as the exact solution above). When $a_2 = 0$ or $\hat{\beta} \downarrow 1$ this is trivial, for then (2.6) reduces to the nonlinear inviscid Burgers equations:

$$q_{t*} + (a_1 q^2)_\xi = 0 \quad \text{or} \quad q_{t*} + \{(a_1 - \frac{1}{2}a_2)q^2\}_\xi = 0, \qquad (2.11)$$

respectively. For these equations the statement above is well known (see [34]). Notice that the shock jump condition for the second equation in (2.11) does not agree with (2.9): this is because the limit $\hat{\beta} \downarrow 1$ is rather singular. On the other hand, when $a_2 \neq 0$ and $\hat{\beta} > 1$, numerical computations in [38] show that breaking occurs. Recently R. Gardner [56] has proven (rigorously) that this breaking does indeed occur.

Remark 2.6 The preceding analysis shows that spontaneously emitted radiating boundary waves break, forming Mach stem structures. The questions of growth and settling to finite amplitude are not resolved. For this, consideration of the actual chemical reactions in operation, with somewhat more detail than the theory being used here allows, is needed.[21] On the other hand, the computations of [38] show that, for certain choices of the parameters a_1, a_2 and $\hat{\beta}$ in (2.6), growth[22] is possible.

2.3. STEADY WAVE BIFURCATION. An alternative approach to the question of formation of transversal Mach stem structures is to ask:

> Under which conditions is there a "path" for a possible bifurcation of a plane steady detonation wave into a triple shock, slip-line configuration (without causality being violated)?

Remark 2.7 The requirement of causality is so that the bifurcation can occur spontaneously.[23]

Specifically: when is it that a triple shock-slip line configuration as the one in Fig. 8, with the angles θ_1 and θ_2 arbitrarily small, exist?

This question was dealt with in [38]. To answer it one must pose the, purely algebraic, equations describing triple shock slip-line configurations

[21] Hopefully without having to go all the way to a theory as complicated as ZND.

[22] Even violent instability.

[23] The issue of causality here is the same as for linear waves, even though we are taking of the energy carried by the trailing shock and contact discontinuity in Fig. 8.

(as obtained from the appropriate Rankine-Hugoniot and entropy conditions) and investigate conditions for a solution in the required regime. The answer is exactly that given by (2.2), including the statement that the angle between S_1 and S_2 must be less than π (see remark 2.5)!

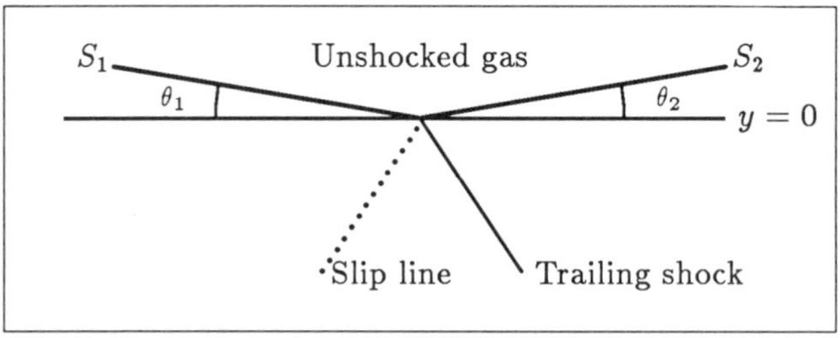

Figure 8: Steady wave bifurcation. S_1 and S_2 form the main wave.

3. SIMPLIFIED MODEL EQUATIONS.

In this section we present simplified model equations for reacting shock fronts (detonation waves), which we use to illustrate the computations in reference [37]. We assume the simplest (CJ) theory of detonations and (for the purpose of mathematical simplicity, as explained in the introduction) assume isentropic gas flow. Thus, the pressure will be a definite function of the specific volume $p = P(\tau)$. The P.D.E's for the flow are then simplified, for one need not consider the equation for the conservation of energy (1.14). As for the Rankine-Hugoniot jump conditions: the role of the Hugoniot line (see (1.2) and (1.8)) is now taken over by the "equation of state" $p = P(\tau)$. We will, of course, assume the equivalent of (1.9) and (1.10):

$$P'(\tau) < 0 \quad \text{and} \quad P''(\tau) > 0. \tag{3.1}$$

Remark 3.1 As we will be taking different equations of state on each side of the detonation wave, the assumption of isentropic flow need not apply across it. Since we will only be considering small perturbations of a *plane* front, the assumption of isentropic gas flow is not as drastic and unphysical as it may look on a first impression.

The plan of this section is as follows: first, we present the differential equations governing the flow, together with the appropriate Rankine-Hugoniot jump conditions. Secondly, we consider the basic detonation wave whose stability we want to consider.

3.1. THE EQUATIONS. Let u and v be the flow velocities in the x and y coordinate directions, respectively. Further, let

$$p = P_0(\tau) \quad \text{and} \quad p = P_1(\tau) \tag{3.2}$$

in the unshocked and shocked regions, respectively; here both P_0 and P_1 obey (3.1) and, for an exothermic reaction, we assume $P_1 > P_0$. The equations governing the flow have the general conservation form:

$$F^t(Z)_t + F^x(Z)_x + F^y(Z)_y = 0, \tag{3.3}$$

where $Z = (\rho, u, v)^T$, $F^t = (\rho, \rho u, \rho v)^T$, $F^x = (\rho u, \rho u^2 + p, \rho v)^T$ and $F^y = (\rho v, \rho uv, \rho v^2 + p)^T$.

We will choose the coordinates so that the unperturbed front is at $y = 0$, with $y > 0$ in the unshocked gas and $y < 0$ in the shocked gas. The equation for the perturbed front will then be of the form:

$$y = \sigma(x, t), \tag{3.4}$$

where σ is small. Introducing the new coordinate $z = y - \sigma$, we have:

$$F^t(Z)_t + F^x(Z)_x + \left\{F^y(Z) - \sigma_t F^t(Z) - \sigma_x F^x(Z)\right\}_z = 0, \tag{3.5}$$

where now the detonation wave is at $z = 0$. Quite clearly then the *Rankine-Hugoniot* conditions at the front are:

$$F^y(Z) - \sigma_t F^t(Z) - \sigma_x F^x(Z) \tag{3.6}$$

is *continuous across* $z = 0$. For an overcompressed detonation we have:

$$M_0 = \left\{(\sigma_t + \sigma_x u - v)/c\sqrt{1 + \sigma_x^2}\right\}_{z=+0} > 1, \tag{3.7a}$$

$$0 < M_1 = \left\{(\sigma_t + \sigma_x u - v)/c\sqrt{1 + \sigma_x^2}\right\}_{z=-0} < 1, \tag{3.7b}$$

where $c = \sqrt{dp/d\rho}$ is the sound speed.

3.2. PLANE STEADY DETONATION WAVE. Consider now a solution of the equations of (§ 3.1) where $\sigma \equiv 0, u \equiv 0$ and:

$$\rho \equiv \bar{\rho}_0,\; v \equiv \bar{v}_0 \text{ for } z > 0 \text{ and } \rho \equiv \bar{\rho}_1, v \equiv \bar{v}_1 \text{ for } z < 0, \tag{3.8}$$

where the quantities with bars are constants. Clearly we must have:

$$\bar{\rho}_1\bar{v}_1 = \bar{\rho}_0\bar{v}_0 < 0 < \bar{p}_1 + \bar{\rho}_1\bar{v}_1^2 = \bar{p}_0 + \bar{\rho}_0\bar{v}_0^2, \tag{3.9a}$$

$$0 < \bar{M}_1 = -\bar{v}_1/\bar{c}_1 < 1 < \bar{M}_0 = -\bar{v}_0/\bar{c}_0 \tag{3.9b}$$

and

$$\mu = \bar{\rho}_1/\bar{\rho}_0 = \bar{\tau}_0/\bar{\tau}_1 = \bar{v}_0/\bar{v}_1 > 1, \tag{3.9c}$$

where the notation $\bar{p}_0 = P_0(\bar{\tau}_0), \bar{\tau}_0 = (\bar{\rho}_0)^{-1}$, etc. is used.

Let now $\bar{Z} = (\bar{\rho}, 0, \bar{v})^T$, i.e.: $\bar{Z} = \bar{Z}_0$ for $z > 0$ and $\bar{Z} = \bar{Z}_1$ for $z < 0$, so that $Z = \bar{Z}$, $\sigma = 0$ solves the equations of (§ 3.1). This is our *unperturbed plane, steady, detonation wave.* We notice that, for $0 < \varepsilon << 1, Z = \bar{Z} + \varepsilon Y$ and $Y = (\rho', u', v')$ arbitrary, we have:

$$F^t(Z) = F^t(\bar{Z}) + \varepsilon A^t Y + \varepsilon^2 B^t(Y, Y), \tag{3.10a}$$

$$F^x(Z) = F^x(\bar{Z}) + \varepsilon A^x Y + \varepsilon^2 B^x(Y, Y) + O(\varepsilon^3), \tag{3.10b}$$

$$F^y(Z) = F^y(\bar{Z}) + \varepsilon A^y Y + \varepsilon^2 B^y(Y, Y) + O(\varepsilon^3), \tag{3.10c}$$

where the A's are 3×3 matrices and the B's are vector valued symmetric quadratic forms (easily expressible in terms of the parameters of the unperturbed wave). Clearly we have $A^t = A_0^t$ for $z > 0$, $A^t = A_1^t$ for $z < 0$, etc. Note that $\det A^y = -(\bar{\rho}\bar{c})^2\bar{v}(1-\bar{M}^2)$, so that A^y is nonsingular.

4. LINEAR ANALYSIS.

Here we consider the linear stability analysis of the plane, steady detonation wave introduced in (§ 3.2). First, we introduce the linearized equations and separate variables, to obtain the normal mode equations. Then, causality is discussed. Next, the results of the linear analysis are presented. Finally, expressions for the radiating boundary waves are displayed.

4.1. THE LINEARIZED EQUATIONS. Writing $Z = \bar{Z} + \varepsilon Y, \sigma = \varepsilon\eta$ (where $\bar{Z}$ is as in (§ 3.2)), substituting into the equations of (§ 3.1), taking ε infinitesimal and using (3.10) we obtain:

$$A^t Y_t + A^x Y_x + A^y Y_z = 0 \tag{4.1}$$

for $z \neq 0$ and continuity across $z = 0$ of

$$A^y Y - \eta_t F^t(\bar{Z}) - \eta_x F^x(\bar{Z}). \tag{4.2}$$

These are the *linearized equations of the stability theory* for the plane steady detonation of (§ 3.1). We now look for the fundamental normal modes of this system. For this we first separate variables:

$$Y = \hat{Y}(z)e^{ikx-i\omega t} \quad \text{and} \quad \eta = \hat{\eta}e^{ikx-i\omega t}, \tag{4.3}$$

where $\hat{\eta}$, k and ω are constants, with k real (for boundedness) and

$$\hat{Y}(z) = \begin{cases} \hat{Y}_0(z) & \text{for } z > 0, \\ \hat{Y}_1(z) & \text{for } z < 0. \end{cases} \tag{4.4}$$

Substituting in (4.1) and (4.2), we obtain the *normal mode equations*:

$$A^y \hat{Y}' = i(\omega A^t - kA^x)\hat{Y} \quad \text{for } z \neq 0, \tag{4.5}$$

where $' = \frac{\partial}{\partial z}$ and

$$i\hat{\eta}\left\{\omega F^t(\bar{Z}) - kF^x(\bar{Z})\right\} + A^y\hat{Y} \tag{4.6}$$

is *continuous across* $z = 0$. We look for solutions of these equations that are bounded as $|z| \to \infty$ and satisfy causality (see (§ 4.2)).

Remark 4.1 For any k and ω these equations can be solved explicitly. The general solution of (4.5) is a linear combination of solutions of the form $\hat{Y} = Xe^{imz}$ where $(\omega A^t - kA^x - mA^y)X = 0$. One then solves (4.5) for $z > 0$ and $z < 0$ separately and matches the solutions so that (4.6) applies. This gives the general solution of (4.5) and (4.6), which must then be restricted so that boundedness and causality applies. The results of this analysis are listed in (§ 4.3) and (§ 4.4).

4.2. CAUSALITY. Consider a bounded solution of (4.5) and (4.6) that does not decay as $z \to \infty$ (or $z \to -\infty$). Then, as $z \to \infty$, the corresponding solution of (4.1) and (4.2) is a linear combination of waves of the form:

$$Y_p = Xe^{i(kx+mz-\omega t)}, \tag{4.7}$$

where $(kA^x + mA^y - \omega A^t)X = 0$ and, it can be shown, m and ω are real,[24] so that (4.7) is either an acoustical or a shear wave.

The flow of energy associated with the wave (4.7) is not normal to the wave fronts, for the wave is propagating in a nonstationary media. The transformation $z' = z - \bar{v}t$ eliminates the background velocity and we have:

$$Y_p = Xe^{i(kx+mz'-\omega' t)}, \tag{4.8}$$

where $\omega' = \omega - m\bar{v}$ is the Doppler shifted wave number. Now energy propagates normal to the wave fronts, with velocity vector:

$$V_e' = (k^2 + m^2)^{-1}\omega'(k, m)^T. \tag{4.9}$$

This vector transforms back to the detonation wave frame of reference as

$$V_e = V_e' + (0, \bar{v})^T = (k^2 + m^2)^{-1}(\omega k - mk\bar{v}, \omega m + \bar{v}k^2)^T, \tag{4.10}$$

which is the *energy propagation velocity vector* for waves as in (4.7).

The *causality principle*:

> No energy propagates from infinity towards the perturbed detonation wave, all energy propagates away from it, (4.11)

is then equivalent to:

$$\omega m + \bar{v}_0 k^2 > 0 \text{ for } z > 0 \quad \text{and} \quad \omega m + \bar{v}_1 k^2 < 0 \text{ for } z < 0. \tag{4.12}$$

The condition $\bar{M}_0 > 1$ implies that no wave in $z > 0$ can satisfy causality.

4.3. RESULTS OF THE LINEAR STABILITY ANALYSIS. In this subsection we summarize the results of the linear stability analysis posed in (§ 4.1):

[24] m must be real for otherwise the wave would be either unbounded or vanishing. An inspection of the equation $\det(kA^x + mA^y - \omega A^t) = 0$ shows then ω is also real.

(i) No exponentially growing modes (i.e.: $\operatorname{Im}\omega > 0$) exist (see (2.1)).

(ii) If $(\mu - 1)\bar{M}_1^2 \leq 1$ then no radiating modes (i.e.: $\operatorname{Im}\omega = 0$) exist either. In this case all linear modes decay exponentially.

(iii) If $(\mu - 1)\bar{M}_1^2 > 1$ radiating modes (i.e.: $\operatorname{Im}\omega = 0$) exist. They are displayed next in (§ 4.4).

4.4. RADIATING BOUNDARY WAVES. Assume $(\mu - 1)\bar{M}_1^2 > 1$. Then the radiating boundary waves of (§ 4.3) (iii) are given by (we display them in terms of the solutions Y of (4.1) and (4.2)):

$$Y \equiv 0 \text{ for } z > 0, \tag{4.13a}$$

$$Y = ik\hat{\eta}\left\{\ell_1 X_1 e^{-ik\lambda_1 z} + \ell_2 X_2 e^{-ik\lambda_2 z}\right\} e^{ik(x-st)} \text{ for } z \leq 0 \tag{4.13b}$$

and

$$\eta = \hat{\eta} e^{ik(x-st)}, \tag{4.13c}$$

where k is real, $\hat{\eta}$ is arbitrary and $\ell_1, \ell_2, \lambda_1, \lambda_2, s, X_1$ and X_2 are as follows:

Let $\beta_2, \beta_3, \alpha_2, \alpha_3$ be defined by:

$$-1 < \beta_3 = -\sqrt{(\mu\bar{M}_1^2 - 1)/(\mu - 1)\bar{M}_1^2} < -\bar{M}_1, \tag{4.14a}$$

$$0 < \alpha_3 = \sqrt{1 - \beta_3^2} < \sqrt{1 - \bar{M}_1^2}, \tag{4.14b}$$

$$-\bar{M}_1 < \beta_2 = -\frac{2\bar{M}_1 + (1 + \bar{M}_1^2)\beta_3}{1 + 2\bar{M}_1\beta_3 + \bar{M}_1^2} < 1 \tag{4.14c}$$

and

$$0 < \alpha_2 = \sqrt{1 - \beta_2^2} < \sqrt{1 - \bar{M}_1^2}. \tag{4.14d}$$

Then we have:

$$\bar{c}_1\sqrt{1 - \bar{M}_1^2} < s = \bar{c}_1(1 + \bar{M}_1\beta_2)\alpha_2^{-1} = \bar{c}_1(1 + \bar{M}_1\beta_3)\alpha_3^{-1}, \tag{4.15a}$$

$$\lambda_1 = -s/\bar{v}_1 > \lambda_2 = \beta_2\alpha_2^{-1} > \lambda_3 = \beta_3\alpha_3^{-1},$$

$$X_1 = (0, \lambda_1, 1)^T, \tag{4.15c}$$

$$X_j = (\bar{\rho}_1, \bar{c}_1\alpha_j, -\bar{c}_1\beta_j)^T \text{ for } j = 2 \text{ or } 3, \tag{4.15d}$$

$$\ell_1 = -\bar{c}_1(\mu - 1)(1 - \bar{M}_1^2)(1 + \bar{M}_1\beta_3)/\mu\alpha_3(1 + 2\bar{M}_1\beta_3 + \bar{M}_1^2) < 0 \tag{4.15e}$$

and

$$\ell_2 = -2(\mu - 1)\bar{M}_1(1 + \bar{M}_1\beta_3)/\mu\alpha_3(1 - \bar{M}_1^2) < 0. \tag{4.15f}$$

Remark 4.2 (α_2, β_2) and (α_3, β_3) are the direction cosines of the reflected and incident[25] acoustical waves associated with the boundary waves.

Remark 4.3 The relationship with the separation of variables notation of (§ 4.1) and (§ 4.2) is $\omega = ks$ and $m = -k\lambda$.

It is convenient to introduce also the "left eigenvectors":

$$X_1^\dagger = \bar{r}_1 \bar{v}_1 (s^2 + \bar{v}_1^2)^{-1}(-\bar{v}_1, \lambda_1, 1), \tag{4.16a}$$

$$X_j^\dagger = -\left[2\bar{\rho}_1 \bar{c}_1^2 (\bar{M}_1 + \beta_j)\right]^{-1} (\bar{c}_1(1 - \bar{M}_1\beta_j), \alpha_j, -\beta_j) \text{ for } j = 2, 3. \tag{4.16b}$$

The objects introduced above in (4.14) through (4.16) are defined[26] by the equations:

$$(A_1^x - sA_1^t - \lambda_n A_1^y)X_n = 0, (n = 1, 2, 3), \tag{4.17a}$$

$$X_n^\dagger (A_1^x - sA_1^t - \lambda_n A_1^y) = 0, (n = 1, 2, 3), \tag{4.17b}$$

$$X_{n_1}^\dagger A_1^y X_{n_2} = \delta_{n_1 n_2}, (n_1, n_2 = 1, 2, 3), \tag{4.17c}$$

$$\ell_1 = X_1^\dagger J(s), \tag{4.17d}$$

$$\ell_2 = X_2^\dagger J(s), \tag{4.17e}$$

$$0 = X_3^\dagger J(s), \tag{4.17f}$$

where $J(s) = \left\{sF^t(\bar{Z}) - F^x(\bar{Z})\right\}\big|_1^0$. The relationship of these equations with the system (4.5), (4.6) is quite obvious. (4.17f) insures causality.

Remark 4.4 There is a second set of radiating boundary waves, moving to the left instead of to the right as the one displayed here. It can be obtained simply by observing that the equations are invariant under the transformation $x \to -x$, $u' \to -u'$ (u' is the second component of Y).

5. WEAKLY NONLINEAR THEORY.

The objective of this section is to illustrate the derivation of equation (2.6) in the context of the isentropic model. We also present a wave interaction interpretation of the integral term there.

[25] Although there is no incident wave at the linear level we introduce here the parameters of this wave anyway, for we will need them in the nonlinear expansion of (§ 5).

[26] Uniquely, except for trivial scalings of the eigenvectors.

5.1. THE NEAR LINEAR EXPANSION. The objective is to calculate nonlinear corrections for the radiating boundary waves of (§ 4.4).[27] We write:

$$\sigma = \varepsilon\eta(\xi,t^*) + \varepsilon^2\sigma_1(\xi,t^*) + \cdots, \tag{5.1a}$$

$$Z = \bar{Z}_0 \quad \text{for } z > 0, \tag{5.1b}$$

$$Z = \bar{Z}_1 + \varepsilon\sum_{j=1}^{2}\eta_\xi(\xi-\lambda_j z,t^*)\ell_j X_j + \varepsilon^2\sum_{j=1}^{3} w_j(\xi,z,t^*)X_j + \cdots \text{ for } z<0, \tag{5.1c}$$

where $\xi = x - st, t^* = \varepsilon t$, we assume $(\mu-1)\bar{M}_1^2 > 1, 0 < \varepsilon \ll 1$ and we follow the notation of (§ 4.4). Again we impose causality (see (2.4)) - which, in particular, immediately implies (5.1b).

Remark 5.1 As pointed out before, the waves of type "3" do not satisfy causality (they are incident acoustical waves). We must, however, include them because they are generated by the nonlinear interactions of the other waves. This, of course, is not a violation of causality.

Substituting these expressions in the equations of (§ 3), using (3.10) and collecting equal powers of ε,[28] one finds, at $O(\varepsilon^2)$:
From (3.8):

$$\begin{aligned} 0 = {} & \lambda_n(w_n)_\xi + (w_n)_z + \sum_1^2 \mu_{nj}\dot{q}(\xi-\lambda_j z,t^*) \\ & + \sum_1^2 \ell_j\lambda_j^2\delta_{nj} q(\xi,t^*)q'(\xi-\lambda_j z,t^*) \\ & + \sum_1^2 \nu_{nj}\left[q(\xi-\lambda_j z,t^*)^2\right]_\xi + \sum_{\substack{1\\ i\neq j}}^{2}\Gamma_{ij}^n q(\xi-\lambda_i z,t^*)q'(\xi-\lambda_j z,t^*), \end{aligned} \tag{5.2}$$

where $n = 1, 2$ or 3, $\cdot = \frac{\partial}{\partial t^*}, \prime = \frac{\partial}{\partial \xi}, q = \eta'$ and

$$\mu_{nj} = \ell_j X_n^\dagger A_1^t X_j, \tag{5.3a}$$

$$\nu_{nj} = \frac{1}{2}\Gamma_{jj}^n, \tag{5.3b}$$

$$\Gamma_{ij}^n = 2\ell_i\ell_j X_n^\dagger\left\{B_1^x(X_i,X_j) - sB_1^t(X_i,X_j) - \lambda_j B_1^y(X_i,X_j)\right\}. \tag{5.3c}$$

[27] Rather than use the Fourier representation of (4.13), it is convenient here to think of $\hat{\eta} = \hat{\eta}(k)$ and integrate over k, to obtain the waves in terms of $\eta(x-st)$.

[28] One must use (4.17) heavily here.

From (3.9):

$$0 = -w_n(\xi, 0, t^*) + \Delta_n q^2(\xi, t^*) + \Pi_n \dot{\eta}(\xi, t^*) + \ell_n \sigma_1'(\xi, t^*), \tag{5.4}$$

where $n = 1, 2$ or 3, $\ell_3 = 0$ (see (4.17 f)) and

$$\Delta_n = -\sum_1^2 \ell_i \ell_j X_n^\dagger B_1^y(X_i, X_j) + \sum_1^2 \ell_j \lambda_n \delta_{nj}, \tag{5.5a}$$

$$\Pi_n = -X_n^\dagger F^t(\bar{Z})|_1^0. \tag{5.5b}$$

Now w_1 and w_2 can include an arbitrary contribution from the homogeneous solution of (5.2).[29] Thus one can always satisfy (5.4) for them. On the other hand, the homogeneous part of w_3 is restricted by the causality principle so that then (5.4) will impose restrictions in the expansion. Thus we solve first for w_3 from (5.2):

$$w_3 = W_3(\xi - \lambda_3 z, t^*) + \sum_1^2 (\lambda_j - \lambda_3)^{-1} \mu_{nj} \dot{\eta}(\xi - \lambda_j z, t^*)$$

$$+ \sum_1^2 (\lambda_j - \lambda_3)^{-1} \nu_{3j} q^2(\xi - \lambda_j z, t^*)$$

$$+ \sum_{\substack{1 \\ i \neq j}}^2 \Gamma_{ij}^3 \int_z^0 q(\xi - \lambda_3 z - (\lambda_i - \lambda_3)y, t^*) q'(\xi - \lambda_3 z - (\lambda_j - \lambda_3)y, t^*) dy, \tag{5.6}$$

where W_3 is arbitrary so far as (5.2) is concerned.

Assume now a localized disturbance, with η vanishing a $|\eta| \to \infty$. (5.7)

Then causality requires that, as $z \downarrow -\infty$, disturbances on the "3" wave should vanish, i.e.:

$$\text{For } \phi = \xi - \lambda_3 z = \text{ const.}, \lim_{z \to -\infty} w_3 = 0. \tag{5.8}$$

There is no question then that we must take:

$$W_3(\phi, t^*) = -\sum_{\substack{1 \\ i \neq j}}^2 \Gamma_{ij}^3 \int_{-\infty}^0 q(\phi - (\lambda_i - \lambda_3)y, t^*) q'(\phi - (\lambda_j - \lambda_3)y, t^*) dy =$$

[29] Since both these waves satisfy causality.

$$= (\lambda_1 - \lambda_3)^{-1}\Gamma^3_{21}q^2(\phi, t^*) - \kappa_3 \int_0^\infty q(\phi + \hat{\beta}y, t^*)q'(\phi + y, t^*)dy, \quad (5.9)$$

where $\hat{\beta} = \frac{(\lambda_1-\lambda_3)}{(\lambda_2-\lambda_3)} > 1$ and $\kappa_3 = (\lambda_2 - \lambda_3)^{-1}\Gamma^3_{12} - (\lambda_1 - \lambda_3)^{-1}\Gamma^3_{21}$.

Substituting now (5.6) and (5.9) into (5.4) we obtain:

$$\kappa_1 \dot{\eta} + \kappa_2 q^2 + \kappa_3 \int_0^\infty q(\xi + \hat{\beta}y, t^*)q'(\xi + y, t^*)dy = 0, \quad (5.10)$$

where:

$$\kappa_1 = \Pi_3 - \sum_1^2 \frac{\mu_{3j}}{\lambda_j - \lambda_3} > 0 \text{ and } \kappa_2 = \left\{ \Delta_3 - \frac{\Gamma^3_{21}}{\lambda_1 - \lambda_3} - \sum_1^2 \frac{\nu_{3j}}{\lambda_j - \lambda_3} \right\}.$$

Taking $\frac{\partial}{\partial \xi}$ of this equation and dividing by κ_1 we obtain (2.6), with $a_1 = \kappa_2/\kappa_1$ and $a_2 = \kappa_3/\kappa_1$. This completes the derivation.

Remark 5.2 The way causality determines (5.10) at second order is exactly analogous to the way it determines s at the linear level.

5.2. WAVE INTERACTION. We now give a geometrical wave interaction interpretation of the integral term in (2.6) and (5.10).

We recall that the propagation vectors for waves of Type 1, 2 and 3 are, respectively (see (4.10)):

$$V_1 = (0, \bar{v}_1)^T, V_2 = \bar{c}_1(\alpha_2, -\beta_2 - \bar{M}_1)^T, V_3 = \bar{c}_1(\alpha_3, -\beta_3 - \bar{M}_1)^T. \quad (5.11)$$

An alternative form for V_2 is:

$$V_2 = \bar{c}_1(1 - \bar{M}_1^2)(1 + 2\bar{M}_1\beta_3 + \bar{M}_1^2)^{-1}(\alpha_3, \bar{M}_1 + \beta_3)^T. \quad (5.12)$$

We refer to Fig. 9. Consider an arbitrary point, A, along the unperturbed wave front at an arbitrary time. W.L.O.G. we can take its coordinates to be $x_A = 0, z_A = 0$ and $t_A = 0$, in particular $\xi_A = 0$. This point can be influenced by incident acoustical waves emitted along the line DA at earlier times. Such emissions would have occurred at:

$$x_D = \bar{c}_1\alpha_3 t', \; z_D = -(\bar{M}_1 + \beta_3)\bar{c}_1 t', \; t = t', \quad (5.13)$$

where $t' < 0$ is arbitrary. Such emissions can only happen if produced by the nonlinear interaction of a wave of Type 1 and a wave of Type 2. The points of emission of these two earlier waves are, clearly:

$$x_B = \alpha_3\bar{c}_1 t', z_D = 0, t_B = -(\beta_3/\bar{M}_1)t' \Rightarrow \xi_B = (\bar{c}_1/\alpha_3\bar{M}_1)(\bar{M}_1 + \beta_3)t' > 0, \quad (5.14)$$

$$x_C = 2\bar{c}_1\alpha_3 t', z_C = 0, t_C = 2(1+\bar{M}_1\beta_3)t'/(1-\bar{M}_1^2) \Rightarrow$$
$$\xi_C = \frac{-2\bar{c}_1}{\alpha_3(1-\bar{M}_1^2)}(\bar{M}_1+\beta_3)^2 t' > 0. \tag{5.15}$$

The strength of the interaction must be a quadratic function of the strengths at ξ_B and ξ_C. To obtain the total effect at $\xi = 0$ we must integrate over all $t' < 0$ (alternatively all $\xi_C > 0$). One can show $\xi_B = \hat{\beta}\xi_C$, so that this argument is in total agreement with the form of the integral term in (2.6) and (5.10). If one were to carry this argument further, even the details of the integral term could be understood in this fashion. For an intuitive picture of how things operate, this seems enough however.

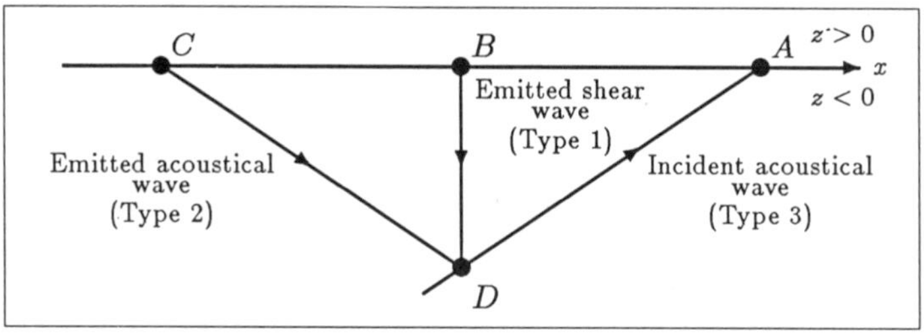

Figure 9: Wave interactions.

6. APPENDIX.

Here we show that, for an overcompressed detonation in the CJ theory of (§ 1.2), *if the reacted gas satisfies a polytropic gas equation of state* then:

$$(\Gamma+1)(\mu-1)M_1^2 < 1. \tag{6.1}$$

Thus (2.2) does not apply and the situation in remark 2.3 holds, so that *spontaneous Mach stem formation is not possible.*

To show this let the state of the unreacted gas: p_0, τ_0 and $e_0 = E_0(p_0, \tau_0)$ be given. Furthermore, let the equation of state for the burnt gas be:

$$e_1 = E_1(p_1, \tau_1) = \frac{1}{\Gamma}p_1\tau_1 + \beta, \tag{6.2}$$

where β is a constant, $\Gamma = \gamma - 1 > 0$ is the Gruneisen coefficient and γ is the ratio of the specific heats for the burnt gas. We assume that E_0

satisfies (1.9) and (1.10). Assume also an exothermic reaction, so that the Hugoniot curve for the reactants, $H_0 = 0$, is below the Hugoniot curve for the products, $H_1 = 0$. Then, in particular:

$$q = -H_1(p_0, \tau_0) = e_0 - \beta - \frac{1}{\Gamma} p_0 \tau_0 > 0, \tag{6.3}$$

where q is a measure of the heat released.

We have then the nondimensional equations (see (1.1) and (1.2)):

$$(\bar{p} - 1) + \varsigma(\bar{\tau} - 1) = 0 \quad \text{and} \quad (\bar{p} + \xi)(\bar{\tau} - \xi) = \bar{q} + (1 - \xi^2), \tag{6.4}$$

where $\bar{p} = p_1/p_0, \bar{\tau} = \tau_1/\tau_0, \varsigma = \rho_0 v_0^2/p_0 = \gamma \bar{p} M_1^2/\bar{\tau}, \xi = (\gamma - 1)/(\gamma + 1)$ and $\bar{q} = 2(\gamma - 1)q/(\gamma + 1)p_0\tau_0$. We note that $0 < \xi < 1$ and $\bar{q} > 0$. Thus, for a strong detonation:

$$\xi < \bar{\tau} < \bar{\tau}_j \quad , \quad \bar{p} = (1 + \bar{q} - \bar{\tau}\xi)(\bar{\tau} - \xi)^{-1} > 1 \tag{6.5}$$

and

$$0 < M_1^2 = \frac{\bar{\tau}(\bar{q} + (1 - \bar{\tau})(1 + \xi))}{\gamma(1 - \bar{\tau})(1 + \bar{q} - \bar{\tau}\xi)} < 1 \,, \tag{6.6}$$

where $\bar{\tau}_j$ is the nondimensional specific volume corresponding to the CJ detonation:

$$\bar{\tau}_j = 1 + (1 + \xi)^{-1} \left(\bar{q} - \sqrt{\bar{q}^2 + \bar{q}(1 - \xi^2)} \right) < 1. \tag{6.7}$$

Then, since $\mu = \bar{\tau}^{-1}$, we have:

$$(\Gamma + 1)(\mu - 1)M_1^2 = (\bar{q} + (1 - \bar{\tau})(1 + \xi))(1 + \bar{q} - \bar{\tau}\xi)^{-1}. \tag{6.8}$$

The right hand side of (6.8) is a decreasing function of $\bar{\tau}$. Thus, evaluating at $\bar{\tau} = \xi$ and $\bar{\tau} = \bar{\tau}_j$, we obtain:

$$(\Gamma + 1)(\mu_j - 1) < (\Gamma + 1)(\mu - 1)M_1^2 < 1, \tag{6.9}$$

where μ_j is the compression ratio of the CJ detonation. This proves (6.1).

BIBLIOGRAPHY

1. Alpert, R. L., and Toong, T. Y., "Periodicity in exothermic hypersonic flows about blunt projectiles", *Acta Astronaut.*, **17**, (1972), pp. 539-560.

2. Abouseif, G. E., and Toong T. Y., "Theory of unstable one-dimensional detonations", *Combust. Flame*, **45**, (1982), pp. 67-94.

3. Barthel, H. O., "Reaction zone-shock front coupling in detonations", *Phys. Fluids*, **15**, (1972), pp. 43-50.

4. Barthel, H. O., "Predicted spacings in hydrogen-oxygen-argon detonations", *Phys. Fluids*, **17**, (1974), pp. 1547-1553.

5. Barthel, H. O., and Strehlow, R. A. , "Wave propagation in one dimensional reactive flows", *Phys. Fluids*, **9**, (1966), pp. 1896-1907.

6. Bethe, H. A., "The theory of shock waves for an arbitrary equation of state", Office of Scientific Research and Development, Rept. 445, (1942).

7. Chapman, D. L., "On the rate of explosion in gases", *Philos. Mag.*, **47**, (1899), pp. 90-104.

8. Chernyi, G. G., "On problems involving gas flow associated with gasdynamics", *Acta Astronaut.*, **2**, (1975), pp. 839-965.

9. Chernyi, G. G., and Medvedev, S. A., "Development of oscillations associated with attenuation of detonation waves", *Acta Astronaut.*, **15**, (1970), pp. 371-375.

10. Chiu, K. W., and Lee, J. H., "A simplified version of the Barthel model for transverse wave spacings in gaseous detonations", *Combust. Flame*, **26**, (1976), pp. 353-361.

11. Courant, R. and Friedrichs, K. O., *Supersonic flow and shock waves*, Springer-Verlag, New York, (1948).

12. Cowperthwaite, M., "Properties of some Hugoniot curves associated with shock instability", *J. Franklin Inst.*, **285**, (1968), pp. 275-284.

13. Doering, W., "On detonation processes in gases", *Ann. Phys.*, **43**, (1943), pp. 421-436.

14. Duvall, G.E., "Shock wave stability in solids", in *Ondes de Detonation*, pp. 337-352. Paris: Editions du Centre National de La Rechenche Scientifique, (1962).

15. D'yakov, S. P., "The stability of shock waves: investigation of the problem of the stability of shock waves in arbitrary media", *Zh. Eksp. Teor. Fiz.*, **27**, (1954), pp. 288-295. [English translation: Atomic Energy Research Establishment, A.E.R.E. Lib./Trans. 648, (1956)].

16. Erpenbeck, J. J., "Stability of steady-state equilibrium detonations", *Phys. Fluids*, **5**, (1962), pp. 604-614.

17. Erpenbeck, J. J., "Stability of step shocks", *Phys. Fluids*, **5**, (1962), pp. 1181-1187.

18. Erpenbeck, J. J., "Structure and stability of the square-wave detonation". In *Ninth Symposium (International) on Combustion*, pp. 442-453. Academic Press, New York, (1963).

19. Erpenbeck, J. J., "Stability of idealized one-reaction detonations", *Phys. Fluids*, **7**, (1964), pp. 684-696.

20. Erpenbeck, J. J., "Stability of idealized one-reaction detonations: zero activation energy", *Phys. Fluids*, **8**, (1965), pp. 1192-1193.

21. Erpenbeck, J. J., "Detonation stability for disturbances of small transverse wavelength", *Phys. Fluids*, **9**, (1966), pp. 1293-1306.

22. Erpenbeck, J. J., "Nonlinear theory of unstable one-dimesional detonations", *Phys. Fluids*, **10**, (1967), pp. 274-288.

23. Erpenbeck, J. J., "Theory of detonation stability", In *Twelfth Symposium (International) on Combustion*, pp. 711-721. The Combustion Institute, Pittsburgh, (1969).

24. Erpenbeck, J. J., "Nonlinear theory of two-dimensional detonation", *Phys. Fluids*, **13**, (1970), pp. 2007-2026.

25. Fickett, W., Jacobson, J. D., and Schott, G. L., "Calculated pulsating one-dimensional detonations with induction-zone kinetics", *AIAA J.*, **10**, (1972), pp. 514-516.

26. Fickett, W., and Wood, W. W., "Flow calculations for pulsating one-dimensional detonations", *Phys. Fluids*, **9**,(1966), pp. 903-916.

27. Fickett, W., Jacobson, J. D., and Wood, W. W., "The method of characteristics for one-dimensional flow with chemical reaction", Los Alamos Scientific Laboratory Report LA-4269, (1970).

28. Fickett, W., and Davis, W., *Detonation*, University of California Press, Berkeley, CA, (1979).

29. Fickett, W., "Stability of the square wave detonation", preprint, (1984).

30. Fowles, G. R., "Conditional stability of shock waves - a criterion for detonation", *Phys. Fluids*, **19**, (1976), pp. 227-238.

31. Fowles, G. R., "Stimulated and spontaneous emission of acoustic waves from shock fronts", *Phys. Fluids*, **24**, (1981), pp. 220-227.

32. Jouguet, E., "On the propagation of chemical reactions in gases", *J. de Mathematiques Pure at Appliquees*, **1**, (1905), pp. 347-425.

33. Kontorovich, V. M., "Concerning the stability of shock waves", *Zh. Eksp. Teor. Fiz.*, **33**, (1957) pp. 1525-1526. [English translation: *Sov. Phys. JETP*, **6**, (1958), pp. 1179-1180.]

34. Lax, P. D., "Hyperbolic systems of conservation laws and the mathematical theory of shock waves", *SIAM* Regional Conf. Series in Appl. Math., No. 11, (1973).

35. Majda, A., "The stability of multi-dimensional shock fronts", *Memoirs Amer. Math. Soc.*, **41**, No. 275, (1983).

36. Majda, A., "The existence of multidimensional shock fronts", *Memoirs Amer. Math. Soc.*, **43**, No. 281, (1983).

37. Majda, A. and Rosales, R., "A theory for spontaneous Mach Stem formation in reacting shock fronts I: the basic perturbation analysis", *SIAM J. Appl. Math.*, **43**, (1983), pp. 1310-1334.

38. Majda, A. and Rosales, R., "A theory for spontaneous Mach Stem formation in reacting shock fronts, II: steady wave bifurcation and the evidence for breakdown", *Studies in Applied Math*, **71**, (1984), pp. 117-148.

39. Oppenheim, A., *Introduction to gas dynamics of explosions*, CISM lectures No. 48, Udine, 1970.

40. Pukhnachev, "The stability of Chapman-Jouguet detonations", *Dokl. Akad. Nauk. SSSR (Phys. Sect.)*, **149**, (1963), pp. 798-801. [English translation: *Soviet Physics - Doklady*, **8**, (1963), pp. 338-340].

41. Rosales, R., "Catastrophic instabilities in square wave models of detonations: a possible explanation", in preparation, (1985).

42. Shchelkin, K. I., "Two cases of unstable combustion", *Zh. Eksp. Teor. Fiz.*, **36**, (1959), pp. 600-606. [English translation: *Soviet Physics JETP*, **36**, (1959), pp. 416-420].

43. Shchelkin, K. I., "Unidimensional instability of detonation", *Dokl Akad. Nauk. SSSR (Phys. Chem. Sect.)*, **160**, (1965), pp. 1144-1146.

44. Shchelkin, K. I., and Troshin, Ya., *Gasdynamics and combustion*, Mono Book Corp., Baltimore, MD, (1965).

45. Sattinger, D. H., *Topics in stability and bifurcation theory*, Springer-Verlag, New York, (1973).

46. Strehlow, R., *Fundamentals of Combustion*, Kreiger Publishing, New York, (1979).

47. Strehlow, R., "Multidimensional detonation wave structure", *Acta Astronaut.*, **15**, (1970), pp. 345-357.

48. Strehlow, R., and Fernandes, F. D., "Transverse waves in detonations", *Combust. Flame*, **9**, (1965), pp. 109-119.

49. Strehlow, R. A., Maurer, R. E., and Rajan, S., "Transverse waves in detonations I: spacing in the hydrogen-oxygen system", *AIAA J.*, **7**, (1969), pp. 323-328.

50. Swan, G. W., and Fowles, G. R., "Shock wave stability", *Phys. Fluids*, **18**, (1975), pp. 28-35.

51. Von Neuman, J., "Theory of detonation waves". In John von Neuman, collected works, vol. **6**, ed. A. J. Taub, MacMillan, New York, (1942).

52. Wood, W. W., and Salsburg, Z. W., "Analysis of steady-state supported one dimensional detonations and shocks", *Phys. Fluids*, **3**, (1960), pp. 549-566.

53. Zaidel, R. M., "The stability of detonation waves in gaseous mixtures", *Dokl. Akad. Nauk SSSR (Phys. Chem. Sect.)*, **136**, (1961), pp. 1142-1145.

54. Zaidel, R. M. and Zeldovich, Ya. B., "One dimensional instability and attenuation of detonation", *Zh. Prikl. Mekh. Tekh. Fiz.*, **6**, (1963), pp. 59-65.

55. Zeldovich, Ya. B., "On the theory of the propagation of detonation in gaseous systems", *Zh. Eksp. Teor. Fiz.*, **10**, (1940), pp. 542-568.

56. Gardner, R., "Solutions of a non-local conservation law arising in combustion theory", preprint, (1985).

DEPARTMENT OF MATHEMATICS
2-337
MASSACHUSETTS INSTITUTE OF TECHNOLOGY
CAMBRIDGE, MA 02139

Lectures in Applied Mathematics
Volume 24, 1986

ROLES OF PERTURBATION METHODS FOR TURBULENT DIFFUSION FLAME

F. A. Williams[1]

ABSTRACT. Currently outstanding problems of turbulent diffusion flames are considered, with emphasis on differences occurring in different combustion regimes. Ways in which perturbation methods have contributed to progress in understanding of turbulent diffusion-flame combustion are indicated, and some prospects for future advances are assessed.

1. INTRODUCTION. Turbulent combustion often tacitly is viewed as a process of flame propagation into a gaseous fuel under conditions of turbulent flow. The "propagation" aspect of this point of view excludes an important class of turbulent combustion problem - turbulent diffusion flames. In diffusion flames the fuel and oxidizer enter the combustion region in different streams and therefore must mix as they burn. This necessity of mixing prompts the label "diffusion". Diffusion flames do not propagate; they may be characterized by rates of heat release or of reactant consumption but not by burning velocities. Therefore turbulent diffusion flames differ in many respects from turbulent flame propagation through homogeneous fuel mixtures.

There are many reasons for interest in turbulent diffusion flames. These flames occur in various combustion chambers (e.g., in diesel engines), in some industrial furnaces and flaring operations and in most fires, for example. Some of the specific ques-

[1]Supported in part by the National Science Foundation.

0075-8485/86 $1.00 + $.25 per page

tions asked are: What is the flame length or flame height? What is the average local volumetric rate of heat release? What are the critical conditions for a turbulent-jet diffusion flame to be lifted off the fuel duct? What are the radiant energy fluxes emitted by the flame? How complete is the combustion? What is the rate of production of oxides of nitrogen? What is the rate of production and liberation of smoke and of unburnt hydrocarbons? What are the properties of noise emission from the flame? A number of these questions can be approached by procedures that make use of perturbation methods. Some illustrations will be given here.

Optimum methods of attack on turbulent combustion problems depend strongly on the combustion regime. Therefore a review of combustion regimes for turbulent diffusion flames is presented first, to provide a necessary background. Methods of analysis will then be addressed.

2. COMBUSTION REGIMES. Regimes of turbulent diffusion flames are conveniently identified in terms of appropriate Reynolds numbers and Damköhler numbers [1]. A helpful graph is shown in Fig. 1. The Reynolds number $R_\ell \equiv \sqrt{2k}\ \ell/\nu$ (where ν is the kinematic viscosity of the gas) is based on the integral scale ℓ and kinetic energy k of the turbulence. Knowledge of turbulence relates R_ℓ to other Reynolds numbers; e.g., that based on the Taylor scale is $R_t = \sqrt{R_\ell}$. The Damköhler number $D_\ell \equiv (\ell/\sqrt{2k})/\tau_c$ is the ratio of a large-eddy turnover time to a representative chemical time τ_c for the overall conversion of reactants to products. Length and time scales associated with the smallest eddies, the Kolmogorov eddies, are $\ell_k = \ell/R_\ell^{3/4}$ and $\tau_k = \ell_k^2/\nu$, and the Damköhler number based on a Kolmogorov scale is $D_k = \tau_k/\tau_c$. Lines for constant values of D_k are shown in Fig. 1, as are lines for constant values of two other parameters, one being $\sqrt{R_\ell D_\ell}$, the ratio of ℓ to the chemical-diffusive length $\sqrt{\nu\tau_c}$, and the other $\sqrt{R_\ell/D_\ell}$, the ratio of a root-mean-square velocity fluctuation $\sqrt{2k}$ to the chemical-diffusive velocity $\sqrt{\nu/\tau_c}$. Different parts of Fig. 1 correspond to different

combustion regimes.

There are a number of sources of ambiguity and uncertainty in this simple picture. There are many chemical times in turbulent combustion; a judicious combination of them relevant to the overall rate of heat release must be selected to draw the graph. Even after the selection is made, the resulting τ_c, as well as ν, and in fact even ℓ and k, vary significantly with temperature, thereby raising further questions as to where any given turbulent diffusion flame should be placed in the diagram. Since most of the chemistry occurs in the hottest parts of the flow it can be reasoned that ν and τ_c should be evaluated near the adiabatic flame temperature and with at least partially depleted reactant concentrations for closest correspondence to what is really happening. While it would appear to be also desirable to evaluate ℓ and k in the hot regions, this is more difficult because the changes in the turbulence are less well known. There are additional uncertainties as to whether the classical relationships among the various turbulence scales continue to apply in the presence of the density variations associated with combustion. Nevertheless the figure at least provides a starting point for discussion.

At sufficiently small values of D_ℓ the chemistry is slow compared with the rates of turbulent mixing and the reactions occur slowly and are broadly distributed throughout the flow. Thus there is a regime of distributed reactions indicated in the figure. At sufficiently large values of D_ℓ the chemistry is fast, where it occurs, and therefore moving and distorted reaction sheets exist in the flow, with fuel and oxidizer diffusing into them from opposite sides. Thus there is also a reaction-sheet regime in the figure. When D_ℓ is extremely large and R_ℓ is not too large the turbulence has little effect on the reaction sheets and the situation is identified as one of weak turbulence, an extreme limit of the reaction-sheet regime. If R_ℓ is large enough the sheets are highly convoluted and fold back on themselves (multiple sheets), while at smaller R_ℓ there is a single well-identifiable reaction sheet in the flow. The transition between these two types of behavior in the reaction-sheet regime is not well characterized, although one suggestion for the boundary, $\sqrt{2k}/\sqrt{\nu/\tau_c}=1$, is indicated in the figure. It is thus seen that there may be a variety of types of behavior in the reaction-sheet regime.

Where transition from reaction sheets to distributed reactions occurs also is somewhat a matter of speculation. It appears that $D_k \gg 1$ is a sufficient criterion for reaction sheets, and it has been suggested that $D_\ell \ll 1$ is a sufficient criterion for distributed reactions, although the more conservative criterion, $\ell/\sqrt{\nu\tau_c} \ll 1$, may be required. Irrespective of these uncertainties, it is seen from Fig. 1 that at large R_ℓ there is an intermediate region that corresponds neither entirely to reaction sheets nor entirely to distributed reactions. This intermediate region has no name, and the character of the combustion there is not understood. There may be more than one intermediate regime at large R_ℓ, or there may be more than one regime at the same values of R_ℓ and D_ℓ in the figure, with the selection being determined by how the combustion is initiated.

Where do practical turbulent diffusion flames lie in Fig. 1? There are intensely stirred chemical reactors with weakly reacting fuel and oxidizer that almost certainly lie in the distributed-reaction regime. However, it appears that in the majority of the real problems, e.g. for open fires or large jet flames employing usual fuels, the combustion occurs in the reaction-sheet regime. Perturbation methods have found their strongest applications in problems that arise in the reaction-sheet regime and the remainder of the discussion is restricted to this regime.

3. APPROACHES TO THE ANALYSIS OF TURBULENT DIFFUSION FLAMES.

Methods of analysis of turbulent combustion can be subdivided into (A) methods based on heuristic conservation equations, (B) methods based on kinetic-theory conservation equation with chemical reactions, and (C) methods based on chemically reacting Navier-Stokes equations. Age theories of chemical reactors provide examples that fall within the first category. There has been relatively little rigorous work in the second category despite indications that in certain very special circumstances kinetic theory may be the proper starting place [2]. A number of different approaches fall within the third category, which is generally accepted to provide the relevant underlying conservation equations for turbulent diffusion flames. In all of these approaches probabilistic ideas are involved in one way or another.

The methods appealing to chemically reacting Navier-Stokes turbulence can be identified as (1) moment methods that give no special thought to probability-density functions, (2) methods involving approximations of probability-density functions through use of selected moment methods, (3) methods based on calculating evolutions of probability-density functions from their underlying conservation equations, and (4) direct numerical simulations for realizations of turbulent reacting flows. These methods have been listed roughly in the order of increasing demand for computational power. The methods (1) encounter difficulties of countergradient diffusion in reaction-sheet regimes and will not be considered further here. Methods (3) and (4) are interesting for future research but as yet have provided relatively few results for turbulent diffusion flames. Method (2) has proven most useful for the reaction-sheet regime and is the main approach to be considered here. This method makes use of coupling functions and introduces perturbation methods for addressing reaction-sheet structures. Advantages of coupling functions and perturbations also can benefit methods (3) and (4); for example, there has been some success by (4) in observing baroclinic vorticity generation by combustion in turbulent diffusion flames [3]. The essential ideas concerning rolls of perturbation methods nevertheless are demonstrated most easily by restricting attention to method (2).

In method (2) attention is focused largely on a mixture fraction Z defined to be unity in the fuel stream and zero in the oxidizer stream. One definition of Z is the ratio of the fuel mass (irrespective of the molecules in which the fuel atoms may be contained) to the total mass in a local volume element; if Soret, pressure-gradient and body-force diffusion are negligible, and if all binary diffusion coefficients are equal to ν, then the conservation equation for Z is simply

$$\partial(\rho Z)/\partial t + \underset{\sim}{\nabla}\cdot(\rho \underset{\sim}{v} Z) = \underset{\sim}{\nabla}\cdot(\rho\nu\underset{\sim}{\nabla} Z) \quad , \tag{1}$$

where ρ is density and $\underset{\sim}{v}$ velocity. In the reaction-sheet regime the chemistry occurs at a particular value of Z, the stoichiometric value, Z_c in Fig. 2, at which the adiabatic flame temperature is T_c; the temperature T as well as mass fractions of fuel Y_F and of oxidizer Y_0 are uniquely related to Z as illustrated by the solid lines in Fig. 2. Also shown in the figure as dashed lines for comparison are the corresponding relationships for frozen flow (identified by the subscript f) i.e., for mixing without chemical reaction. In each of these limits T, Y_F and Y_0 are known functions of Z which is seen from (1) not to involve a chemical source term. This "conserved-scalar" aspect of Z is what makes it most useful for turbulent diffusion flames. It aids in reducing the turbulent combustion problem to one of turbulent mixing.

In method (2) a probability-density function P(Z) is parameterized with from 2 to 4 parameters. Representative shapes that the parameterizations are designed to reproduce are shown in Fig. 3, where the vertical arrows represent delta functions accounting for intermittency. The evolutions of the parameters in the turbulent flow are then calculated by moment methods to obtain the evolution of P(Z). Current uncertainties in method (2) concern the fidelity with which the turbulence modeling in these moment methods accounts for the variable-density aspects of the derived turbulent mixing problem. Continuing research on modeling variable-density mixing flows is needed for reducing these uncertainties. As yet there is no firm indication - either theoretical or experimental - that the countergradient diffusion problems, plaguing moment modeling of fields with chemical sources, extent to the conserved scalars like Z. Therefore successive refinements in selection of parameterized functions P(Z) and in modeling to obtain the evolutions of the parameters may provide accurate bases for calculating many aspects of turbulent diffusion flames of interest.

From probability density functions for Z can be calculated probability density functions for species mass fractions and for temperature, as illustrated in Fig. 4. The methods are straightforward and enable averages such as $\overline{T}$ or average square fluctuations such as $\overline{T'^2}$, where $T' \equiv T-\overline{T}$, to be obtained at any given point in the flow. Thus perturbation methods have relatively little to do with finding these primitive turbulence properties by method (2). Questions concerning more interesting properties, such as average local rates of heat release, however encounter strong needs for perturbation methods.

4. AVERAGE RATES OF HEAT RELEASE. Under the approximations leading to (1) the mass rate of production of chemical species i is

$$w_i = \partial(\rho Y_i)/\partial t + \underset{\sim}{\nabla}\cdot(\rho \underset{\sim}{v} Y_i) - \underset{\sim}{\nabla}\cdot(\rho \nu \underset{\sim}{\nabla} Y_i) \quad , \tag{2}$$

where Y_i is the mass fraction of species i. If H_i is the enthalpy of formation per unit mass for species i then the rate of heat release per unit volume is

$$Q = -\sum_i H_i w_i \tag{3}$$

Finding the local average $\overline{Q}$ therefore entails finding $\overline{w}_i$, which may be thought to be straightforward from chemical-kinetic expressions for w_i if P(Z) and results like Fig. 2 are available. However, when it is realized that typically w_i is proportional to the product $Y_F Y_O$ and that Fig. 2 shows this product to be always zero, a difficulty becomes apparent. This difficulty concerns the approach to chemical equilibrium - a condition under which w_i really is the difference between large terms that nearly cancel. The difficulty calls for a perturbation approach in which the right-hand side of (2) is employed with $Y_i=Y_i(Z)$ in the first approximation.

With this assumption (2) may be written as

$$w_i = \frac{dY_i}{dZ}\left[\frac{\partial(\rho Z)}{\partial t} + \underset{\sim}{\nabla}\cdot(\rho \underset{\sim}{v} Z) - \underset{\sim}{\nabla}\cdot(\rho\nu\underset{\sim}{\nabla} Z)\right] - \rho\nu|\underset{\sim}{\nabla} Z|^2 \frac{d^2Y_i}{dZ^2} \quad . \tag{4}$$

In view of (1), the quantity in the square brackets in (4) vanishes. Therefore in a turbulent flow under the given approximations, the local, instantaneous rate of production of species i per unit volume is [4]

$$w_i = -\rho\nu(d^2Y_i/dZ^2)(\nabla Z)\cdot(\nabla Z) \quad . \tag{5}$$

Multiplication by heats of formation, and summation over i, provides a similar expression for the local, instantaneous rate of heat release from (3).

Although d^2Y_i/dZ^2 is a function of Z in (5), the dot product $(\nabla Z)\cdot(\nabla Z)$ is not. Therefore knowledge of P(Z) is insufficient for obtaining the average $\overline{w}_i$ from (5); a joint probability-density function for Z and for $\chi\equiv 2\nu\nabla Z\cdot\nabla Z$ must be known. The non-negative quantity χ is closely related to the instantaneous rate of scalar dissipation which is similarly defined in terms of the fluctuation Z' instead of Z. For high turbulence Reynolds numbers these two dissipations almost always are practically equal because gradients of fluctuations then are much larger than gradients of averages. The occurrence of χ in (5) underscores the importance of a joint probability-density function, $P(\chi,Z)$, for the mixture fraction Z and its dissipation rate χ. In terms of $P(\chi,Z)$, the average of (5) is

$$\overline{w}_i = -\frac{1}{2}\int_0^1\int_0^\infty \rho\,(d^2Y_i/dZ^2)\chi P(\chi,Z)\,d\chi dZ \quad , \tag{6}$$

and from (3) then

$$\overline{Q} = \frac{1}{2}\int_0^1\int_0^\infty \rho\,[\sum_i H_i(d^2Y_i/dZ^2)]\,\chi P(\chi,Z)\,d\chi dZ \quad . \tag{7}$$

These formulas show that knowledge of the average production rates of near-equilibrium species and of heat relies on knowledge of the joint probability-density function for Z and χ. Here χ appears because the magnitude of the gradient is a measure of the rate of diffusion of reactants into the reaction sheet.

Thermochemical calculations enable d^2Y_i/dZ^2 at equilibrium to be obtained in general cases. From Fig. 3 it may be seen that d^2Y_i/dZ^2 are delta functions located at the reaction sheet, $Z=Z_c$. In this approximation $P(\chi,Z)$ need not be known for all values of

Z to obtain $\overline{w}_i$ and $\overline{Q}$; it is sufficient to know $P(\chi,Z)|_{Z=Z_c} \equiv P(\chi,Z_c)$. Since $P(\chi,Z)=P(\chi|Z)\ P(Z)$, it is seen that the conditioned-average rate of dissipation, $\overline{\chi}_c \equiv \int_0^\infty \chi P(\chi|Z_c)d\chi$ must be known to obtain $\overline{w}_i$ or $\overline{Q}$. Since knowledge of the joint or conditioned functions is practically absent, often statistical independence is hypothesized, or else it is merely assumed, less restrictively, that the conditioned-mean dissipation equals the unconditioned mean. Some measurements of this important conditioned mean have been made, and others are in progress.

After $\overline{\chi}_c$ is obtained, the local average rates of fuel consumption and of heat release may be calculated, in the flame-sheet approximation, from the formulas

$$\overline{w}_F = -\frac{1}{2}\rho_c\overline{\chi}_c P(Z_c)\ Y_{F,0}/(1-Z_c) \tag{8}$$

and

$$\overline{Q} = \frac{1}{2}\rho_c\overline{\chi}_c P(Z_c)[(qY_{F,0})/(W_F\nu_F)]/(1-Z_c) \quad , \tag{9}$$

which are derived by using Fig. 2 to obtain the delta functions. Here q is the standard heat of reaction at temperature T_c, so that $q/(W_F\nu_F)$ is the heat released per unit mass of fuel consumed. The density ρ_c is the function $\rho(Z)$ evaluated at $Z=Z_c$, i.e., the density at the relevant adiabatic flame temperature. The influence of the fluid mechanics emerges in the proportionality of the local, average rates to the product $\overline{\chi}_c P(Z_c)$; in various flows $\overline{\chi}_c$ may range perhaps from $10^{-1}s^{-1}$ to $10^4 s^{-1}$, while $P(Z_c)$ generally is somewhere between 10^{-3} and 10.

Equation (5) may be treated as the first term in a perturbation about equilibrium flow. Given a scheme for the chemical kinetics, an expansion of w_i may be generated in the departures of T and of Y_i from their chemical-equilibrium dependences on Z. From (2), a general equation for the departures from equilibrium may then be derived, and suitable modeling in this equation then may provide the first correction to $\overline{w}_i$. The procedure, never simple, is least complicated if the departure from chemical equilibrium can be characterized fully in terms of a single variable. More research remains to be done on the subject.

5. PRODUCTION OF TRACE SPECIES. Oxides of nitrogen and soot-related species are examples of chemical components present in very low concentrations in turbulent diffusion flames; they are trace species. The trace species of interest often are far from equilibrium, as are soot-related species always and oxides of nitrogen almost always. The fact that the concentrations of these species are low means that they affect the thermochemistry to a negligible extent and that therefore finite-rate effects for them can be analyzed more easily than those for major species. Methods of analysis have been developed in the literature [1]. The local average rates of production of nonequilibrium trace species generally are of the greatest concern. Their local average concentrations are of lesser interest. This is fortunate because the production rates often can be obtained without introducing a conservation equation for the concentration of the species of interest. It is difficult to analyze the conservation equations for species experiencing finite-rate chemistry in turbulent flows, irrespective of whether the species are present in trace amounts.

Let there be T trace species out of equilibrium, identified by subscript i=1,...,T, and assume (quite reasonably) that there chemistry involves at most binary collisions between trace species. The equilibrium diffusion flame is to be analyzed first, by the methods indicated above, with the trace species neglected. This gives the temperature and the concentrations of all other species in terms of Z. The mass rates of production w_i of the nonequilibrium trace species then may be expressed in terms of Z and the mass fractions Y_i of the trace species in the general form

$$w_i = w_{io}(Z) + \sum_{j=1}^{T} w_{ij}(Z)Y_j + \sum_{j=1}^{T}\sum_{k=1}^{T} w_{ijk}(Z)Y_jY_k, \quad i=1,\dots T, \quad (10)$$

in view of the assumed chemistry. Here the first term arises from steps in which none of the reactants are nonequilibrium trace species, the second from steps involving one such trace species as a reactant and the third from steps involving two. For purposes of characterizing the chemistry, the reasonable approximation of constant pressure has been employed here. In view of (10), the local average production rate evidently is

$$\bar{w}_i = \int_0^1 w_{i0}(Z)\,P(Z)dZ + \sum_{j=1}^{T} \int_0^1\int_0^1 Y_j w_{ij}(Z)\,P(Z,Y_j)dY_j dZ$$

$$+ \sum_{j=1}^{T}\sum_{k=1}^{T} \int_0^1\int_0^1\int_0^1 Y_j Y_k w_{ijk}(Z)\,P(Z,Y_j,Y_k)\,dY_k dY_j dZ,$$

$$i=1,\ldots,T, \qquad (11)$$

where $P(Z,Y_j)$ and $P(Z,Y_j,Y_k)$ are joint probability-density functions. Thus, $\bar{w}_i$ may be evaluated directly from knowledge of the appropriate probability-density functions.

Given the above methods, calculation of the first term of (11) by numerical integration poses no difficulty in principle. The second and third terms can be evaluated only if the appropriate joint probability-denisty functions can be obtained. This would entail either measurement or theoretical consideration of the conservation equations for the nonequilibrium trace species. These are challenging problems; typical interactions through chemical kinetics are strong enough, for example, to cause hypotheses of statistical independence for $P(Z,Y_j)$ or for $P(Z,Y_j,Y_k)$ to be questionable. Therefore well-justified approaches for obtaining $\bar{w}_i$ are available today only if the second and third terms are negligible, i.e., only if the nonequilibrium trace species participate to a negligible extent in the net production of the species of interest. This is often true, as the following example will illustrate. Therefore useful methods are available for obtaining the average rates of production in some interesting cases.

First consider production of NO by the Zel'dovich mechanism, $O+N_2 \rightarrow NO+N$, $N+O_2 \rightarrow NO+O$. The molar rate of production of NO is $\omega_{NO} = 2k_f c_{N_2} c_O$, where k_f is the specific reaction-rate constant for the first step, since the second is fast. If equilibrium is maintained for O_2 dissociation, then the concentration of O is related to that of O_2 by $c_O^2 = K_c c_{O_2}$, where K_c is the equilibrium constant for concentrations in the dissociation reaction. Both k and K_c depend strongly on the temperature T since $k_f=7\times10^{13}\exp(-37,750/T)\mathrm{cm}^3/$ mol s and $\sqrt{K_c}=4.1\ \exp(-29,150/T)(\mathrm{mol/cm}^3)^{1/2}$, with T in °K. Thus,

$$\omega_{NO} = 5.7 \times 10^{14}\ c_{N_2}\sqrt{c_{O_2}}\ e^{-66,900/T} \tag{12}$$

Since N_2 and O_2 are major species, the first of which is practically inert and the second of which usually nearly maintains chemical equilibrium, the equilibrium chemistry for the diffusion flame gives $\omega_{NO}(Z)$ from (12). Then, as in (11)

$$\overline{\omega}_{NO} = \int_0^1 \omega_{NO}(Z)\ P(Z)\ dZ \tag{13}$$

Equation (13) provides an estimate of the average rate of production of NO.

Because of the strong dependence of ω_{NO} on T in (12), evaluation of $\overline{\omega}_{NO}$ from (13) by numerical integration requires very fine step sizes. However, the integral may readily by evaluated as an asymptotic expansion for large values of T_a/T_c, where T_a=66,900°K. If a flame-sheet approximation for the diffusion-flame structure is adopted, then $c_{O_2}=0$ for $Z\geq Z_c$, so the integral in (13) extends only from 0 to Z_c, and near $Z=Z_c$, c_{O_2} is linear in Z_c-Z, so that $c_{O_2}=c(Z_c-Z)$, where c is constant. On the oxygen side of the flame sheet T is also linear in Z_c-Z, say $T=T_c[1-b(Z_c-Z)]$, where b is constant. Employing the expansion $[1-b(Z_c-Z)]^{-1}\approx1+b(Z_c-Z)$ in the argument of the exponential, the asymptotic expansion

$$\overline{\omega}_{NO} = 5.7 \times 10^{14} e^{-T_a/T_c}\ \frac{c_{N_{2,c}}\sqrt{c}P(Z_c)}{[(T_a/T_c)b]^{3/2}} \int_0^\infty \sqrt{\xi}\ e^{-\xi}d\xi \quad , \tag{14}$$

is obtained, where the substitution $\xi=(T_a/T_c)\ b(Z_c-Z)$ has been made. The integral in (14) is a gamma function with value $\sqrt{\pi}/2$. Equation (14) shows that the average production rate is closely proportional to the value of the probability-density function for Z at the stoichiometric point Z_c. Experimentally, the production rate has a maximum on the fuel-rich side of the average stoichiometric surface in a turbulent-jet diffusion flame; this is readily explained if for the profiles illustrated in Fig. 3, $P(Z_c)$ is greater at a fuel-rich point that at a point on the average stoichiometric surface.

The calculation outlined here is inaccurate in that equilibrium dissociation of oxygen is not attained in most diffusion flames. It is known that superequilibrium concentrations of oxygen atoms occur; these may lead to appreciably increased values of $\overline{\omega}_{NO}$. An approach to take this into account without changing the formulation is to employ an empirical correlation for $c_O(Z)$. It might be thought that to achieve better accuracy O atoms could be treated in a framework similar to that of NO, since they also are nonequilibrium trace species. However, this approach encounters difficulties because reactions such as $O+H_2 \rightarrow OH+H$, involving O atoms and major species, are important, so that a joint probability-density function may have to be considered in (11); moreover, accounting for three-body recombinations, such as $O+OH+M \rightarrow HO_2+M$, brings the three-variable (trivariate) joint probability-density functions into (11). The ultimate utility of the general formulation for these more complicated processes is unclear because the approach that has been presented is relatively unexplored.

6. EFFECTS OF REACTION-SHEET STRUCTURES. From these considerations it is seen that structures of reaction sheets have important bearings on turbulent diffusion flames. Perturbation methods are quite useful in investigating these structures. The studies need not consider turbulence at all since the same types of problems arise in laminar flows. However, for application to turbulence problems it is desirable to employ formulations that are relatively independent of specific geometrical configurations. The mixture fraction Z is a helpful variable for achieving this independence [5].

It is possible to treat Z as an independent variable in analyzing reaction-sheet structures. To make a formal transformation introduce an orthogonal coordinate system with Z being one of the coordinates and with x and y, distances along surfaces of constant

Z, being the other two. In this system let u and v be velocity components in the x and y directions. It may then be shown from equations (1) and (2) that

$$\rho\frac{\partial Y_i}{\partial t} + \rho \underset{\sim}{v}_{\perp}\cdot\underset{\sim}{\nabla}_{\perp} Y_i = w_i + \rho\nu|\underset{\sim}{\nabla}Z|^2 \frac{\partial^2 Y_i}{\partial Z^2} + \underset{\sim}{\nabla}_{\perp}\cdot(\rho\nu\underset{\sim}{\nabla}_{\perp} Y_i)$$

$$-\rho\nu\underset{\sim}{\nabla}_{\perp}(\ln|\underset{\sim}{\nabla}Z|)\cdot\underset{\sim}{\nabla}_{\perp} Y_i \qquad (15)$$

where the x and y components of the transverse velocity vector $\underset{\sim}{v}_{\perp}$ are u and v, and where $\underset{\sim}{\nabla}_{\perp}$ is the two-dimensional gradient involving x and y.

In equation (15), the convective term that remains involves velocities parallel to surfaces of constant Z, i.e., transverse to reaction sheets. Convection in the normal direction is now implicit in the term involving $\partial^2 Y_i/\partial Z^2$, which superficially exhibits a purely diffusive character. In analyzing structures of thin sheets a stretching of Z about a fixed value, usually the stoichiometric value, is often useful. Unless the time-derivative or transverse-derivative terms in (15) become large, the stretching in Z approximately reduces this equation to the simplified equation

$$\partial^2 Y_i/\partial Z^2 = -w_i/(\rho\nu|\underset{\sim}{\nabla}Z|^2) \qquad (16)$$

which exhibits an apparent diffusive-reactive balance in the new coordinate. Equation (16) turns out to be quite useful for analyses of flame structure.

As an example consider a one-step reaction written as

$$\nu_F F + \nu_O O \rightarrow \text{Products} \quad , \qquad (17)$$

where ν_i is a stoichiometric coefficient for species i, so that in (16) for fuel

$$w_F = -\left[\frac{W_F \nu_F B T^{\alpha}}{W_F^{n_F} W_O^{n_O}}\right]\rho^{n_F+n_O} Y_F^{n_F} Y_O^{n_O} e^{-E/R^\circ T} \quad , \qquad (18)$$

in which W_i denotes the molecular weight of species i. In this one-step approximation the empirical reaction orders n_F and n_O have been introduced, since the overall stoichiometry seldom provides the correct dependence of the rate on concentrations. The overall activation energy is E, and the overall frequency factor for the rate of fuel consumption is $A_F=\nu_F BT^{\alpha}$. The justification for the stretching in (16) then arises by considering $E/R^\circ T_c$ to be a large parameter. Detailed understanding of diffusion-flame structure was obtained by perturbation methods of this type [6], termed activation-energy asymptotics [7].

To carry out an analysis of the reaction zone by activation-energy asymptotics, Z may be stretched about Z_c and T about T_c. As an independent variable introduce $\eta=(Z-Z_c)/\varepsilon$ and as a dependent variable $\phi=(T_c-T)E/R^\circ T_c^2-\gamma\eta$, where

$$\varepsilon = 2R^\circ T_c^2 Z_c(1-Z_c)/[E(T_c-T_{f,c})] \tag{19}$$

and

$$\gamma = 2Z_c-1-2Z_c(1-Z_c)(T_{F,0}-T_{0,0})/(T_c-T_{f,c}) \quad , \tag{20}$$

in which $T_{f,c}$ is T_f evaluated at $Z=Z_c$. It can be shown that ε is the formal small parameter of expansion for activation-energy asymptotics applied to the diffusion flame and that $1-\gamma$ is twice the ratio of the heat conducted from the flame toward negative values of $Z-Z_c$ to the total chemical heat released at the flame. The quantity $1/\varepsilon$ is essentially the relevant Zel'dovich number. By introducing ϕ and η into (16) treating these variables as being of order unity and expanding in ε, in the first approximation (if $n_F=n_0=1$) the differential equation

$$d^2\phi/d\eta^2 = \delta(\phi^2-\eta^2)e^{-(\phi+\gamma\eta)} \tag{21}$$

is obtained, where a reduced Damköhler number is

$$\delta = \left(\frac{Z_c\nu_0 Y_{F,0}\rho A_F e^{-E/R^\circ T_c}}{W_F\nu_F\nu|\nabla Z|^2}\right)\left[\frac{R^\circ T_c^2}{E(T_c-T_{f,c})}\right]^3 [2Z_c(1-Z_c)]^2 \quad , \tag{22}$$

with $A_F\rho/(\nu|\nabla Z|^2)$ understood to be evaluated at $Z=Z_c$. The boundary conditions are found from matching to be $d\phi/d\eta\to\pm 1$ as $\eta\to\pm\infty$. Solutions to (21) that satisfy these boundary conditions exist for a range of values of δ. These solutions show that the maximum temperature is reduced below T_c by a fractional amount, proportional to ε, that increases as δ decreases. They also show, through evaluation of $\lim_{\eta\to\pm\infty}(\phi\mp\eta)$, that either fuel or oxidizer or both leak through the reaction zone in a fractional amount, of order ε, that increases as δ decreases. The thickness of the reaction zone corresponds to $Z-Z_c$ of order ε.

Solution to the problem leads to results of the type illustrated in Fig. 5. There are critical values of reduced Damköhler numbers corresponding to ignition (δ_I) and extinction (δ_E). Analysis shows that the minimum value of δ below which solutions to (21) cease to exist is

$$\delta_E = [(1-|\gamma|)-(1-|\gamma|)^2 + 0.26(1-|\gamma|)^3 + 0.055(1-|\gamma|)^4]e \tag{23}$$

within one percent for $|\gamma|<1$. For $|\gamma|>1$, one of the two boundary temperatures exceeds the flame temperature, and an abrupt extinction event does not exist for the diffusion flame. Use of (23) in (22) provides an explicit expression for conditions of extinction of the diffusion flame.

The studies thus far by perturbation methods mainly have been based on one-step chemistry. Current research is beginning to address influences of more complicated chemical kinetics. There are numerical solutions of the full equations for hydrocarbon-air flames that suggest useful simplified kinetic schemes amenable to treatment by perturbations. An example of studies of more detailed kinetics is a recent consideration of influences of dissociation [8]. These further studies still seem best based on activation-energy asymptotics in one form or another. Methods of this type also can address effects of radiant energy transfer [9]. Conditions for extinction still arise; a review on extinction is available [10].

From (22) and (23) it is seen that if $|\underset{\sim}{\nabla}Z|^2$ becomes too large at any given position and time, at a point on the reaction sheet in a turbulent diffusion flame, then extinction occurs at that point. An increase in $|\underset{\sim}{\nabla}Z|^2$ produces an increase in the magnitude of the local, instantaneous, diffusion-controlled reaction rate, but within the context of activation-energy asymptotics, this increase terminates abruptly with extinction at a critical maximum value beyond which the chemistry is not fast enough to keep pace with the rate at which the reactants are diffusing into the reaction zone. Since $|\nabla Z|^2$ is proportional to the local, instantaneous rate of dissipation of the mixture fraction, which in turn is proportional to the square of a strain rate (a velocity gradient) in the turbulent flow, sufficiently large strain rates cause local extinctions.

Criteria for local extinction are readily stated from (22) and (23). These criteria provide restrictions that must be satisfied if the turbulent flame is to lie within the reaction-sheet regime. What happens at higher strain rates is not known today. There are many studies that could help to improve our understanding of processes that may occur subsequent to local extinction. For example, time-dependent analyses of the dynamics of extinction and analyses of the evolution of a hole in a continuous, planar, diffusion-flame sheet would provide helpful information.

A working hypothesis that seems reasonable for many turbulent diffusion flames, notably for those in open environments, with cold reactants that release large amounts of energy over narrow ranges of stoichiometry at rates strongly dependent on temperature, is that combustion is not re-established after a local extinction. This hypothesis can be employed to investigate liftoff and blowoff of turbulent-jet diffusion flames [11-13]. In the turbulent range, as the velocity of the fuel issuing from the jet is increased, the average flame height changes little. At sufficiently low fuel velocities, the base of the flame extends to the mouth of the duct, where it is stabilized. When a critical exit velocity, the liftoff velocity, is exceeded, the flame abruptly is detached from the duct and acquires a new configuration of stabilization in which combustion begins a number of duct diameters downstream. The axial distance from the duct exit to the plane at which the flame begins, the liftoff height of the lifted diffusion flame, increases as the exit velocity is increased further. When the exit velocity exceeds a second critical value, the blowoff velocity, the flame no longer can be stabilized in the mixing region, and combustion ceases (blowoff occurs). Liftoff heights and blowoff velocities are of both fundamental and applied interest, and a number of measurements of them may be found in the literature.

The occurrence of liftoff may be understood in terms of critical strain rates for sheet extinction [11]. As the exit velocity is increased, the average local strain rate increases, thus causing an increase in $\overline{\chi}_c$. An increasing fraction of reaction sheets then encounters extinction conditions, $\delta \leq \delta_E$. When too many of the sheets are extinguished, a network of diffusion-flame sheets no longer is connected to the burner, and liftoff must occur. In a first, rough approximation, this may be assumed to happen when $\overline{\chi}_c \geq \chi_E$, where χ_E is the value of $2\nu|\underset{\sim}{\nabla} Z|^2$ obtained from (22) by putting $\delta = \delta_E$. Since this value is calculable from thermochemical and chemical-kinetic parameters alone, the liftoff criterion expresses a comparison between a fluid-mechanical quantity ($\overline{\chi}_c$) and a chemical property of the system. Thus, $\overline{\chi}_c$ is relevant not only to average rates of heat release (9) but also to liftoff phenomena.

After liftoff occurs, the lower temperatures at the duct exit produce lower molecular diffusivities, i.e. higher Reynolds numbers, so that $\overline{\chi}_c$ is increased appreciably in the vicinity of the exit. Farther downstream, as the jet spreads, $\overline{\chi}_c$ is reduced. Beyond some distance from the duct exit, the condition $\overline{\chi}_c \leq \chi_E$ again is established, and the turbulent diffusion flame may exist. The liftoff height then may be determined as the height at which $\overline{\chi}_c = \chi_E$ [11]. This height increases as the jet velocity increases because at any fixed height, $\overline{\chi}_c$ increases with jet velocity. Although there are significant uncertainties in estimating $\overline{\chi}_c$ for turbulent jets, reasonable agreements between measured and estimated liftoff heights have been obtained. However, this is not the only view of liftoff, and the subject remains an active topic of research. This research may employ turbulence modeling with perturbation methods applied to analysis of the reaction sheets.

7. CONCLUDING REMARKS. From these discussions it is seen that for analyses of turbulent diffusion flames perturbation methods are attractive mainly in studying the structures and dynamics of reaction sheets. Perturbations have played other roles, e.g. in the contexts of expansions for low turbulence intensities or large turbulence scales. However, in the past as well as the future, careful investigations of reaction sheets by use of perturbations appear to contribute most to enhancing understanding of turbulent diffusion flames.

BIBLIOGRAPHY

1. P. A. Libby and F. A. Williams, editors, Turbulent Reacting Flows, Springer-Verlag, Berlin (1980).

2. F. A. Williams, Combustion Theory, 2nd ed., Benjamin/Cummings, Menlo Park, CA (1985).

3. P. A. McMurthy, W. H. Jou, J. J. Riley and R. W. Metcalf, AIAA Paper No. 85-0143 (1985).

4. R. W. Bilger, Prog. Energy Combust. Sci., **1**, 87 (1976).

5. F. A. Williams, in Recent Advances in Aerospace Sciences, C. Casci, ed. Plenum, New York (1985), 415-421.

6. A. Liñán, Acta Astronautica, **1**, 1007 (1979).

7. J. D. Buckmaster and G. S. S. Ludford, Theory of Laminar Flames, Cambridge University Press, Cambridge (1982).

8. N. Peters and F. A. Williams, in Dynamics of Flames and Reactive Systems, J. R. Bowen, N. Manson, A. K. Oppenheim and R. I. Soloukhin, eds., American Institute of Aeronautics and Astronautics, New York (1985), 37-60.

9. S. H. Sohrab, A. Liñán and F. A. Williams, Combust. Sci. Tech. **27**, 143 (1982).

10. F. A. Williams, Fire Safety Journal **3**, 163 (1981).

11. N. Peters and F. A. Williams, AIAA Journal **21**, 423 (1983).

12. N. Peters, Combust. Sci. Tech. **30**, 1 (1983).

13. J. Janicka and N. Peters, Ninteenth Symposium (International) on Combustion, The Combustion Institute, Pittsburgh (1982), 367-374.

MECHANICAL & AEROSPACE ENGINEERING DEPARTMENT
D-325 ENGINEERING QUADRANGLE
PRINCETON UNIVERSITY
PRINCETON, NEW JERSEY 08544

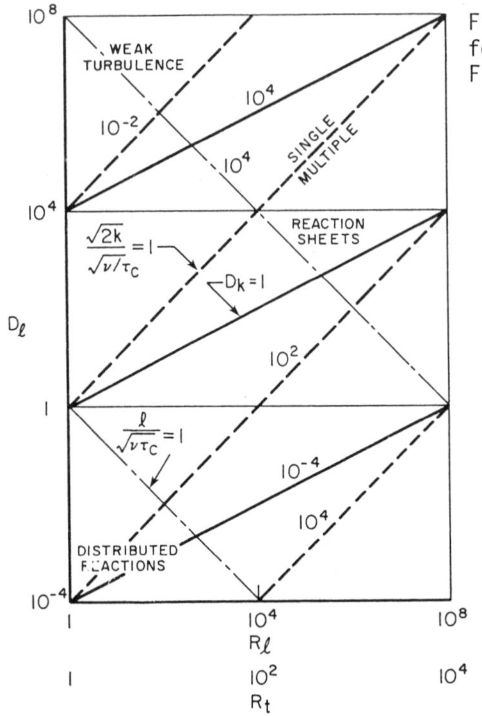

Fig. 1. Combustion Regimes for Turbulent Diffusion Flames.

Fig. 2. Dependences of the Concentrations and Temperature on the Mixture Fraction in the Frozen-Flow and Reaction-Sheet Limits.

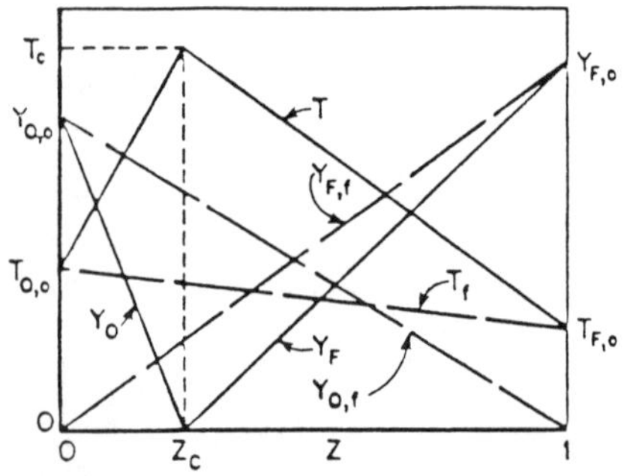

Fig. 3. Representative Shapes for Probability Density Functions for the Mixture Fraction.

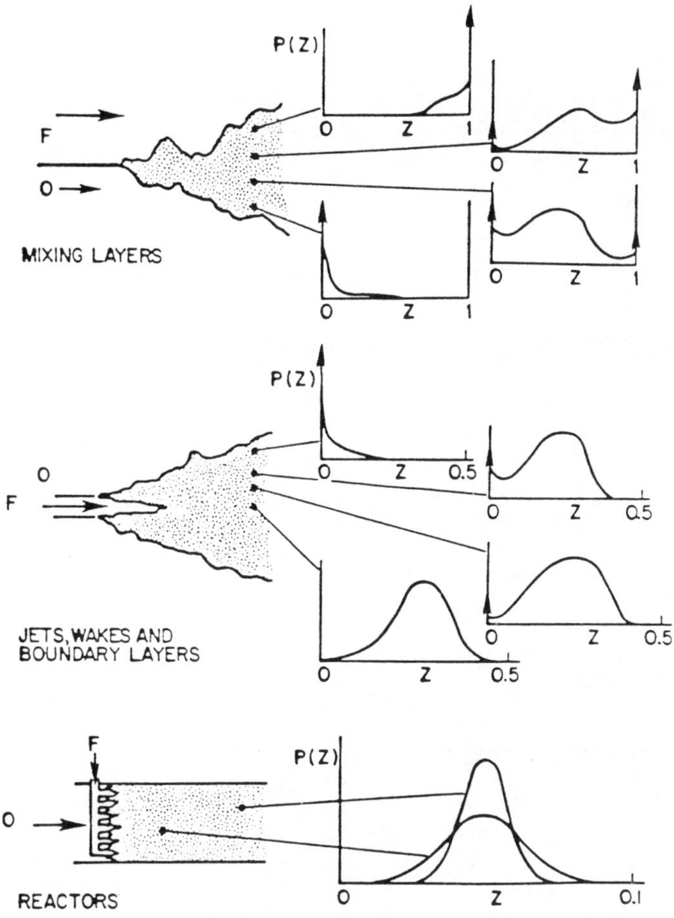

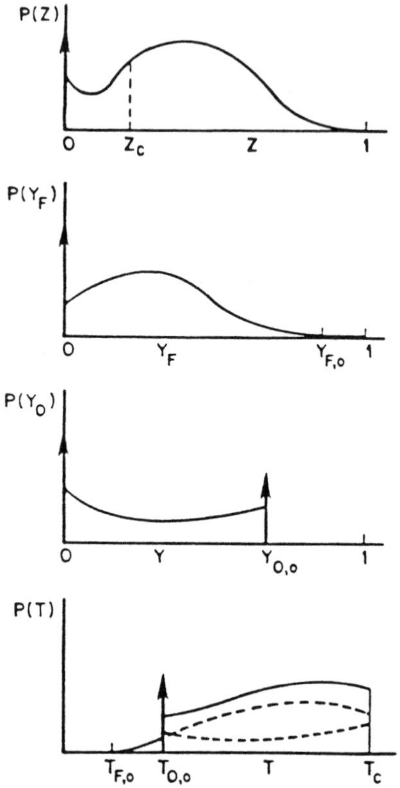

Fig. 4. Generation of Probability Density Functions for Reactant Mass Fractions and Temperature from that for Mixture Fraction.

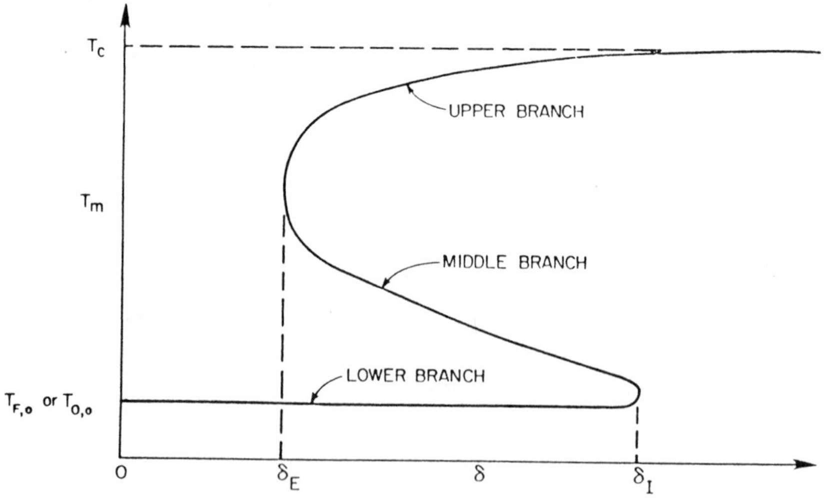

Fig. 5. Dependence of Maximum Temperature on Reduced Damköhler Number for Diffusion Flames.

Lectures in Applied Mathematics
Volume 24, 1986

THE LIFTING OF TURBULENT DIFFUSION FLAMES AND THE NOISE THEY MAKE

J. Buckmaster & N. Peters

ABSTRACT. Peters' theory of lifted turbulent diffusion flames is briefly reviewed, and some preliminary results on noise generation outlined.

INTRODUCTION. Turbulent diffusion flames will lift from a burner if the efflux is strong enough, stabilizing at a finite distance from the exit plane, and a satisfactory explanation of this phenomenon provides important insight into the processes that occur in these flames. In a sequences of paper, e.g. [1], N. Peters has developed a simple theory based on an extinction analysis for laminar diffusion flames, and this theory is briefly summarized in these pages.

The mechanism responsible for lifting is local quenching of the flame sheet. Buckmaster, Stewart, Ignatiadis and Williams [2] have discussed the nature of the flow field that is induced when a diffusion flame is quenched, and the implications of these results for lifted flames are briefly discussed.

We shall consider a simple kinetic scheme of the form

$$\nu_F F + \nu_0 0 \rightarrow \text{Products} \tag{1}$$

at a rate

1980 Mathematics Subject Classification.

$$w = B\,\rho^2 \frac{Y_0\,Y_f}{M_0\,M_f}\,e^{-E/RT}, \tag{2}$$

where Y_0, Y_F are mass fractions, M_0, M_F are molecular weights, ρ the density, and T the temperature. We shall suppose that the mass and heat diffusion coefficients are equal so that if the differential operator L is defined by

$$L \equiv \rho \frac{D}{Dt} - \nabla \cdot (\rho D \nabla), \tag{3}$$

the equations governing Y_0, Y_F and T are

$$LT = \frac{Qw}{C_p},\ L\,Y_0 = -\nu_0 M_0 w,\ L\,Y_F = -\nu_F M_F w. \tag{4}$$

Q is the heat released due to reaction.

For a burner flame we have an oxygen supply, a domain where

$$Y_F = 0,\ Y_0 = Y_{0_S},\ T = T_{S_0}, \tag{5}$$

and a fuel supply where

$$Y_F = Y_{F_S},\ Y_0 = 0,\ T = T_{S_F}, \tag{6}$$

Schvab-Zeldovich variables may be defined in the usual way, i.e.

$$\begin{aligned} S_0 &= Y_0 + T\,\frac{C_p}{Q}\,\nu_0 M_0, \\ S_F &= Y_F + T\,\frac{C_p}{Q}\,\nu_F M_F, \end{aligned} \tag{7}$$

andthese satisfy reaction free equations, i.e.

$$L(S_i) = 0,\ i = 0, F. \tag{8}$$

For a large class of problems, and consistent with (5),(6), it follows that S_0 and S_F are linearly related, the situation to which we shall restrict ourselves.

SURFACE OF STOICHIOMETRIC MIXTURE AND QUENCHING. Any linear combination of S_0 and S_F satisfies equation (8), and an important example is the conserved scalar, Z, defined by

$$Z\left(\frac{Y_{0_S}}{\nu_0 M_0}+\frac{Y_{F_S}}{\nu_F M_F}\right)=\frac{Y_{0_S}}{\nu_0 M_0}+\frac{Y_F}{\nu_F M_F}-\frac{Y_0}{\nu_0 M_0}\ . \tag{9}$$

This vanishes in the oxygen supply and equals 1 in the fuel supply. At intermediate points in the combustion field it takes on the stoichiometric value Z_{st}, identified by the condition

$$\frac{Y_F}{\nu_F M_F}=\frac{Y_0}{\nu_0 M_0}\ , \tag{10}$$

so that

$$Z_{st}=\frac{Y_{0_S}}{Y_{0_S}+\frac{\nu_0 M_0}{\nu_F M_F}Y_{F_S}}\ . \tag{11}$$

The surface of stoichiometric mixture defined by

$$Z=Z_{st} \tag{12}$$

plays an important role in the structure of the flame when the activation energy E is large. For then reaction is confined to a small neighborhood of this surface, a flame-sheet which divides the combustion field into two portions. In one of them, including the oxygen supply, Y_F is sensibly zero in the other, including the fueld supply, Y_0 is negligably small.

The flame-sheet structure is well understood, [3]. Since it is thin, the operator L may be approximated within it by

$$L\cong-(\rho D)_{st}\frac{\partial^2}{\partial n^2}+\rho_{st}\frac{\partial}{\partial t} \tag{13}$$

where n is the distance measured normal to the sheet, and the time derivative is only important if temporal changes are rapid, a situation that we shall discuss later.

A crucial step in the theory is to use Z rather than n as independent variable, so that

$$L\cong-(\rho D)_{st}|\nabla Z|^2_{st}\frac{\partial^2}{\partial z^2}+\rho_{st}\frac{\partial}{\partial t}. \tag{14}$$

In order to write down the structure equation we introduce the expressions

$$\bar{T} + \bar{T}_{st} + \frac{\bar{T}_{st}}{\theta}\,\phi(\xi),$$
$$Z = Z_{st} + \frac{1}{\theta}\xi. \qquad (15)$$

where

$$\bar{T} = \frac{C_p T}{Q}, \qquad (16)$$

and

$$\theta = \frac{EC_p}{QR} \qquad (17)$$

is the large nondimensional activation energy. Moreover, because of the linear connection between the Schvab-Zeldovich variables, Y_0 and Y_F are linear function of Z and T (the coefficients are represented by the constants α and β in equation (18) below) so that the structure equation is

$$\frac{1}{\theta^2}\frac{\partial\phi}{\partial t} - \frac{1}{2}X_{st}\frac{\partial^2\phi}{\partial\xi^2} = \frac{(\phi-\alpha\xi)(\phi-\beta\xi)e^{\phi}}{\tau_{chem}}. \qquad (18)$$

X_{st} is the <u>scalar disruption rate</u>, defined by

$$X_{st} = 2D_{st}|\nabla Z|^2_{st}, \qquad (19)$$

and τ_{chem} is a chemical time that depends, amongst other things on B (cf. (2)).

To obtain boundary conditions for (18) we note that in any region where Y_0 and Y_F are essentially zero, there is a linear relation between Z and T. Indeed, we have

$$\lim_{\xi\to\infty}\frac{\partial\phi}{\partial\xi} \to \beta(>0) \qquad \lim_{\xi\to-\infty}\frac{\partial\phi}{\partial\xi} \to \alpha(<0). \qquad (20)$$

α and β are known constants, a consequence of using Z as an independent variable. A formulation in the physical space would require the specification of $\frac{\partial T}{\partial n}$ on each side of the sheet, a quantity which is not known in the absence of a description of the combustion field beyond the reaction zone.

The stationary problem defined by (18) and (20)(with the time-derivative dropped) is Linan's structure problem [4] and is context free. The specific problem at hand only plays a role through the specification of the parameter X_{st}, which is an inverse Damköhler number. A solution only exists if the Damköhler number is large enough, i.e.

$$X_{st} < (X_{st})_{crit}, \tag{21}$$

the critical value being identified with static extinction. $(X_{st})_{crit}$ can be calculated (by numerical integration of the structure problem) or better - in view of the crude kinetic model - obtained from experimental data. Experimental results obtained for the counterflow flame are sufficient for the latter purpose, [5].

THE LIFTING HYPOTHESIS AND ITS SUBSTANTIATION. The essential idea is that if the turbulent intensity, and therefore X_{st} at the burner mouth, is large enough, the flame will be locally quenched and will establish itself further downstrem where X_{st} is smaller. However, in a turbulent flame X_{st} is a stochastic variable so that at any instance there will be a fraction of the surface $Z = Z_{st}$ for which X_{st} is not so large as to suppress combustion. This fraction will decrease as the mean value $\overline{X_{st}}$ increases. Thus in order to have a well-defined lift-off criterion it is necessary to argue that if the fraction is less than some critical value, there can be no local burning.

This can be justified by an abstract appeal to percolation theory. As an example of percolation theory, suppose that an electric current is passed between two parallel sides of a rectangular conducting sheet. If we now cut holes in this sheet, in a random fashion, there will be a well-defined average area of sheet that must be removed in order to destroy the connectivity and stop the current flow. By analogy, if a sufficiently large fraction of the flame sheet is destroyed, it can no longer define a flame.

The physical underpinnings that could justify this point of view are concerned with the problem of reignition. A well defined flame can only exist if a portion of the sheet that is quenched when X_{st} exceeds the critical value is reignited when X_{st} subsequently drops below the critical. This ignition comes about because of the proximity of portions of burning sheet, and if there are not enough of these it may plausibly be argued that reignition is not possible. Autoignition will play no role since this can only occur for values of X_{st} that are orders of magnitude smaller than $(X_{st})_{crit}$.

The key evidence to substantiate the lifting hypothesis may be found in the experimental data of [6]. It the non-dimensional lift-off height H/d (H = lift-off height, d = burner diameter) is plotted against the nondimensional residence time $(X_{st})_{crit}d/u_0$ (u_0 = burner gas speed), the data for a wide range of mixture strengths collapses to a single curve.

WIND GENERATION AND ACOUSTICS. If X_{st} exceeds the critical value, a steady solution of (18), (20) does not exist. The unsteady collapse of the flame is described in terms of the fast-time τ, where

$$\tau = \theta^2 t. \tag{22}$$

As the temperature drops the reaction rate becomes small and

$$\phi \to -\infty \quad \text{as } \tau \to \infty \tag{23}$$

so that the asymptotic behavior is governed by the equation

$$\frac{\partial \phi}{\partial \tau} - \frac{1}{2} X_{st} \frac{\partial^2 \phi}{\partial \xi^2} = 0, \tag{24}$$

which has a similarity solution of the form

$$\phi = \sqrt{\sigma}\, f(\xi/\sqrt{\sigma}), \sigma = \tfrac{1}{2} X_{st}\tau. \tag{25}$$

This solution satisfies the gradient conditions (20).

The drop in temperature is associated with an increase of density and therefore of mass within the flame-sheet, and the net mass-flux that provides this increase can be explicitly calculated (it is proportional to $\int_{-\infty}^{\infty} \frac{\partial \phi}{\partial \tau} d\xi$). The jump in velocity across the reaction zone, in the limit $\tau \to \infty$, is given by the approximate formula

$$[v_n]_{st} \cong \frac{1}{2} \frac{X_{st}}{|\nabla Z|_{st}} \cdot \frac{1}{Z_{st}(1-Z_{st})} \tag{26}$$

and has values which are typically .2-.3 m/s. This is a significant wind induced by the collapsing flame. The temperature change which causes this wind, and the domain in which the change occurs, are small. Nevertheless the wind is large because the change is so rapid.

This induced wind has important implications for the noise generated by a lifted flame. The continuity equation, when globally integrated, is

$$\rho_\infty \iint_S v_n dS = -\iiint_V \frac{\partial \rho}{\partial t} dV, \tag{27}$$

provided S, the surface of the volume V, lies in the far-field where the density is constant (ρ_∞). The volume integral can be written as

$$\iiint_V \frac{\partial \rho}{\partial t} dV = \iiint_{V_r} \frac{\partial \rho}{\partial t} dV + \iiint_{V_f} \frac{\partial \rho}{\partial t} dV \tag{28}$$

where V_r is the portion of V in which significant reaction occurs, and V_f is the remainder, where the reaction is frozen. If L is a length that characterizes the size of the combustion field, we have the estimates

$$V \sim V_f \sim L^3, \quad V_r \sim L^3/\theta. \tag{29}$$

It follows that, if quenching is negligable, so is the contribution to the monopolefield from V_r; the acoustic field is due simply to turbulent fluctuations of a nonuniform temperature field. If quenching is significant, however, $\frac{\partial \rho}{\partial t}$ within V_r is large and the contribution from V_r is comparable to that from V_f. Thus we would expect lifted flames to be noisy, and indeed they are.

ACKNOWLEDGEMENTS. The work of John Buckmaster was supported by the National Science Foundation and the Army Research Office. He was the recipient of a Humboldt Foundation U.S. Senior Scientist Award which made this collaboration possible, and this is acknowledged with gratitude.

BIBLIOGRAPHY

1. N. Peters, Combustion Science and Technology, 30, 1-17, 1983.

2. J. Buckmaster, D.S. Stewart, A. Ignatiadis and M. Williams, 'On the wind generated by a collapsing diffusion flame', to appear in Combustion Science and Technology.

3. J.D. Buckmaster and G.S.S. Ludford, Theory of Laminar Flames, Cambridge University Press, NY, 1982, Ch. 6.

4. A. Linan, Acta Astronautica, 1, 1007-39, 1974.

5. S. Ishizuka and H. Tsuji, 18th Symposium (International) on Combustion, The Combustion Institute, Pittsburgh, 695-703, 1981.

6. S. Donnerhack and N. Peters, Combustion Science and Technology, 41, 101-8, 1984.

J.D. Buckmaster
Aero-Astro Engineering
University of Illinois
Urbana, IL 61801

N. Peters
Lehrgebiet für Mechanik
RWTH, D-5100 Aachen
W. Germany

ABCDEFGHIJ-89876